Rapidly Quenched Metals

Rapidly Quenched Metals

Second International Conference
Section I

Edited by
N. J. Grant and B. C. Giessen

ACKNOWLEDGMENT

This conference was supported by grants from the following U.S.
Government agencies:

   Army Research Office

   Energy Research and Development Administration

   National Aeronautics and Space Administration

   National Science Foundation

   Office of Naval Research

PUBLISHER'S NOTE

This format is intended to reduce the cost of publishing certain
works in book form and to shorten the gap between editorial preparation and final publication. The time and expense of detailed ed-
iting and composition in print have been avoided by photographing
the text of this book directly from the editors' typescript.

Printed and bound in the United States of America

Library of Congress Cataloging in Publication Data

International Conference on Rapidly-Quenched Metals,
    2d, Massachusetts Institute of Technology, 1975.
    Rapidly Quenched metals.

    Includes bibliographical references and index.
    1. Metals--Quenching--Congresses. I. Grant,
Nicholas John. II. Giessen, B. C., 1932-
III. Massachusetts Institute of Technology. IV. Title.
TN672.I54 1975   669'.94   76-43340
ISBN 0-262-07066-9

TABLE OF CONTENTS

E. MECHANICAL PROPERTIES

F. PHYSICAL PROPERTIES: MAGNETIC, ELECTRICAL AND ELECTROCHEMICAL CHARACTERISTICS

CHAIRMEN AND COMMITTEE MEMBERS

Conference Co-Chairmen

B.C. Giessen
N.J. Grant

International Steering Committee

T. R. Anantharaman	India
R.W. Cahn	Great Britain
J. Dixmier	France
B.C. Giessen	U.S.A.
N.J. Grant	U.S.A.
H. Jones	Great Britain
B. Leontić	Yugoslavia
R. Maddin	U.S.A.
T. Masumoto	Japan
H. Matyja	Poland
H. Warlimont	Switzerland

Session Chairmen

Session A: R.W. Cahn
         P. Duwez

Session B: T.R. Anantharaman
         B.C. Giessen

Session C: J.J. Gilman
         D.E. Polk

Session D: J. Dixmier
         J. Vander Sande

Session E: N.J. Grant
         R. Maddin

Session F: F. Gardner
         B. Leontić

Publication Review Committee

B.C. Giessen	W.B. Nowak
N.J. Grant	D.E. Polk
H. Jones	C.C. Tsuei
F. Luborsky	D. Turnbull
J. Vander Sande	

Local Committee

F. Gardner	N.J. Grant
B.C. Giessen	D.E. Polk
J. Vander Sande	

Affiliations of Chairmen and Committee Members

THE SECOND INTERNATIONAL CONFERENCE ON
RAPIDLY-QUENCHED METALS

November 17, 18, 19, 1975

Massachusetts Institute of Technology
Cambridge, Massachusetts, U.S.A.

Conference Dedicated to

PROFESSOR POL DUWEZ

California Institute of Technology

For awakening the technical and scientific world to the potential and
excitement of metastable metallic structures produced by rapid quenching from the melt; for developing the gun technique for rapid quenching;
for his excellent publications on the subject; and for the many fine
scientists and engineers who obtained their training under his guidance.

In 1970, the (First) International Conference on Metastable Metallic
Alloys was held at Brela, Yugoslavia. A total of 46 papers was presented, and the Proceedings were published in Fizika, Vol. 2 — Sup-
plement 2, shortly thereafter. A search of the literature produced
about 200 papers, starting from 1960, following Professor Pol Duwez's
first reporting of unusual structures produced by rapid quenching
from the melt by splat cooling. The Brela Conference was primarily concerned with the preparation of materials and early reporting of "structural, electrical, and mechanical properties of these new, metastable
metallic materials". Dr. M. Paić in his foreword then stated,

> "These efforts are at their very beginning and the entire
> field is yet to be explored. Nevertheless, one feels confi-
> dent that by systematic and careful observations and measure-
> ments, on judiciously chosen series of these new alloys, im-
> portant regularities and, possibly, fundamental laws will ul-
> timately be discovered".

A short five years later and over 2000 publications downstream of Brela,
the Second International Conference on Rapidly Quenched Metals was held
at the Massachusetts Institute of Technology on November 17, 18, 19,
1975. In response to requests for abstracts of papers to be presented,
more than 120 were submitted; many more were offered after the dead-
line. Authors from 17 countries responded. Efforts to keep the attendance level within the bounds of the meeting room (170) were quite unsuccessful since 210 registered; many last minute requests to attend,
unfortunately, had to be denied.

Out of the papers presented and submitted, a total of 97 papers was
accepted for inclusion into the Conference Proceedings. For publication purposes, these papers were divided roughly equally into two sections, each containing papers from all six of the topical areas A-F
treated at the Conference. Section I, which includes the six invited
introductory papers to the topical areas, is being published by the
M.I.T. Press; Section II appears as a special issue of Materials Science and Engineering. The publications, appearing in summer and
fall of 1976, will have been edited by the conference co-chair-

man acting with the assistance of an editorial advisory board.

The two publications will be complementary, and each will contain an index with the titles of the papers appearing in the other volume to achieve a complete listing of each.

The coverage of the field in the two volumes is almost encyclopedic; the compilation can serve as a standard reference, bringing together material otherwise dispersed over many journals in various fields (metallurgy, solid state physics, magnetism, thin films, corrosion, to name but a few) and several countries.

The present volumes will also enable the Conference participants to peruse the proceedings with the leisure regrettably often not to be found during the hectic sessions, especially in the poster sessions where many papers of interest could not be studied to the degree desirable in the allotted time.

The editors feel that through the present conference the field of metastable alloys has reached the stature of a well-defined and rapidly maturing field. The large attendance and enthusiastic response of attendees of the Conference, which have resulted in the double publication, have also stimulated a decision of a Third Conference to be held in 1978, in England; announcements and details regarding the Third Conference will be forthcoming within six to nine months.

Summarizing the Proceedings, one notes the following:

At this stage of progress on rapid quenching from the melt (or vapor) there were a number of new thoughts on materials preparation techniques; emphasis was on technique improvements rather than on novel ideas.

The Conference showed tremendous interest in glassy metals, from fundamental concern over structure and occurrence to preparation and characterization, and extending to basic mechanical magnetic behavior: features which make amorphous alloys such fascinating materials with respect to their engineering properties. Among properties investigated, interest focused primarily on mechanical behavior, although not to the degree anticipated; magnetic properties received the second broadest coverage. For the first time, a group of papers appeared on corrosion and stress corrosion cracking behavior of these new materials. Clearly there is an intense and immediate interest in assessing these materials in terms of their engineering contributions.

[The magnitude of the interest in amorphous metals stimulated a one and a half day ONR Materials Workshop on the State of the Art and Applications of Amorphous Metals, held immediately following the present conference at Northeastern University in Boston. A synopsis of this Workshop will appear in Section II of these Proceedings.]

It is of merit to observe areas of omission or neglect in currently reported programs. Such areas include: lower than expected interest in metastable solid solutions, both as materials in their own right and as starting materials for precipitation strengthening; almost no work on microcrystalline alloys; a low level of alloy development research; limited study of the fundamental nature of "deformation" processes in amorphous metals, with indecisive and indiscriminate use of terms such as dislocations, slip, shear bands, deformation bands, plasticity, viscosity, etc. It appears that this latter area could stand much more thoughtful work; fracture behavior, too, deserves more attention. Interest in alloy phases is almost totally absent, and, finally, the respective merits of various liquid and vapor quenching techniques are not clear at this time and more comparative study seems called for.

The Chairmen acknowledge with deep thanks the contributions to this successful meeting made by the various committees listed overleaf. They thank the International Steering Committee for its effective handling of local publicity, inquiries, guidance, and advice.

Thanks are due to the session chairmen who contributed so greatly to the (almost) frictionless progress of the conference.

Special thanks are due to the members of the Publication Review Committee who handled many papers in the short time available for their task between the conference and the publication dates, to Mr. R. Dey for dedicated typing work, and to Drs. J. Megusar, M. Madhava, and Ms. Pat Corcoran for invaluable help in proofreading and other contributions.

Appreciation is also extended to the local committee for their valuable services, with a special acknowledgment to Dr. D.E. Polk; to

Ms. Anne Wallace of the Presentation Company for able and effective day-by-day operations; and to Dianne Banda of M.I.T. and Barry Ann Swanson of Northeastern University for extra hours and efficient coverage of thousands of other details; and to Lydia White who helped in many ways from start to the end of this entire operation.

Finally, we are particularly appreciative of the financial and moral support of the following groups, since it was this support which made possible the smooth and efficient operation of the Conference:

Army Research Office (Durham)
Energy Research and Development Administration
NASA
The National Science Foundation
Office of Naval Research

Nicholas J. Grant
Cambridge, Massachusetts

Bill C. Giessen
Boston, Massachusetts

EFFECTS OF EXPERIMENTAL VARIABLES IN RAPID QUENCHING FROM THE MELT

H. Jones

Department of Metallurgy
University of Sheffield
Sheffield, U.K.

1.0 INTRODUCTION

Techniques for producing nonequilibrium effects in solids divide into
three categories. Quenching involves changing from one set of conditions (of temperature and pressure, in particular) to another rapidly
enough to limit the amount of transformation towards equilibrium for
the new conditions. It includes conventional quenching entirely within
the solid state and quenching through the liquid to solid transformation. Molecular deposition involves growth from the vapour phase by
thermal evaporation, sputtering or chemical reaction, or from salt
solution by electroless displacement or by electrolysis. These differ
from normal quenching in that the deposited species are well-separated
or in chemical combination until the instant of deposition. External
action, for example, on a solid by deformation, irradiation or chemical attack, forms the third category.

   The limitations of the standard method of achieving large depar-
tures from equilibrium in solids by quenching from one temperature to
another, both within the solid state, include:

   (i)   Restriction of the initial state to what happens to be
          stable in the solid state at a given temperature and
          composition.
   (ii)   A cooling rate normally not exceeding 10^5K/s (1) be-
          cause of the problem of achieving sudden close contact
          with an efficient heat sink.
   (iii) Restricted possibilities for deliberately changing the
          shape and dimensions of the specimen during the quench.

The crucial steps in extending the potential of quenching to include
a liquid initial state were taken by Duwez, et al.[2,3] They reported
in 1960 that complete nonequilibrium solid solubility and new non-
equilibrium solid phases (one [3] of them non-crystalline) could be

1

made in certain alloy systems by quenching the melt at cooling rates estimated (e.g., Ref. 4) to exceed 10^6K/s. The gun/inclined substrate arrangement first described by Duwez and Willens[5] has established itself as the basic splat cooling technique as well as being capable of the highest cooling rates, estimated (e.g., Refs. 6,7) to reach 10^{10}K/s in the thinnest areas.

The present selective survey is limited to what is known about the mechanisms of quenching from the melt, the effect of process variables and the status of monitoring and control of the process. A more comprehensive review[8] and bibliography[9] can be consulted for further references to work published before 1972.

2.0 CHARACTERISTICS OF METHODS OF QUENCHING FROM THE MELT

The earlier review classified methods of forming and quenching a thin element of liquid in terms of whether a double, single, or no separate heat sink was used. An alternative approach[10] is to group according to how the _melt_ finds itself in contact with the heat sink, i.e., by _spraying_, _injection_, or _discharge_.

2.1 Spraying

Metal deposition by spraying liquid from a heated source has been in use for more than 50 years for protective coating, hardfacing, or reclamation of worn machine parts (see, e.g., Ref. 11). Duwez' _gun_ technique of splat cooling achieved production of new nonequilibrium effects by (a) limiting the metal charge to $\sim$0.1g and (b) expelling it with a _shock-wave_. This atomizes the molten charge[12] forming the resulting flake-like splat by impact, spreading and coalescence of individual atomized droplets. Since the distributions of droplet size, position and velocity in the spray are not even, typical gun splats are individually nonuniform in the following respects:

    (i)   In _thickness_, varying from up to $\sim$100μm in central areas to less than 0.1μm, particularly near extremities.

    (ii) In _thermal contact_ with the substrate, sometimes

2

approaching ideality on initial impact but de-
clining subsequently due to entrapment of ambient gas, overlapping of adjacent drops and
action of thermal stresses leading to incorporation of porosity and surface films and formation of 'lift-off' areas. [13]

(iii) In thermal history, varying from single-stage
freezing and cooling of single or coalesced
impacted droplets to multistage freezing resulting from solidification being interrupted
or succeeded by arrival and freezing of further material. Examples of the heavily-layered
structure resulting when good thermal contact
is combined with a low enough deposition rate
are shown in Fig. 1 for the gun technique [14]
and for spray rolling [15] (see below).

(iv) In growth direction, varying from essentially
normal to the substrate in areas of better
contact and higher nucleation rate to radial
and parallel to the substrate elsewhere.

(v) In macrostructure, from fully columnar in thin
areas to partially equiaxed and from relatively sound in single-stage areas to significant
levels of included porosity and non-metallics
in multistage areas.

(vi) In microstructure, for example, from featureless single-phase solid solution to two-phase
cellular or dendritic areas of spacing variable between 0.01 and 0.3μm within single Al-
Fe alloy gun splats. [14,16]

These features are useful in allowing a continuous range of solidification structures to be identified for a given alloy (e.g., Ref.
7) but can be a drawback when controlled and uniform structures
are required for subsequent studies. The widespread use of the
technique is attributable to its simplicity and its capability of
producing the highest estimated cooling rates of 10^{10}K/s referred to
earlier. The only major modifications to have been introduced are to
alloys melting of refractory allows and materials and to control atmospheres (see references in Ref. 8).

Continuous or batch versions employ fluid atomization (e.g., Ref.
15), centrifugal atomisation (e.g., Ref. 17) or plasma jet or arc
spraying (e.g., Ref. 18) instead of a shock tube and normally require relative motion between source and substrate. Macrostructures

3

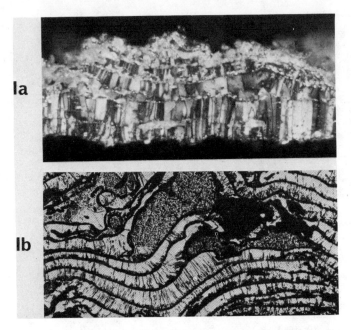

la

lb

Figure 1. Section macrostructures of (a) aluminium gun splat x 5300[14] and (b) duralumin
spray deposit x 350[15]

(Fig. 1), microstructures[19] and properties[20] equivalent to those of
gun splats can be obtained. Cooling rates are by implication considerably higher than for free-fall atomisation. Lebo and Grant,[21] ,
for example, reported dendrite spacings of 0.7μm for splat-cooled
flakes of 2024 aluminium alloy compared with 7μm for atomised powder[22] and solid solubility extension, as indicated by smaller lattice
parameter, was enhanced more by splat cooling.[23]   These differences
were manifest also in the superior <u>mechanical</u> properties of extrusions from splat-cooled compared with atomised material as measured
at room temperature for 2024[23] and at elevated temperature for
Al-8wt%Fe.[20]   It is also notable that techniques such as <u>spray roll-</u>
<u>ing</u>[15] were developed primarily with a view to improved production
economics for <u>conventional</u> tonnage products, e.g., thin strip. This

4

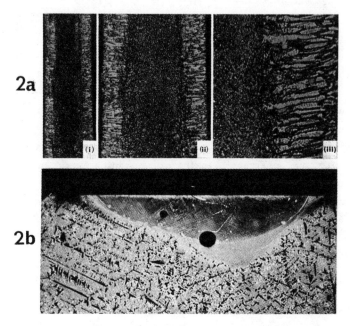

Figure 2. Section macrostructures of Ag-Cu alloy
(a) piston-and-anvil splat (i) x180,
(ii) x 390, (iii) x1060[28]
(b) laser-melted zone x640 [43]

capability as a more direct route to an established product is quite apart from their ability to generate desirable microstructures which then become a bonus allowing improved properties to be developed as well as process economy.

2.2 Injection

This feature characterizes methods in which the melt contacts the substrate in <u>undivided</u> form. They can be regarded as developments of the injection chill mould technique used industrially for sampling liquid metals and employed by, for example, Falkenhagen and Hoffmann[24] to generate nonequilibrium freezing effects in a number of alloys. A variety of <u>piston-and-anvil</u>, <u>two-piston</u> and <u>hammer-and-anvil</u> tech-

5

niques have been developed (see Refs. 8 and 9 for references) to make flat specimens 5 to 300μm thick at measured cooling rates of 10^4 to 10^6 K/s (e.g., Refs. 25-27). Dendrite spacings are typically 0.5 to 3μm compared with 0.01 to 0.5 m for the gun technique. Although cooling rates cannot match the highest attainable with gun splats, generally improved uniformity of specimen thickness, macrostructure (Fig. 2a) and microstructure[25-28] are distinct advantages for property studies. Non-uniformities that do arise are attributable to effects such as prior contact with one substrate allowing significant penetration of freezing before contact is made with the other one.[25,26] Pond and Maddin[29] showed that lengths of ribbon could be made by injecting liquid metal onto the inside surface of a rotating cup translating along its axis of rotation and lengths of sheet have been made in twin-roll devices.[30-32] Fully continuous versions include melt drag, melt extrusion, melt spinning[33,34] and melt extraction.[35] Melt drag employs a rotating drum or moving belt to drag molten metal through an orifice at which it has formed a stable meniscus. Cooling rates as high as 10^4K/s have been estimated[33] for tape as thin as 0.2mm made at up to 2m/s. Melt extrusion involves pressure-expulsion of a strand of melt through an orifice into a cooling fluid under conditions that ensure progress of solidification faster than instabilities can develop to break-up the jet. Thorne[36] has produced stable jets of tin 1m long and 0.1mm diameter by this method and the alternative Taylor process (e.g., Ref. 37) allows feedstock sheathed in glass to be drawn to even finer sizes through a heating coil. Estimates of cooling rate have not been traced for these methods, but absence of heat extraction by a solid substrate must be a limitation in that respect. Melt spinning involves melt extrusion onto a rotating substrate such as a cooled disc that shapes and solidifies the jet into ribbons 2 to 30μm thick at 4 to 40m/s[34] and at cooling rates estimated[38] at $\sim 2 \times 10^6$K/s for 25μm thick ribbons of tin. Melt extraction occurs by contact between the periphery of a rotating disc and the surface of melt either enclosed in a crucible or pendant at the end of otherwise solid rod feedstock. Section sizes as small as 25μm can be cast at cooling rates

6

estimated[35] as exceeding 10^6K/s. Improved tensile properties at room
temperature[39] as well as superplastic behaviour[40] have been reported
for extrusions of melt-spun commercial aluminium alloys and production of amorphous solid alloy by melt extraction has been claimed.[35]
Again it is notable that these methods have a wider potential for
more economical manufacture of products presently made by multistage
reduction from ingot or billet, or which are too brittle for that
route.

2.3 Discharge

These methods employ some form of intense energy discharge to rapidly
melt an area of the surface of a block of feedstock material, to be
followed immediately by rapid freezing when the discharge stops or has
moved onto a neighbouring area. Capacitance discharge, laser and electron beam sources (see Refs. 8,9) as well as explosive welding[41] have
been successfully employed to produce nonequilibrium effects associated with splat cooling. Heat flow theory and measurements[14] indicate
that cooling rates in excess of 10^6K/s can be achieved in melted zones
$\sim$0.1mm deep. The energy density applied and pulse duration are critical to achieve adequate penetration of melting while minimizing vaporization   (drilling). Jones[14] employing capacitance discharge and
Elliott, et al.[42,43] using a laser have shown that 2 to 4 kW/mm^2 for 4
to 10ms produces melted zones $\sim$0.3mm deep in Al-Fe and Ag-Cu alloys
respectively with negligible drilling. The dendrite arm spacing of
0.35μm obtained for 2024 aluminium alloy[43] is characteristic of gun
splats much thinner than 0.3mm (except where they have frozen in successive layers), reflecting the ideal contact between melt and substrate assured in discharge methods. Melted zones produced without
drilling characteristically contain spherical pores as large as 0.1mm
in diameter (Fig. 2b). The mechanism by which these defects grow to
such a size in a few milliseconds has yet to be defined. Possibilities
include trapped gas and metal vapour. Elliott, et al.[43] attribute
banding they observe in fusion zones to energy spikes $\sim$2μs duration
in the 10ms laser pulse they employed. A similar effect occurs, on a

7

larger scale, in conventional weld fusion zones (e.g., see Reference 44).

Lux and Hiller[45] showed how the entire surface of a block of material can be progressively melted and rapidly resolidified by traversing an electron beam. More conventional welding equipment has been employed in a number of investigations to produce continuous lengths of rapidly solidified material with properties markedly superior to the conventionally processed parent material (see, e.g., Ref. 44). Such methods can be regarded as a form of surface treatment without deposition of new material or change of composition, analagous, for example, to induction hardening of steels. Alternatively, continuous fibres could be produced by following the translating heat source with a cutting tool to separate newly solidified material from the parent material.

3.0 THEORY OF THE EFFECT OF PROCESS VARIABLES

The central feature in splat cooling and associated techniques is the ability to achieve a high cooling rate $\dot{T} > 10^6$K/s which may promote sufficient supercooling ΔT_- ($\sim$100K) to allow formation of new non-equilibrium phases observed and produce a high enough solidification front velocity $\dot{x}$ ($\sim$10m/s) in certain cases to exceed the expected rate of crystal growth and form metallic glasses (e.g., Ref. 46). All three parameters can be determined in principle from cooling curves[24,25] but records of these tend to be incomplete at the highest cooling rates (§5.1). Estimates from microstructure (§5.3,§5.4) are secondary standards at best so recourse to the predictions of heat conduction/crystal growth theory has been necessary to identify key variables.

3.1 Models of cooling and freezing

Both spray and injection methods involve cooling by sudden contact between melt and substrate. Analytical[12,25,47] and numerical[6,46] calculations assume one-dimensional parallel heat-flow normal to the substrate and a planar solidification front, although solidification can be locally radial and parallel to the substrate, particularly in 'lift-off' areas,

8

TABLE. Estimates of Cooling Rate from Microstructure for
Electron-Transparent Areas of Gun Splats

Cooling rate K/s	Method	Alloy and Conditions	Reference
10^{10}	Dendrite spacing (0.025µm)	Al–Ge on Cu	Ramachandrarao, et al.[7]
5×10^9	Eutectic spacing ((0.01µm)	Al–Cu on Cu in Ar	Davies and Hull[48]
2×10^8	Dendrite spacing (0.1µm)	Al alloys on Cu	Matyja, et al.[49]
10^8*	Dendrite spacing (0.14µm)	Al–Ag on Cu	Roberge and Herman[50]
3×10^6	Width of zones denuded of vacancy loops (0.1µm)	Al on Cu	McComb, et al.[51]
2×10^6	Dendrite spacing (0.53µm)	Al–Ge on Cu	Suryanarayana and Anantharaman[52]

*presumed misprint corrected.

and is frequently dendritic rather than planar. Of ten experimental variables investigated, Ruhl[6] concluded that splat thickness z and heat transfer coefficient h at the interface with the substrate would be the most significant in practice in affecting average cooling rate, predicted (e.g., in Refs. 8,47) to vary between inverse parabolic to inverse linear with thickness between the extremes of ideal contact and Newtonian conditions. Available direct measurements for a range of thicknesses between 1µm and 1mm agree to within an order of magnitude with predictions for h equal to 0.1 W/mm^{2}K.[8] Much larger apparent variations in h seem to be applicable to estimates of cooling rate from microstructure, particularly (see Table) for electron-transparent areas.

The range of these variations, equivalent to as much as four ordered in magnitude in h, is too large to be attributable to thickness variations below 0.01μm or failings in the method of estimation and so presumably does reflect real differences in thermal contact.

While cooling rate is relatively insensitive to the level of supercooling to which it applies,[6] rate of crystal growth is strongly dependent on ΔT_-. (see, e.g., Ref. 53) and may achieve some independence of rate of heat extraction by rejecting latent heat into the supercooled liquid (recalescence). Shingu and Ozaki[46] predicted the level of undercooling achieved in quenching by equating the solidification front velocity $\dot{x}$ that could be sustained by external heat extraction to the rate of crystal growth as governed by a simple atomistic law. For one example, they showed that R never reached $\dot{x}$ at any undercooling for splats thinner than 0.5μm so that a glass would then be formed. The magnitude of h required by their calculations to achieve undercoolings of hundreds K believed to be necessary to form certain metal glasses and other nonequilibrium phases, is, however, greater than $100W/mm^2K$. This value is substantially larger than the highest estimates of $\sim 8W/mm^2K$[48,54] for splat cooling, and Shingu and Ozaki concede that even a trace of adsorbed gas on the substrate would prevent this level of h from being achieved. Some possible explanations for this discrepancy are:

(i) Dendritic solidification would produce more rapid penetration of the melt than planar growth for the same conditions of heat extraction.

(ii) The crystal growth law employed is but one of a number[53] that could govern atomic attachment and additional, and generally larger, contributions to ΔT_- arise for dendritic growth to support interface curvature and solute partitioning.

(iii) Any role of nucleation is not considered.

3.2 Factors governing h and z

Heat transfer coefficient h is undoubtably the most elusive variable in quenching from the melt. Observations on surface topography and microstructure of chill cast and splat-cooled specimens suggest that good contact is highly localized with intervening lift-off areas[8,55,56].

10

Conduction through films is likely to control h at all points but the controlling film may range from adsorbed gas or thin oxide film at points of good contact to a layer of trapped gas. The average h of $0.1W/mm^2K$ characteristic of cooling rate measurements in splat cooling is equivalent to that given by conduction through a layer of air $0.5\mu m$ thick. This would represent an average value for areas of good and bad contact and seems reasonable for splats $\sim 10\mu m$ thick. Prates and Biloni[55] showed that the spacing between predendritic nuclei at the surface of a chill cast Al-Cu alloy was identical to the spacing of roughness peaks on the mould surface. Casting fluidity measurements of h were shown to be proportional to $\sqrt{N}$,* where N is the number of predendritic nuclei per unit area. This relationship predicts h ~ 0.01 and 1 W/mm^2K respectively for the predendrite spacings ~ 80 and $0.4\mu m$ observed for thicker[57] and electron-transparent[7] areas of Al-Ge alloy gun splats. Although these values of h are within the scatter of values derived from cooling rate measurements and estimates, the general applicability of this approach to estimating h remains to be explored.

Splat thickness z can be preselected with some precision only in a limited number of techniques that do not normally extend to thicknesses below $\sim 50\mu m$ (§4.3). In other cases, terminal thickness may be controlled by conversion of impact energy into spreading of liquid against viscous and capillary restraints, by advancing solidification or by alignment and flatness of two substrates.[8] For the gun technique, Davies and Hull[48] have predicted from impact energy that smaller droplets, diameter $\sim 0.3\mu m$, and larger ones, diameter $\sim 3\mu m$ of Al-Cu alloy should spread to limiting thicknesses ~ 0.04 and $0.3\mu m$ respectively, considered to be in reasonable agreement with observations. Blétry[58], on the other hand, reported good agreement between measured splat thicknesses of ~ 30 and $>5\mu m$ for an aluminium alloy using copper and steel pistons respectively and predictions based on thickness limitation by advancing solidification. The agreement he reported on the

---

* $h/\sqrt{N} \sim 0.77$ W/mK for Al-5wt%Cu alloy chill cast into a copper mould.

same basis for 1μm thick gun splats obtained by Predecki et al.[12] is
subject to the assumption of an initial drop size of 5μm found only
for bismuth of the metals studied. The marked effect of substrate material in the two-piston apparatus is, however, not readily explica-
ble on the basis of energetics. Measurements of the effect of increasing melt superheat and impact velocity in decreasing splat thickness
in a hammer-and-anvil apparatus were also interpreted in terms of
limitation by solidification but assuming that contact was imperfect
rather than ideal[54] . Limitation of thickness by energetic factors is
in fact normally competitive with limitation by solidification only in
spray techniques where impact velocities can be much higher than solidification front velocities and kinetic energy is confined to the relatively small mass of incident droplets.

4.0 RESULTS OF CHANGES IN PROCESS VARIABLES

A feature of splat cooling is that thermal contact and sample thickness
expected to have the main effect on cooling rate, are not normally under _direct_ control. Variables under more direct control, such as melt
superheat and substrate material, however, have greater effect than
expected simply from heat flow models, due to their important _indirect_
effects on, for example, degree of spreading on the substrate and extent of thermal contact with it.[6] The present approach will be initially in terms fo these variables under more direct control, prior to
considering effects of thickness and thermal contact.

4.1 Variables Associated with the Melt

Of the prime variables melt _identity_, _mass_, and _temperature_ only the
latter has been widely studied.

Cooling rate measurements indicate that aluminium cooled up to an
order of magnitude more slowly than silver in a gun technique[12] and
than lead in a piston-and--anvil.[25] This difference is mainly attributable to the lower specific heat of lead in the latter case and to a
higher heat transfer coefficient for silver in the gun, possibly be-

12

cause of less interference from oxide films for silver. The effect of
alloying lead with tin was small in the piston-and-anvil but it is not
clear whether a factor of a hundred decrease in cooling rate for
Au-14at%Sb compared with silver for the gun technique is attributable
to alloying or not. Predel and Schluckebier[59] reported a minimum splat
thickness at the eutectic composition in Ag-Cu alloys but any effect
on cooling rate was not recorded.

It is normal to establish an optimum <u>charge</u> and operating condition
for a given alloy in given apparatus, usually within the range 0.1 to
1g for noncontinuous equipment. In the gun technique the minimum
charge ∿20mg will maximise the proportion of electron-transparent areas
while charges of nearer 200mg will ensure continuity. Increased shock
pressure has been reported to displace equilibrium by nonequilibrium
phases in Al-Ge[60] and decreased orifice size would be expected to
form thinner layers on the substrate by increasing atomization.[60] Specific momentum can also be adjusted by changing the combination of
driver and driven gases.[61]

As for other casting processes, a minimum superheat ∿50K is usually
necessary to make satisfactory specimens.[61] Studies of the effects of
superheat above such thresholds have been reported. Russian workers[62-66]
in particular, reported increased terminal solid solubility extension
as indicated both by lattice parameter (Fig. 3) and physical property

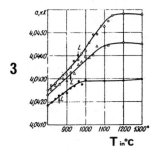

measurements, with increasing superheat.
Shingu et al.,[67] on the other hand, found
no detectable effect of superheat on solubility extension for Al-Si alloys. Again,
others[60, 68-69] have reported the formation
of increasingly metastable phases with
increasing superheat, but Scott[70] was unable to confirm such an effect reported
earlier[60] for Au-Ge alloys. Reports of
effects on <u>morphology</u> as opposed to <u>con-
stitution</u> are fewer. Suryanarayana[71] reported increased truncation of pyramidal

3

Figure 3
Lattice parameter 'a' as a
function of melt-tempera-
ture 'T' for Al-Zr alloys[64]
● 0.8   Δ 1.6   o 2.2%Zr

13

growth of a metastable phase in Al-Ge alloys with increasing superheat and Scott[72] found increased amounts of degenerate as opposed to
lamellar eutectic in Al-Cu eutectic alloy. Scott also found that the
effects of increasing superheat in increasing solidification front
velocity, as indicated by decreased eutectic spacing, and in decreasing splat thickness,[54] were small in his experiments. Likewise, some
workers[73-74] found increased quenched-in vacancy concentration with increasing superheat while others[51,75] could not detect any effect. Thus,
there is a body of evidence that superheat can affect constitution
and microstructure but the extent to which this is attributable to its
effects on, for example, splat thickness and thermal contact, is not
at all clear. Prior supercooling of an Fe-Ni alloy has been shown[26]
to partially replace the rold-like dendritic structure typical of
superheating by a highly segregated structure typical of slower solidification following recalescence.

4.2 Variables Associated with the Substrate

Major factors are substrate environment and surface condition as well
as substrate identity and temperature.

Effects of quenching atmosphere have been reported by Jansen[61] and
by Davies and Hull.[48] Jansen found that aluminium alloy gun splats
made in air contained a range of two-phase microstructures while those
made in argon were uniformly single-phased up to the limit of solid
solubility extension. Davies and Hull reported that nominal increases
in oxygen partial pressure in argon systematically decreased the proportion of noncrystalline areas in Al-Cu alloy gun splats. Similar effects to those of Jansen were also obtained[14] by grit-blasting the substrate to improve thermal contact, also producing decreases in Al-Cu
eutectic spacing by a factor of four equivalent to at least an order
of magnitude increase in h and cooling rate.[76] Shingu et al.[77] reported using a rubber flap on the substrate until the moment of splatting
in an attempt to deal with adsorption of residual gases at liquid nitrogen temperature, though it is not clear to what extent the formation of the glassy Fe-C alloy obtained is dependent on this measure.

14

General experience records little difference in effectiveness between the various high conductivity metals as quenching substrates.[6] Little difference was found even between copper and stainless steel substrates in one study,[60] while (§3.2) Blétry[58] obtained thinner splats in his two-piston apparatus when alloy steel rather than copper pistons were used. An earlier study,[25] also referred to in §2.2, found that piston material only affected splat thickness when one piston contacts the substrate before the other one. Diamond at liquid nitrogen temperature, at which it has 20 times the conductivity of copper at room temperature, was found to be more effective both in supersaturating Al-Cu and in forming noncrystalline Al-Ge alloys.[78] While Burov and Yakunin[79] showed that glass was less effective than copper as a quenching substrate in their experiments, as expected on conductivity grounds, Scott,[80] in contrast, found the reverse effect attributable to the more intimate thermal contact afforded by surface melting of the glass by the incident droplets. The explanation is that the expected effectiveness of a high conductivity substrate can be realised only if thermal contact and other conditions are at least equal to those of lower conductivity materials. Thus, glass may be more effective than copper if thermal contact with the copper is poor and diamond at liquid nitrogen temperature could be less effective than copper at room temperature if condensed films impair the effectiveness of the diamond. The deliberate use of insulating inserts to obtain thinner splats and controlled cooling rates has been reported.[25, 81] Substrate temperature as such is expected to have little effect[6] provided it is below the critical temperatures both for formation of the nonequilibirum effect (e.g., below the glass transition temperature T_g) and for its subsequent decomposition (below the recrystallization temperature T_x).

4.3 Effects of Splat Thickness

A range of splat thickness can be obtained by <u>presetting</u>, <u>selection</u>, or by employing <u>different techniques</u>.

Presetting of thickness has been achieved by injection into a wedge-shaped[82-86], stepped[87], or variable section[88] mould cavity. Results, for

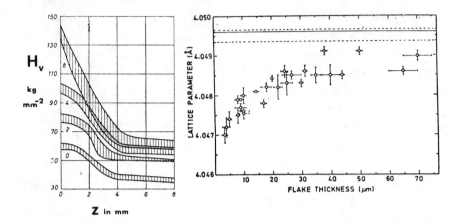

Figure 4. Hardness H_V as a function of wedge section thickness z for Al-Si-5wt%Cr alloy.[82] Figures by curves indicate wt%Si.

Figure 5. Lattice parameter as a function of splat thickness for aluminium.[96]

thicknesses down to $\sim$200μm, include:

(i) Systematic changes with wedge thickness of hardness (Fig. 4) and of diffracted x-ray intensity for Al-Si-Cr alloys.[82]

(ii) Critical thickness, and composition-dependence of coupled eutectic growth in Al-Cu alloys.[83]

(iii) Decreased thickness increasing the proportion of αAl dendritic growth for Al-Si eutectic[88] and increasing supercooling for Al-4wt%Mn.[84]

(iv) Critical cooling rates for solid solubility extension as a function of super-saturation for Al-Mn, -Cr, and -Zr alloys.[85]

(v) Increased hardness with increasing cooling rate for Al and Al-Fe alloys.[86]

Selection of thicknesses from the range produced by a single technique has yielded the following effects:

(i) Although variable thermal contact ensures that thinner specimens are not inevitably the most rapidly quenched[21,89,90], it is commonly observed that particular non-equilibrium effects do not occur above a critical thickness, typically 10-20μm[91~94], although values do

16

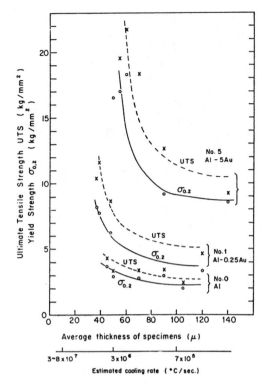

Figure 6. Strength as a function of splat thickness for aluminium and Al-Au alloys.[99]

range more widely (e.g., from 0.1μm to 2mm[95] for certain metallic glasses).

(ii) This is reflected in the effect of <u>position</u> within the area of a splat on constitution and microstructure leading, for example, to construction of nonequilibrium constitution diagrams.[7]

(iii) Decreasing <u>lattice parameter</u> with decreasing thickness (Fig. 5) for aluminium splats made by a piston-and-anvil technique.[97] Such a reduction in lattice parameter on splat-cooling has also been reported for cadmium[98] but Suryanarayana was unable to confirm it for aluminium[75] and Agarwal et al.[98] reported that lattice parameter was independent of thickness in a given splay of Al-Zn alloy. This apparent conflict may relate to the question of whether or not vacancy concentration can be increased by quenching the melt (§4.1) as well as to the variability or otherwise of thermal contact.

(iv) Increased <u>microhardness</u>[92-93],[99] and (Fig. 6)
      <u>strength</u>[99] with decreasing thickness for Al
      alloys, attributable to increasingly effec-
      tive solid solution or dispersion hardening.
      Any effect in pure Al, due, for example, to
      decreased grain size, was not detectable ac-
      cording to Tonejc et al.,[100] although Toda and
      Maddin[99] detected some effect on strength
      (Fig. 6).

A <u>range of techniques</u> has been employed most frequently to obtain
a wide range of thicknesses and cooling rates. Cooling rates have
been either calculated retrospectively[101-102] from theoretical relationships or, less commonly, measured.[103] Reports of lattice parameter
indications of increased solid solubility extension with decreasing
thickness or increased cooling rate are particularly numerous,[102],[104-108]
and an increased proportion of a nonequilibrium phase with decreasing
thickness has been detected.[109] Although the danger of misjudging the
difference in cooling rate between different thicknesses is greater
when they are made by different methods, the effects reported (Fig. 7)
are nevertheless significant.

In summary, despite some conflicting findings, it has been shown
that decreasing splat thickness can lead to increased cooling rate,
supercooling, solid solubility extension, nonequilibrium phase formation, microstructural refinement and strength properties. The results
of Lebo and Grant[21] for example, illustrate that other effects can
override the effect of thickness. They obtained for 2024 aluminium
alloy splats higher cooling rate, as indicated by smaller dendrite
spacings, for <u>thicker</u> flakes, leading to enhanced fatigue life in extrusions from thicker than from thinner flake.

5.0 MONITORING THE EFFECTIVENESS OF QUENCHING

The significant effect of some process variables underlines the need
for reliable methods of assessing the efficiency of a given quench.
Available and possible methods range from direct recording of cooling
curves to measurement of physical properties and few have been widely
tested.

18

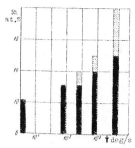

Figure 7. Extended
limit of solid solu-
bility as a function
of cooling rate for
Sn-Sb alloys

■ single-phase range

▨ β' phase also
present

Relatively complete cooling curves showing cooling of the liquid, supercooling below the equilibrium liquidus temperature, a solidification
arrest and cooling of the solid, have been recorded for injection mould,[24,85] piston-and-anvil[25,110] and two-piston[81] methods. Methods of
measurement include optical pyrometry[110] or a
thermocouple in the substrate[25] or in the melt.[110]
Cooling rates as high as 10^6K/s were recorded
for the liquid decreasing by as much as an order
of magnitude for the solid.[25] Supercoolings as
large as 330K have been recorded[81] and solidification arrests as short as 0.25ms,[25] the latter indicating an average solidification velocity $\dot{x}$ of 200mm/s for a piston-and-anvil splat
88μm thick. Other estimates of $\dot{x}$[76,112] for comparable specimens are an
order of magnitude lower.

Cooling curves reported for the gun technique[12,67] and attributed to
impact of individual droplets are less complete and not so reproducible.
This is attributable to limitations in response time of the thermocouple detector, for the cooling rates as high as 10^8K/s recorded, and
the effect of neighbouring impact events not coincident in time. Thus
only the lower part of the cooling curve, more representative of cooling of the solid, is recorded, and cooling rates of the liquid could
be an order of magnitude higher.

5.2 Splat Dimensions

Some of the earliest estimates of cooling rate and solidification velocity were from the length of slivers[112] or the contact time with a
moving piston[112-114] or with rotating rolls[30,112,115]. An upper limit to the
average cooling rate for ideal cooling can be calculated for a known
thickness z from calculated values tabulated by Ruhl[6] or analytically.[47]

An estimate of the actual cooling rate requires a value for h as well as z. Estimates of h for splat cooling cover a wide range (§3.1) and calculation of h from contact topography (3.2) has yet to be developed. Assumption of the mean value of h of 0.1 W/mm^2K for splat cooling (3.2) could lead to appreciable errors for particular cases.

5.3 Scale of Microstructure

Direct measurements of <u>microstructural</u> parameters such as dendrite arm spacing or GP zone radius characterize <u>local</u> conditions of cooling and are most useful for monitoring variations with <u>position</u> in a splat.

Matyja et al.[49] first suggested that the power correlation between <u>dendrite arm spacing</u> and cooling rate established for aluminium alloys at cooling rates below $10^3K/s$ could be used to estimate cooling rate in splat cooling. Similar correlations are now available for Cu[16] and Fe[110,117] base alloys. Confirmation has been obtained that the power relation changes systematically with alloy composition[103] and that the exponent changes from nearly 1/3 for secondary arms to nearly 1/2 for primary arms.[118] The effect, if any, on the power correlation of the tendency to suppress secondary arms at high cooling rates has yet to be established.

The corresponding power relation between <u>eutectic interphase</u> spacing and solidification velocity $\dot{x}$, first used[76] to estimate $\dot{x}$ and hence $\dot{T}$ for Al-Cu eutectic splats has yet to be applied to splats of other alloys, although the necessary steady state data are available for $\dot{x}$ up to 2mm/s (e.g., Ref. 119). The further relation between spacing and growth temperature (e.g., Ref. 120) allows corresponding values of undercooling to be derived. For example, the estimated undercooling corresponding to the eutectic spacing of 8nm measured by Davies and Hull[48] is 30K (for $\dot{x} \sim 2m/s$ and $\dot{T} \sim 10^{10}K/s$). The undercooling in neighbouring regions would have needed to be some ten times larger however to form the glassy phase observed there.

Other microstructural spacings have had less attention. The <u>width w of grain boundary zones denuded of vacancy loops</u> was employed by McComb, et al.[51] to estimate cooling rate for aluminium gun splats.

20

His estimate (see Table) was based on the normal width of $\sim 1\mu m$ for solid state quenching at 3×10^4K/s and the assumption that vacancy sinking was diffusion-controlled so that w should be proportional to $\dot{T}^{-1/2}$. Dislocation array size χ in Fe-Cr-Ni alloy splats was shown by Wood and Honeycombe[13] to be within a factor of three of that predicted on the theoretical basis[121] that $\chi \propto \dot{T}^{-1/2}$. Possibly GP zone radius, as determined for example by electron microscopy or small angle x-ray scattering, also might be used to estimate cooling rate. It has already been employed qualitatively as an indicator of relative efficiencies of quenching[50,90-92,122]. On a coarser scale, systematic changes of grain size with cooling rate have been reported.[123] This could be useful for estimating cooling rates of pure metals, for example, for which many of the above methods are not applicable.

5.4 Nonequilibrium Constitution

The earliest proving tests for splat cooling equipment involved demonstration of the formation of standard nonequilibrium phases (e.g., β-phase in Ag-26at%Ge), implying that some critical cooling rate had been exceeded. Recent extensions of this approach (e.g., Ref. 95) employ critical cooling rates calculated from phase transformation theory for various glassy metal phases observed in splats. The possibility[124] of employing degree of terminal solid solubility extension, as for example indicated by lattice parameter determinations, has the greater attraction that the actual magnitude of the cooling rate could be obtained, at least in principle. The amount of a nonequilibrium second phase[70] or of degenerate eutectic[72] might be used in the same way, although, in both cases, the possibility of variations in constitution with position in splats would have to be taken into account especially in interpretating x-ray diffraction data.

5.5 Physical Properties

In addition to lattice parameter, measurements of properties such as electrical resistivity, thermoelectric power, magnetic susceptibility

and hardness are all possible means of indicating effectiveness of
quenching. Relationships have been reported between lattice parameter
and residual resistivity,[125] between lattice parameter and hardness,[126]
between superconducting transition temperature T_c and residual resistivity[127] and between temperature T_K of minimum resistivity ratio,[128]
all for single compositions, presumably formed at different cooling
rates. Direct relationships between T_c and cooling rate[129] as well as
between hardness and cooling rate (§4.3) have also been reported. Related possibilities include the determination of T_x and T_g for glassy
phases and corresponding thermal responses[92,130] of nonequilibrium
crystalline phases. Such determinations should at least be a guide to
the equivalence or otherwise of nominally identical products of
quenching made in different laboratories or by different methods.

6.0   PROSPECTS FOR CONTROL OF RAPID QUENCHING FROM THE MELT

The ultimate aims of attempting to understand how methods of rapid
quenching from the melt actually work are to obtain better control and
reproducibility of results from existing methods and to develop superior methods. Unpredictable process conditions for a given alloy as well
as inadequate knowledge of its physical metallurgy at present prevent
reliable forecasting of what will result from quenching it in given
equipment. Progress has been made, however, in identifying the factors
that must be controlled to achieve, for example, the highest cooling
rates or minimization of macrodefects. Reduced thickness z is not sufficient in itself to guarantee higher cooling rates (§3.1) because
thermal contact becomes increasingly important as z decreases. Special
attention to minimizing oxidation during melting and to preparation
and maintenance of a clean substrate are obvious examples of steps
that may be necessary. Such measures, however, would be redundant, for
example, for thicker sections of poorly conducting specimen materials,
for which diffusion of heat through the specimen may control the overall rate of heat flow. While the most fragile gun splats obtained by
minimizing the charge of melt, intensifying the shock wave and reducing the orifice diameter, may give the highest cooling rates and pro-

22

vide the most extensive areas thin enough for electron microscopy without thinning, they are also the most discontinuous. Although consolidation is a practical possibility, spray techniques are at a disadvantage when shape control and uniformity is more important than achieving the highest cooling rates. Methods of making filaments from the melt (§2.2), for example, may achieve a degree of thickness control by adjustments in the rate of feed and rate of substrate rotation but detailed studies have yet to be reported. The possibilities of spray techniques are by no means exhausted, however, as promised by the commercial installation[131-132] of large-scale inert gas atomization equipment for making alloy powder, readily adaptable for making splat-cooled flake, for subsequent compaction into engineering components for high-duty applications. Commercial production by spray rolling or by melt-spinning/extraction, primarily to achieve more economical manufacture of existing products, in a similar way would allow the additional benefits of rapid quenching to be realized if they are of insufficient justification of investment in their own right.

REFERENCES

1. Duwez, P., in 'Techniques of Metals Research', $\underline{1}$(1) Ed. R.F. Bunshah, Interscience, New York (1968), p. 347.
2. Duwez, P., R. H. Willens, and W. Klement, J. Appl. Phys., $\underline{31}$, 1136, 1137, 1150 (1960).
3. Klement, W., R.H. Willens, and P. Duwez, Nature $\underline{187}$, 869 (1960).
4. Duwez, P., in Progr. in Solid State Chem. $\underline{3}$, Ed. H. Reiss, Pergamon, Oxford (1966), p. 377.
5. Duwez, P. and R.H. Willens, Trans. Met. Soc. AIME, $\underline{227}$, 362 (1963).
6. Ruhl, R.C., Mater. Sci. Eng. $\underline{1}$, 313 (1967).
7. Ramachandrarao, P., M.G. Scott, and G.A. Chadwick, Phil. Mag. $\underline{25}$, 961 (1972).
8. Jones, H., Rep. Progr. Phys. $\underline{36}$, 1425 (1973).
9. Jones, H. and C.S. Suryanarayana, J. Mater. Sci. $\underline{8}$, 705 (1973).
10. Pond, R.B., presented at Battelle Dev. Corp. Conf. (june, 1971).
11. Ballard, W.E., Metallurgical Reviews $\underline{7}$, 19 (1962).
12. Predecki, P., A.W. Mullendore, and N.J. Grant, Trans. Met. Soc. AIME $\underline{233}$, 1581 (1965).
13. Wood, J.V. and R.W.K. Honeycombe, J. Mater. Sci. $\underline{9}$, 1183 (1974).
14. Jones, H., Mater. Sci. Eng., $\underline{5}$, 1 (1969).
15. Singer, A.R.E., Metals and Materials $\underline{4}$, 246 (1970); vide: J. Inst. Met. $\underline{100}$, 185 (1972) and Light Metal Age $\underline{32}$ (9,10), 5 (1974).
16. Burden, M.H. and H. Jones, Fizika $\underline{2}$, Suppl. 2, Paper 17 (1970).

17. Jones, H. and M.H. Burden, J. Phys. E:Sci. Instrum. 4, 671 (1971).
18. Kellerer, H. and B. Looman, Metall. 22, 212 (1968).
19. Jacobs, M.H., A.G. Doggett, and M.J. Stowell, J. Mater. Sci. 9, 1631 (1974).
20. Thursfield, G. and M.J. Stowell, J. Mater. Sci. 9, 1644 (1974).
21. Lebo, M. and N.J. Grant, Met. Trans., 5, 1547 (1974).
22. Grant, N.J., 'The quenched powder approach to superior metallic materials', to be publ.
23. Grant, N.J., Fizika, 2, Suppl. 2, Paper 16 (1970).
24. Falkenhagen, G. and W. Hofmann, Z. Metallk. 43, 69 (1952).
25. Harbur, D.R., J.W. Anderson, and W.J. Maraman, Trans. Met. Soc. AIME, 245, 1055 (1969).
26. Kattamis, T.Z., W.F. Brower, and R. Mehrabian, J. Cryst. Growth, 19, 229 (1973).
27. Kattamis, T.Z., Y.V. Murty, and J.A. Reffner, ibid., 19, 237 (1973).
28. Linde, R.K., Trans. Met. Soc., AIME, 236, 58 (1966).
29. Pond, R. and R. Maddin, ibid., 245, 2475 (1969).
30. Chen, H.S. and C.E. Miller, Rev. Sci. Instrum. 41, 1237 (1970).
31. Babić, E., E. Girt, R. Krsnik, and B. Leontić, J. Phys. E: Sci. Instrum. 3, 1014 (1970).
32. Duhaj, V., V. Sladek, and P. Mrafko, Cesko Cas. Fys. A23, 617 (1973).
33. Butler, I.G., W. Kurz, G. Gillot, and B. Lux, Fibre Sci. & Technol. 5, 243 (1972).
34. Hubert, J.C., F. Mollard, and B. Lux, Z. Metallk. 64, 835 (1973).
35. Maringer, R.E., and C.E. Mobley J. Vac. Sci. Technol. 11, 1067 (1974).
36. Thorne, D.J., Fibre Sci. & Technol. 7, 79 (1974).
37. Manfre, G., G. Servi and C. Ruffino, J. Mater. Sci., 9, 74 (1974).
38. Pond, R.B. and R.B. Pond, presented at Battelle Conf., June (1971).
39. Mobley, C.E., Clauer, A.H., and B.A. Wilcox, J. Inst. Met. 100, 142 (1972).
40. Wilcox, B.A. and A.H. Clauer, Proc. Cambridge Conf. on 'Microstructure and Design of Alloys', 1, 227 (1973).
41. Dhir, P., Ph.D. Thesis, Belfast (1973).
42. Elliott, W.A., F.P. Gagliano, and G. Krauss, Appl. Phys. Lett. 21, 23 (1972).
43. Idem, Met. Trans. 4, 2031 (1973).
44. Davies, G.J. and J.G. Garland, Int'l Met. Revs, 20, 83 (1975).
45. Lux, B. and W. Hiller, Prakt. Metallogr. 8, 218 (1971).
46. Shingu, P. and R. Ozaki, Met. Trans. 6A, 33 (1975).
47. Jones, H., Mater. Sci. Eng. 5, 297 (1970).
48. Davies, H.A. and J.B. Hull, J. Mater. Sci. 9, 707 (1974); see also Scripta Met. 6, 241 (1972).
49. Matyja, H., B.C. Giessen, and N.J. Grant, J. Inst. Met. 96, 30 (1968).
50. Roberge, R. and H. Herman, J. Mater. Sci. 8, 1482 (1973).
51. McComb, J.A., S. Nenno, and Meshii, M., J.Phys. Soc. Japan, 19, 1691 (1964).
52. Suryanarayana, C. and Anantharaman TR, J. Mater. Sci. 5, 992 (1970).
53. Jackson, K.A. in Progr. in Solid State Chem. 4, Ed. H. Reiss, Pergamon, Oxford, Oxon. (1967), p. 53.
54. Williams, C.A. and Jones H., Mater. Sci. Eng. 19, 293 (1975).
55. Prates, M. and H. Biloni, Met. Trans. 3, 1501 (1972).
56. Wood, J.V. and I. Sare, submitted to Met. Trans.
57. Ramachandrarao, P. and T.R. Anantharaman, Trans. Met. Soc. AIME, 245, 890 (1969).

24

58. Blétry, J., _J. Phys. D: Appl. Phys._ 6, 256 (1973).
59. Predel, B. and G. Schluckebier, _Metallk._ 63, 782 (1972).
60. Ramachandrarao, P. and T.R. Anantharaman, _Trans. Indian Inst. Met._ 23 (2), 58 (1970).
61. Jansen, C.H., Ph.D. Thesis, Massachusetts Inst. of Tech. (1971).
62. Vengrenovich, R.D. and V. Psarev, _Russ. Met._ (1969) No. 4, p. 103, (1970) No. 5, p. 138; _Phys. Metals Metallogr._ 29 (3) 93 (1970); _Russ. J. Phys. Chem._ 44, 1119 (1970).
63. Varich, N.I. and R.B. Lyukevich, _Russ. Met._ 1970 No. 2, p. 135, 1970 No. 4, p. 58, 1973 No. 1, p. 73; _Russ. J. Phys. Chem._ 47, 592 (1973).
64. N.I. Varich, et al., _Phys. Met. Metallogr._ 27 (2), 176 (1969).
65. Varich, N.I. and A.N. Petrunina, _ibid._, 33 (2), 106 (1972).
66. Kolpachev, A.A., T.V. Vukolova, and V.D. Sharabkova, _Russ. Met._ (1972), No. 4, p. 152.
67. Shingu, P.H., K. Kobayashi, K. Shimomura, and R. Ozaki, _J. Jap. Inst. Met._ 37, 433 (1973).
68. Filonenko, V.A., _Russ. J. Phys. Chem._ 43, 874 (1969).
69. Poleysya, A.F. and V.N. Gudzenko, Izv. Ak. Nauk SSSR, _Neorg. Mat._ 10 (6) , 1011 (1974); Engl. version, p. 869.
70. Scott, M.G., _Mater. Sci. Eng._ 18, 279 (1975).
71. Suryanarayana, C., _Trans. Indian Inst. Met._ 25 (1), 36 (1972); _Metallography_, 4, 79 (1971).
72. Scott, M.G., _J. Mater. Sci._ 10, 269 (1975).
73. Thomas G. and R.H. Willens, _Acta Met._ 12, 191 (1964); 13, 139 (1965); 14, 1385 (1966).
74. Rastogi, P.K. and K. Mukherjee, _Met. Trans._ 1, 2115 (1970).
75. Suryanarayana, C., _Phys. Stat. Sol._ a18, K135 (1973).
76. Burden, M.H. and H. Jones, _J. Inst. Met._, 98, 249 (1970).
77. Shingu, P.H., K. Kobayashi, K. Shimomura, and R. Ozaki, _Scripta Met._ 8, 1317 (1974).
78. Ramachandrarao, P., P. Laridjani, & R.W. Cahn, _Z.Metallk._ 63, 43 (1972).
79. Burov, L.M. and A.A. Yakunin, _Russ. J. Phys. Chem._ 39, 1022 (1965).
80. Scott, M.G., _J. Mater. Sci._ 9, 1372 (1974).
81. Miroshnichenko, I.S. and G.P. Brekharya, _Phys. Metals Metallogr._ 29 (3), 233 (1970); See also _Industr. Lab._ 35, 362 (1969).
82. Esslinger, P., _Z. Metallkunde_ 57, 12, 109 (1966); See also _Z. Wirtsch. Fert._ 60, 449 (1965).
83. Miroschnichenko, I.S. and A. Ya. Andreeva, _Dokl. Akad. Nauk. SSSR_ 186 (5), 1142 (1969), Engl. ver., p. 128. See also _Russ. Met._ (1968), No. 5, p. 128.
84. Pasalski, V.M. A.F. Polesya, and I.I. Chalyi, _Rost. Def. Met. Krist._ (1972), p. 427.
85. Ichikawa, R., T. Ohashi, and T. Ikeda, _Trans. Jap. Inst. Met._ 12, 280 (1971).
86. Furrer, P., Proc. Cambridge Conference on 'Microstructure and Design of Alloys', 1, 227 (1973).
87. Polesya, A.F. and V.V. Kovalenko, _Russ. Met_ (1970), No. 1, p. 114.
88. Serita, Y., T. Ishikawa, and F. Kimura, _J. Jap. Inst. Light Met._ 20 (1), 1 (1970).

89. Tonejc, A., Met. Trans. 2, 437 (1971).

90. Kranjc, K. and M. Stubičar, Met. Trans., 4, 2631 (1973).

91. Tonejc, A.M., A. Tonejc, and A. Bonefačić, J. Mater. Sci. 9, 523 (1974); See also J. Appl. Phys. 40, 419 (1969) and Met. Trans. 2, 437 (1971).

92. Kranjc, K. and Stubičar, M. Fizika 2, Suppl. 2, Paper 31 (1970).

93. Fontaine, A. and A. Guinier, Phil. Mag. 31, 65 (1975). See also Compt. Rend. B271, 231 (1970) and Fizika 2, suppl. 2, paper 23 (1970).

94. Ramachandrarao, P. and M. Laridjani, J. Mater. Sci. 9, 434 (1974).

95. Davies, H.A., J. Aucote, and J.B. Hull, Scripta Met. 8, 1179 (1974).

96. Kirin, A. A. Tonejc, & A. Bonefačić, Scripta Met. 3, 943 (1969).

97. Laine, E., Phys. Stat. Sol. a13, K27 (1972); See also J. Mater. Sci. 6, 1418 (1971).

98. Agarwal, S., M.J. Koczak, & H. Herman, Scripta Met. 7, 401 (1973).

99. Toda, T. and R. Maddin, Trans. Met. Soc. AIME 245, 1045 (1969).

100. Tonejc, A., D. Ročák, and A. Bonefačić, Acta Met., 19, 311 (1971), Fizika 2 Suppl. 2, Paper 7 (1970).

101. Miroshnichenko, I.S., in 'Crystallization Processes', Ed.: N.N. Sirota, et al., Con. Bur., New York (1966), p. 61.

102. Burov, L.M. and A.A. Yakunin, Russ. J. Phys. Chem. 42, 540 (1968).

103. Miroshnickenko, I.S., A.K. Petrov, V.A. Golovko, B.P. Brekharya, and V.I. Novikov, Industr. Lab., 38, 1869 (1972).

104. Miroshnichenko, I.S. in 'Crystallization Processes', Ed. N. N. Sirota, et al., Consultants Bureau, New York (1966), p. 55; Dokl. Phys. Chem. 164, 647 (1965); in 'Growth and Imperfections of Metallic Crystals', Ed. D.E. Ovsienko, Con. Bur., New York (1968), p. 255.

105. N.I. Varich and A.A. Yakunin, Russ. Met. (1968), No. 2, p. 148.

106. Miroshnichenko, I.S. and G.A. Sergeev, in 'Mechanisms and Kinetics of Crystallization', Minsk (1969), p. 40.

107. Polesya, A.F., L.P. Slipchenko, L.M. Burov, V.N. Gudzenko, and V.I. Demeshkin, Izv. VUZ Chern. Met. (1971), No. 9, p. 114.

108. Dobatkin, V.I. and R.H. Sizova, Izv. VUZ Tsvetn. Met. (1974), No. 3, p. 117.

109. Polesya, A.F. and V.N. Gudzenko, Russ. Met (1973), No. 5, p. 152.

110. Brower, W.E., R. Strachan, and M.C. Flemings, Cast. Met. Res. J. 6, 176 (1970).

111. Miroshnichenko, I.S. and I.V. Salli, Industr. Lab. 25, 1463 (1959).

112. Salli, I.V. and L.P. Limina, ibid. 31, 141 (1965).

113. Booth, A.R. and J.A. Charles, Nature 212, 750 (1966).

114. Ohring, M. and A. Haldipur, Rev. Sci. Instrum. 42, 530 (1971).

115. Revyakin, A.V., B.D. Glyuzitsky, and A.M. Samarin, Industr. Lab. 38, 461 (1972).

116. Sarin, V.K. and N.J. Grant, Met. Trans. 3, 875 (1972).

117. Joly, P.A. and R. Mehrabian, J. Mater. Sci. 9, 1446 (1974).

118. Young, K.P. and D.H. Kirkwood, Met. Trans. 6A, 197 (1975).

119. Livingston, J.D., H.E. Cline, E.F. Koch, and R.R. Russell, Acta Met. 18, 399 (1970); See also Trans. Met. Soc. AIME 245, 1987 (1969).

120. Moore, A. and R. Elliott, in 'The Solidification of Metals', Inst. of Metals, London (1968), p. 167.
121. Frank, F.C. in 'Deformation and Flow in Solids', Springer, Berlin (1956), p. 73.
122. Kähkönen, H.A. and M.T. Yli-Penttilä, J. Appl. Cryst. 6, 412 (1973).
123. Savitsky, E.M., A.V. Revyakin, Yu.V. Efimov, B.D. Glyuzitskii, and V.N. Sumarokov, Dokl. Akad. Nauk SSSR 210 (2) 405 (1973), Engl. Vers., p. 73.
124. Vitek, J. and N.J. Grant, J. Mater. Sci. 7, 1343 (1972).
125. Babić, E., E. Girt, R. Krsnik, Leontić and I Zarić, Phys. Lett. 33A, 368 (1970); Fizika 2 Suppl. 2, Paper 30, (1970).
126. Babić, E., E. Girt, R. Krsnik, B. Leontić, M. Očko, Z. Vučić, & I. Zorić, Phys. Stat. Sol. 16, K21 (1973).
127. Babić, E., R. Krsnik, B. Leontić, and I. Zorić, Phys. Rev. B2, 3580 (1970); See also J. Phys. F: Met. Phys. 1, L18 (1971).
128. Lin, S.C.H., J. Appl. Phys. 40, 2175 (1969).
129. Bychkov, Yu.F., D.S. Kamenetskaya, V.S. Kruglov, and B.T. Sizov, Phys. Metals. Metallogr. 36 (4), 72 (1973).
130. Scott, M.G., Z. Metallkunde, 8, 563 (1974).
131. Metallurgia, 83, 63 (1971).
132. Metals and Materials, 7, 389 (1973).

27

AN EXPERIMENTAL METHOD FOR THE CASTING OF
RAPIDLY QUENCHED FILAMENTS AND FIBERS

R.E. Maringer, C.E. Mobley, and E.W. Collings
Battelle Laboratories
Columbus, Ohio

INTRODUCTION

Numerous procedures have been devised for the rapid quenching of metals and alloys from the melt. Most methods yield samples in the form of flakes or pieces of film. However, the "melt extraction" procedure[1,2] developed at Battelle*, and the "melt spinning" process of Pond[3,4] are particularly attractive because of their ability to produce, under steady state operating conditions, rapidly quenched materials in the form of continuous ribbons or filaments.

Melt spinning[3] involves the jetting of molten metal from a small orifice. On the other hand, in melt extraction[1,2] the liquid source is stationary, the rapidly quenched product being the result of the liquid surface being contacted by the edge of a rotating disc. The molten metal solidifies onto this, adheres for a short time, then spontaneously releases as a solid filament. Two variants of the melt extraction process exist—the so-called "crucible melt extraction" (CME) and "pendant drop melt extraction" (PDME). In the former, the edge of the extraction disc touches the surface of a bath of liquid metal contained in a crucible. In PDME the filament is extracted from the molten end of a rod of metal suspended just above the rotating wheel. The great advantage of PDME lies in the elimination of both the orifice and the crucible, and their attendant problems. Its use makes possible the fiberization of virtually any metal or alloy which can exist as a stable droplet.

The objective of this paper is to describe in some detail the application and use of Pendant Drop Melt Extraction for preparing rapidly solidified metals and alloys.

---

* Patents on the melt extraction processes are owned by the Battelle
  Development Corporation.

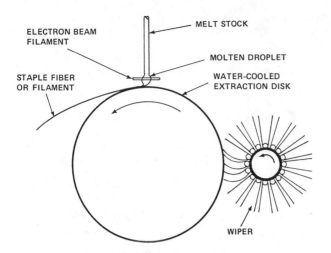

Figure 1. Schematic Diagram of Pendant Drop Melt Extraction

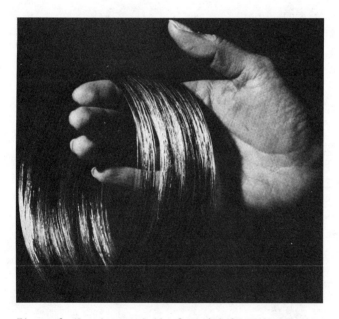

Figure 2. Continuous Coil of Ti-6Al-4V PDME Filament

30

While there are many variants to PDME, its essence is shown schematically in Fig. 1. The water-cooled disc is often cut from 0.04-inch (1mm)
thick copper or brass sheet. The periphery is ground to a point, usually with about a 90° included angle. Copper or brass appears to be adequate for the casting of fine filaments of metals with melting points
up to 2000°C or so, while for higher melting point metals, high-melt-
ing-point disc material is generally necessary. The periphery of the
disc is continuously cleaned by means of a wiper whose flaps (Fig. 1)
are made from fiberglass, cotton, plastic belting, or any other material which will keep the surface in good condition without leaving
a deposit of its own.

The tip of a sample of melt stock, usually in the form of a rod
perhaps 6- to 25-mm in diameter, is melted by an appropriate source.
The sample holder is then lowered until the resulting droplet contacts
the edge of the disc. Once fiber production has begun, the intensity
of heating and the feed rate are adjusted to stabilize a continuous
casting process. By adapting a "carousel" to the system[5], it is possible to collect the filament in a continuous coil. Figure 2 shows a
coil of Ti-6Al-4V filament with an effective diameter of about 0.08mm
and a length of 200 meters or so. This was cast in vacuum (at about
2 meters per second) in an 18-inch (46-cm) diameter bell jar using an
electron beam melting source.

Since the edge of the disc is in fact a mold, an interruption to
the edge interrupts the casting process. Thus an edge notched as in
Fig. 3 will produce staple fiber. These are shown in Fig. 4. Such fiber can be made of almost any length or diameter (L/D) ratio. For L/D
ratios of the order of 10 or less, the product handles very much like
powder. Ti-6Al-4V fiber prepared in this fashion (we call them "L/D
powders") appear to have considerable potential as a substitute for
conventional powder in some powder-metallurgical applications.[6] For
L/D ratios considerably greater than 10, the staple fibers appear to
be ideally suited for certain composite applications, in which case

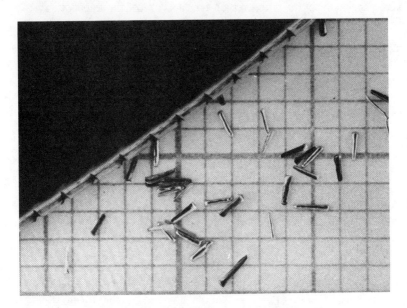

Figure 3. Extraction Disc showing Notches and Product

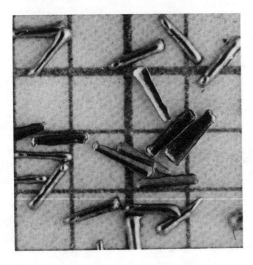

Figure 4. Staple Ti-6Al-4V Fibers against Millimeter-Square Background

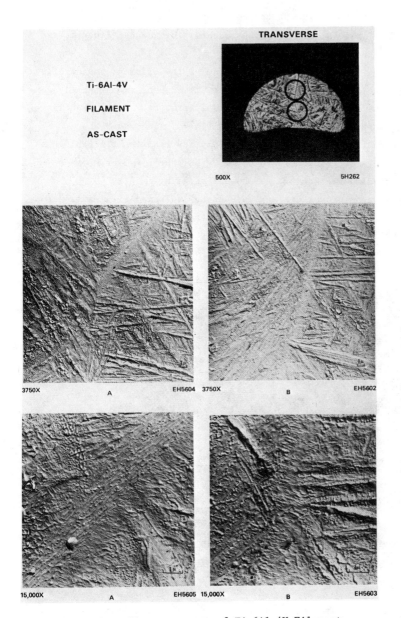

Figure 5. As-Cast Microstructure of Ti-6Al-4V Filament

33

modification of the design of the notch and of the disc edge to produce "dogbone" fibers results in a product with enhanced fiber-matrix
bonding characteristics.

Because the disc is the mold, one side of the filament essentially
replicates the disc surface. As a result, the fibers are generally not
round in cross section (Figure 5). A kidney-shaped cross section is
fairly characteristic of melt-extracted fibers, although "L" or "D"-
shaped, or flat ribbon-type cross sections are readily attainable. As
major diameters decrease below 0.05 mm or so, the filament cross sections become almost circular, as shown in Fig. 6.

In the process of solidification, the newly formed filament adheres
rather strongly to the disc edge surface and is rapidly chilled. Once
the filament spontaneously releases from the disc, the heat transfer
rate drops considerably. Thus the overall cooling of the fiber is a

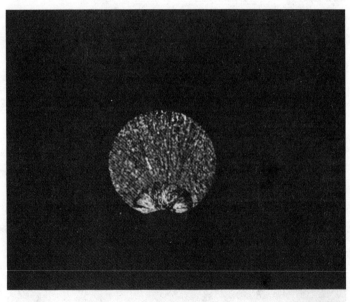

1500 X   OHO34

Figure 6. Metallographic View of PDME Filament of 304 Stainless Steel

two-stage process. Measurement of secondary dendrite arm spacings in PDME filament of 310 stainles steel (such as that shown in Fig. 6) suggests that the primary cooling rate during residence is of the order of 10^{6}°C per second. Observation of the time required for a stainless steel filament to travel from the melt to where it lost incandescence during flight in air, indicated an overall quench rate of about 10^{5}°C per second. Metallographic examination of a PDME filament of $Fe_{.80}P_{.13}C_{.07}$ suggest that it is amorphous through the primary stage of cooling but crystallizes during the secondary (free flight) stage. When secondary-stage oil quenching was introduced, the PDME fibers of this composition remained amorphous, indicating an overall quench rate in excess of 10^{5}°C per second. Some additional observations on quench rates observed in melt extraction are reported by Pond and Winter.[7]

MATERIALS INVESTIGATED BY PENDANT DROP MELT EXTRACTION, SUMMARY

A distinct advantage of PDME is the ease with which diverse metals and alloys may be fiberized. Most metals and alloys available as rod or wire can be utilized directly as melt stock. "Green" powder compacts, either prealloyed or mixed as elemental powders, may also be used. With these, alloying and homogenization occur in situ. Similarly, bundles of rods or wire can be co-melted, with alloying or homogenization occurring within the droplet. Powdered alloying agents (e.g., B, TiC, C, etc.) can be packed into a slot cut into the side of a rod of the base alloy and co-melted. We have, for example, produced Ti alloys containing Cu, Fe, Co, Ni, B, Zr, C, Si, W, and/or Cr using this procedure. Materials we have used in the pendant drop melt extraction of fiber include Bi, Elgiloy, Cr, Nb, Co, Nitinol, Udimet 700, Al_2O_3, TiAl, a Nb/ bronze composite, and a wide variety of stainless steels.

The ability of PDME to fiberize in the laboratory useful quantities of such materials as continuous filament, staple fiber, or powder-equivalent has been demonstrated. The relatively high quench rates associated with PDME coupled with the steady-state operation and the ability of the process to control to some extent the size and shape of

35

the filamentary products, encourage us to believe that the process
will be of considerable value in advancing the science of rapid
quenching and rapidly quenched materials.

ACKNOWLEDGMENTS

The authors are grateful to the Battelle Development Corporation for
their financial support of the Melt Extraction Process, to the Materials Laboratory of the Wright-Patterson Air Force Base (Dr. H.L. Gegel,
Project Monitor) for their financial support of the study of melt-extracted titanium alloys, and to Mr. Oliver Stewart for his continued
assistance in the experimental programs. Figures 2-5 and the materials
shown therein were prepared as part of the Wright-Patterson Air Force
Base-sponsored program under Contract No. F33615-74-C-5179. Figure 6
and the 310 stainless steel fiber shown therein were prepared on a
BCL research program sponsored by H.R. Filters, Pacoima, California.

REFERENCES

1. Maringer, R.E. and C.E. Mobley, "Casting of Metallic Filament and
   Fiber", J. Vac. Sci. Technology, 11, Nov/Dec (1974), pp 1067-1071.
2. The melt extraction process is covered by the following US patents:
   3,838,185   3,904,344   3,871,349   3,812,901   3,861,450.
3. Pond, R.B.,"Metallic Filaments and Method of Making Same," US Patent
   No. 2,825,108 (1958).
4. Pond, R.B., R.E. Maringer, and C.E. Mobley, "High Rate Continuous
   Casting of Metallic Fibers and Filaments". Paper presented at 1974
   ASME Pre-Congress Seminar, entitled New Trends in Materials Fabrication. Paper in press.
5. US Patent No. 3,812,901, Oliver M. Stewart, Robert E. Maringer, and
   Carroll E. Mobley, Jr., inventors. Issued May 28, 1974.
6. Maringer, R.E., C.E. Mobley, and E. Collings, "Preparation and
   Properties of Compacts of Cast (Melt Extracted) Staple Fibers of
   Ti-6Al-4V". Paper presented at 1975 Vacuum Metallurgy Conference[in press]
7. Pond, R.B., Sr., and J.M. Winter, Jr., Proc. Second International
   Conference on Rapidly Quenched Metals, 82, N.J. Grant and B.C.
   Giessen, Eds., Mat. Sci. Eng. 23 (1976) [in print].

36

PREPARATION AND CHARACTERIZATION OF THICK METASTABLE SPUTTER DEPOSITS*

R.P. Allen, S.D. Dahlgren, and M.D. Merz

Battelle Memorial Institute
Pacific Northwest Laboratories
Richland, Washington 99352

INTRODUCTION

Sputtering is a nonevaporative process that uses high energy ion bombardment to detach atoms from the surface of a sputtering target. The sputtered atoms are collected on a substrate to form a deposit that represents a literal atom-by-atom reconstruction of the target material. Thus, although the deposit may have the same composition as the target, the structure and properties of the sputtered material can be significantly different.

Sputtering has traditionally been a slow process and consequently restricted to the preparation of thin films. The recent development of high-rate sputtering techniques has made it possible to produce thick, homogeneous deposits of a variety of materials including metallic amorphous and metastable crystalline alloys. Figure 1, for example, shows a 5.0 mm-thick by 5.5 cm-diameter amorphous Sm_2Co_{17} deposit that was sputtered at a rate of 1μ/min. This paper describes the apparatus and procedures used to produce thick metastable deposits and illustrates some of the applications and advantages of this new metastable materials preparation technique.

DEPOSIT PREPARATION

The metastable deposits discussed in this paper were prepared using a dc supported-discharge sputtering system of the type illustrated in Fig. 2. The essential components consisted of a high-vacuum sputtering chamber that was evacuated to a pressure of less than 5×10^{-9} torr prior to each deposition run, a tungsten filament thermionic cathode,

---

* This paper is based on work performed under US Energy Research and Development Administration Contract No. E(45-1)-1830.

Figure 1. Photograph of a 5.0 mm-thick by 5.5 cm-diameter Sm_2Co_{17} deposit prepared by high rate sputter deposition

a water-cooled anode, a water-cooled target, and a variable-temperature substrate. The sputtering gas was high-purity Kr at a pressure of $\sim 5 \times 10^{-3}$ torr and was ionized by the passage of a high-current, low-voltage arc discharge between the cathode and the anode.

A negative target potential of 1500 Vdc was used to accelerate the Kr ions from the plasma to the target surface. The target current density was normally 10 to 20 mA/cm^2, representing a power input of 15 to 30 W/cm^2 to the target. Deposition rates for the various materials ranged from 0.1 to more than 1.0μ/min. The deposition process was semiautomated by controlling the filament current to maintain a

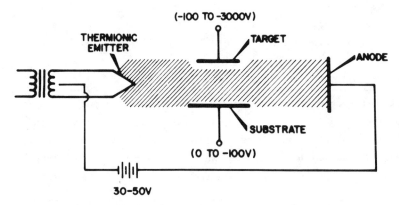

Figure 2. Schematic diagram of a dc supported-discharge sputtering system.

38

constant target current and deposition rate. The sputtering system could thus operate virtually unattended for the time required to produce a deposit of the desired thickness.

The substrates were metallographically-polished Cu, Al, or stainless steel discs and were cooled using either water or chilled nitrogen, or heated by a nichrome heater to permit the preparation of deposits at controlled temperatures from less than -100°C to almost 1000°C. The substrates were ion cleaned and etched at the inception of the deposition run to provide a strong, high-integrity bond between the deposit and substrate.

A Cu disc coated with a low-melting Pb-Sn solder was used as the substrate for free-standing deposits. Either the entire deposit or selected areas could be removed by heating above the melting point of the solder. This type of substrate could be reused simply by recoating and repolishing the surface.

The provision of a suitable target is the most critical and often the most difficult and expensive step in preparing thick sputter deposits. The target must be designed to give a homogeneous deposit of the desired composition and also must be adequately cooled to prevent changes in surface composition due to diffusion or vaporization of target components.

Three types of targets were used to prepare thick metastable sputter deposits. The simplest was a disc fabricated directly from alloy stock of the desired composition. For example, hot rolled steel and malleable cast iron were used as target materials to prepare supersaturated Fe-C solid solution deposits.[1] A homogeneous deposit is obtained from alloy targets even for inhomogeneous, multiphase target material, and the deposit composition is usually close to that of the target.

A compacted powder target was used for metastable compositions that could not be obtained conveniently in allow form. This type of target also produces homogeneous deposits and is made by mixing elemental powders in the desired ratios and using high energy rate forming techniques to simultaneously consolidate and bond the powder compact to a

Cu backing plate. Deposits thus can be prepared representing virtually any combination of elements that can be obtained in powder form and compacted into a mechanically-sound target. The only disadvantage of compacted powder targets is the possibility of significant oxygen contamination when working with reactive powders. This problem can be minimized by using chips or turnings rather than powders of the reactive elements.

A composite target was used to prepare high-purity deposits of reactive alloys. For example, a target consisting of a ring of Sm in a Co disc produced an amorphous Sm-Co deposit that contained less than 200 ppm oxygen.[2] Similarly, targets composed of a hexagonal array of 19 rods of high-purity Pu in Ta, Ag, and Co discs were used to prepare metastable sputter deposits for these alloy systems.[3] Proper design of the composite target is essential, as the composition and homogeneity of the resulting deposit are governed by the relative areas, sputtering rates, and spacing of the target components.[2]

APPLICATIONS AND ADVANTAGES OF HIGH-RATE SPUTTERING

The supported-discharge sputtering techniques described in the preceding section were used to prepare 0.1 mm to 5.0 mm-thick deposits of a number of different metallic amorphous and metastable crystalline alloys. One of the primary advantages of sputtering is the ability to deposit material representing almost any combination of elements that can be fabricated into a target, including compositions that cannot be obtained using conventional alloying techniques because of disparate vapor pressures, immiscibility, or reactivity problems. In addition, the high effective vapor quenching rates associated with high deposition rates promote the formation of metastable structures for a wide range of alloy systems, compositions, and deposition conditions.

The Table, for example, lists some of the amorphous binary and ternary rare-earth-transition metal compositions that have been prepared using high-rate sputtering techniques. It should be noted that these sputtered amorphous alloys do not require the presence of metalloid or "glass former" atoms for their formation or stabilization. Moreover,

TABLE

Amorphous Rare-Earth-Transition Metal Sputter Deposits

$SmCo_5$	Sm_2Co_{17}	$PrFe_2$	$ErFe_2$
$Sm_{.5}Pr_{.5}Co_5$	$Sm_2(Co_{.7}Fe_{.3})_{17}$	YFe_2	$DyFe_2$
$PrCo_5$	$Sm_2(Co_{.5}Fe_{.5})_{17}$	$GdFe_2$	$HoFe_2$
$Pr(Co_{.9}Fe_{.1})_5$	$SmFeCo$	$YbFe_2$	$TbFe_3$
YCo_5	$SmFe_2$	$TmFe_2$	Tb_2Fe_{17}

the disordered structure is obtained over a continuous composition range extending from less than 10 at.% to at least 75 at% of the rare earth component[4,5] irrespective of the nature of the equilibrium liquid-solid and solid-solid phase relationships.

High-rate sputtering can also be used to prepare amorphous alloys over a surprisingly wide temperature range. The deposits listed in the Table were all sputtered at room temperature. Figure 3, however, shows the cross-section of a 0.2 mm-thick $SmCo_5$ deposit sputtered onto an OFHC Cu substrate at 425°C. An amorphous deposit was obtained even at this temperature except for two metallographically distinct bands of crystalline $SmCo_5$ that were formed by brief increases of less than 10°C in the substrate temperature. Conversely, Pu_2Co_{17} sputtered at

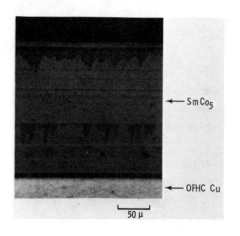

← SmCo5

← OFHC Cu

50 μ

Figure 3. Etched cross-section of $SmCo_5$ sputter deposited onto OFHC Cu at 425°C.

room temperature was microcrystalline, indicating the need for higher Pu contents or lower substrate temperatures to obtain the same
type of amorphous structure observed for the rare-earth-transition
metal alloys.[3]

In addition to amorphous alloys, high-rate sputtering techniques
also have been used to prepare thick metastable solid solution deposits for a number of partial solid solubility and immiscible systems including Fe-C, Ag-Co, Ag-Ni, Ag-Pu, Al-Si, Cu-Ta, and Ta-Pu.
Low substrate temperatures as well as high deposition rates were
required for solid solution formation in the Cu-Ta and Ag-Pu systems. The Fe-C sputter-deposits formed solid solutions containing
up to 5 wt%C and exhibited hardnesses greater than martensite of the
same composition.[1] The Ta-Pu deposits were also of particular interest
as they formed a unique two-phase α-Ta and β-Ta structure that permitted the simultaneous investigation of the metastable solid solution
formation and annealing characteristics of both phases.[3] High-rate
sputtering techniques also were used to produce thick metastable bcc
Nb_3X deposits (X = Al, Sn, Ge, and AlGe) that transformed on annealing to form the superconducting A-15 phase.[6]

The ability to prepare thick, homogeneous deposits has made it possible to investigate the structure, properties, and annealing behavior
of these unique sputtered metastable alloys using techniques that are
difficult or impractical for thin films or limited amounts of test material. For example, use of the massive amorphous sputter deposit shown
in Fig. 1 permitted, for the first time, the measurement of both the
bulk modulus and the shear modulus for a binary metallic amorphous alloy.[7] The changes in the elastic constants and density with crystallization were also determined and correlated with theoretical predictions.

Similarly, the availability of 1 mm-thick sputter deposits permitted an extensive neutron diffraction investigation of the structure and
magnetic properties of amorphous rare-earth-transition metal alloys.[5]
The results of both the neutron diffraction and x-ray scattering[8]
studies show that the thick sputter deposits are crystallographically

amorphous with interatomic distances indicative of a dense random-packed arrangement of the rare earth and transition metal atoms.

The ability to prepare thick sputter deposits has other important advantages in addition to facilitating subsequent experimental studies. Many of the properties of conventional sputter deposits and other thin film specimens are very sensitive both to the nature of the substrate and to film thickness. The anomalous-appearing region nearest the substrate in Fig. 3, for example, is representative of the thin film material formed by conventional deposition techniques. The preparation of thick specimens ensures that the properties of the metastable sputter deposits are representative of "bulk" material and also independent of deposit thickness.[5]

Thick deposits and high deposition rates also enhance the purity of sputtered reactive alloys. The thin film region of the deposit getters impurities in the initial system and the high deposition rates minimize the incorporation of impurities in the remainder of the deposit. The resulting impurity content of thick reactive deposits is usually the same or less than that of the target.[2] High deposition rates also permit the preparation of metastable deposits at higher substrate temperatures by minimizing composition changes due to loss of high vapor pressure elements from the deposit surface.

Another very important advantage of high-rate supported-discharge sputter deposition is the ability to reproducibly prepare metastable materials with the same structure and properties.[3,5] All of the essential inputs for the sputtering process are electrical and can be precisely controlled and duplicated from run to run. This means that metastable material of the same or differing compositions can be prepared using exactly the same deposition conditions.

Conversely, deposition rate, substrate temperature, sub-substrate bias, and other deposition parameters can be varied as illustrated in Fig. 3 to produce material with the same composition but different structure and properties.

REFERENCES

1. Dahlgren, S.D. and M.D. Merz, <u>Met. Trans.</u> <u>2</u> (1971) 1753.
2. Allen, R.P., S.D. Dahlgren, H.W. Arrowsmith, and J.P. Heinrich, AFML-TR-74-87, Air Force Materials Laboratory, Wright-Patterson Air Force Base, Ohio, May 1974.
3. Allen, R.P., S.D. Dahlgren, and R. Wang (to be published in Proc. 5th Int'l Conf. on Plutonium and Other Actinides).
4. Cochrane, R.W., R. Harris, J.O. Strom-Olsen, and M.J. Zuckermann, <u>Low Temperature Physics-LT14</u> <u>3</u> (1975) 118.
5. Rhyne, J.J., S.J. Pickart, and H.A. Alperin, <u>AIP Conf. Proc.</u> <u>18</u> (1974) 563.
6. Wang, R. and S.D. Dahlgren (this volume).
7. Merz, M.D., R.P. Allen, and S.D. Dahlgren, <u>J. Appl. Phys.</u> <u>45</u> (1974) 4126.
8. Cargill, G.S., <u>AIP Conf. Proc.</u> <u>18</u> (1974) 631.

SOLIDIFICATION RATE OF RAPIDLY QUENCHED METALS

P.H. Shingu, K. Shimomura, K. Kobayashi, and R. Ozaki
Department of Metal Science and Technology
Kyoto University, Kyoto, Japan

The formation of amorphous metallic materials by splat-cooling requires a high cooling rate of the liquid specimen so that the rate of
crystallization cannot exceed the rate of solidification. The present
paper extends previous calculations[1] to evaluate the effects of the
thermal properties of the quenched metal, of the form of the viscosity-temperature relation and of alloying on freezing temperature and
viscosity.

1. EFFECT OF THERMAL PROPERTIES OF QUENCHED METAL

Figure 1 shows the calculated results of a comparison of the solidification rate and the crystal growth rate for iron. The model and
the method of calculation are given in the previous work.[1] The
growth rate is calculated using the equation,

$$v = K\frac{L\Delta T}{\eta T_m} \tag{1}$$

where $K = a^2\beta/3\pi\lambda$, a: atomic diameter, β: geometric factor, λ:
thickness of solid-liquid interface, T_m: equilibrium melting temperature, ΔT: undercooling below T_m and η: viscosity of liquid.

The value of K used in the calculation is 6.7×10^{-9}cm.

The solidification rates for curves (1) and (2), which correspond to the cases of near ideal and intermediate cooling are smaller
than the corresponding cases for pure aluminum[1] by a factor of about
3, whereas for the case of curve (3) the solidification rates for
iron and aluminum splats are almost equal. This result confirms that
in the case of near Newtonian cooling, the thermal conductivity of
the splat material does not have much influence on the solidification
rate. Hence the magnitude of the peak in the crystal growth rate is

45

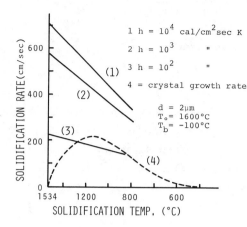

Figure 1. Comparison of solidification rate and crystal growth rate for iron. h: heat transfer coefficient at the splat-substrate interface, d: splat thickness, T_o: initial splat temperature, T_b: initial substrate temperature

more important than this in determining the condition for the formation of amorphous phases.

The peak crystal growth rate for aluminum calculated using the same value of K as for iron (which is nearly an order of magnitude smaller than that used in Reference 1: the absolute value of K is not accurate because of the uncertainties in λ and β) is larger than that for iron by a factor of about 3 because of the smaller value and smaller temperature dependence of viscosity for aluminum compared to iron.

Thus, these calculations indicate that by splat cooling from near the melting point, the maximum of crystal growth rate can be surpassed more easily for iron than for aluminum for the same condition of thermal contact at the splat-substrate interface.

2. EFFECT OF FORM OF VISCOSITY-TEMPERATURE RELATION

In the calculation of crystal growth rate in Fig. 1, the viscosity of undercooled liquid iron is assumed to follow the extrapolation of the Arrhenius relation reported for temperatures above T_m. The viscosity of undercooled metallic liquid should, however, increase more steeply

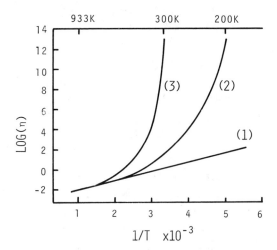

Figure 2. Calculated viscosity of undercooled aluminum liquid. (1): Arrhenius relation, $\eta = 0.1492 \times 10^{-2}\exp(3950/RT)$. (2): Eq. (2), $T_g = 200K$. (3): Eq. (2), $T_g = 300K$.

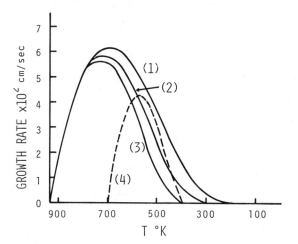

Figure 3. Calculated crystal growth rate for aluminum based on the viscosity curves in Fig. 2. Curve (4) is for Al-Ge eutectic alloy.

with decreasing temperature than the Arrhenius relation suggests[2].

Figure 2 shows the trial calculation of viscosity for under-cooled liquid aluminum. Curve (1) is calculated from the Arrhenius relation and curves (2) and (3) are calculated assuming a Tammann and Hesse type relation[3,4], i.e.,

$$\log (\eta) = A + \frac{B}{T - T_\infty} \qquad (2)$$

where A, B, and T_∞ are constants which are determined assuming,

$\eta = 10^{13}$ poise at $T = T_g$ (glass temperature)

η = value for the Arrhenius relation at $T = T_m$

$d(\log[\eta]) \,/\, d(1/T)$ = slope of the Arrhenius relation at $T = T_m$

Although the apparent differences between these curves are great, the peak crystal growth rates calculated for these curves by Eq. (1) using the same value of K as that used for iron, do not differ so much, as shown in Fig. 3. Numbers (1), (2), and (3) given for curves in Fig. 3 refer to the curves in Fig. 2 used for the calculations.

The results shown in Figures 2 and 3 thus suggest that the high T_g value does not by itself control the ease of amorphous phase forma-tion. It has to be noted, however, that the present analyses apply only for the case of heterogeneous nucleation of crystal nuclei at small undercoolings below T_m. The effect of T_g on homogeneous nuclea-tion has to be considered separately in relation to the condition for the amorphous phase formation.

3. EFFECT OF ALLOYING ON T_m AND ON VISCOSITY

The curve (4) in Fig. 3 is a result of calculation for the case when T_m is lowered to 697K (equal to the eutectic temperature T_e of Al-Ge system) with the other physical properties left equal to these for pure aluminum. The value of viscosity and the slope of $\log(\eta)$ vs $1/T$ plot at T_e (697K) are set equal to these values for aluminum at T_m (933K). Curve (4) suggests that the lowering of melting temperature is more effective in reducing the peak crystal growth rate than the reduction of T_g. When the η-value at T_e is set equal to that for the

Arrhenius relation for aluminum at T_e, the peak height becomes as low as about 1/2 for that for curve (4).

Since in the range of Newtonian cooling, lowering of solidification temperature does not increase the solidification rate, the ease of amorphous phase formation for deep eutectic systems may at least partly be explained by the decrease in crystal growth rate due to the reduction in melting temperature.

The effect of alloying in enhancing the formation of amorphous phases is not limited to the thermal effect such as the reduction of T_m. The chemical interaction between different atomic species which increase the viscosity of liquid should be one of the reasons for the effect of alloying. It is of interest, in this connection, to note the range of compositions for which the amorphous phase is formed for Al-Ge alloys.

Figure 4 is for experimental results showing the range of compositions for which the amorphous phase is formed by the gun method of splat-cooling. The experimental procedure is the same as that reported previously. [5]

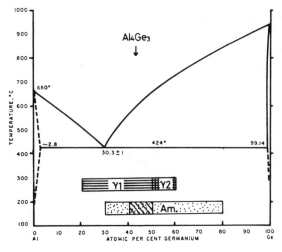

Figure 4. The Al-Ge system, showing the ranges of compositions for which the amorphous phase and the metastable compound phases have been observed to form.

In the composition range of 30 to 80at%Ge for which the amorphous phase is formed, it is only for the 40 to 50at%Ge range that the entire specimen can be made amorphous when the examination by x-ray diffractometer is performed. The phases γ_1 and γ_2 in the figure are the metastable compound phases obtained at lower cooling rates.

The fact that the composition range which is slightly away from the eutectic is more favourable in producing the amorphous phase could be due to the increased liquid viscosity of this alloy in the displaced composition range.

It is interesting to note that the 40 to 50at%Ge range corresponds to the composition for which the tendency of formation of Zintl phase Al_4Ge_3 is reported by Predel[6] from thermodynamic considerations.

REFERENCES

1. Shingu, P.H. and R. Ozaki: Met. Trans. 6A, 33 (1975).
2. Beyer, R.T. and E.M. Ring: "Liquid Metals", S.Z. Beer ed. Marsel Dekker, 451 (1972).
3. Tammann and Hesse: Z.anorg. u. allg. Chem., 156, 245 (1926).
4. Turnbull, D.: "Liquids" ed. T.J. Hughel, Elsevier (1965), 6.
5. Shingu, P.H., Kobayashi, K. Shimomura, and R. Ozaki: Scripta Met. 8 (1974), 1317.
6. Predel, B. and D.W. Stein: "The Properties of Liquid Metals", ed. S. Takeuchi, Taylor & Francis, 495 (1973).

STUDIES OF THE UNDERCOOLING AND NUCLEATION
OF LIQUID METALS AND ALLOYS
UTILIZING THE DROPLET TECHNIQUE

D.H. Rasmussen, J.H. Perepezko, and C.R. Loper, Jr.

Department of Metallurgical and Mineral Engineering
University of Wisconsin at Madison

Rapid cooling from the melt techniques are commonly used to achieve large undercoolings during solidification. Large undercoolings may also be achieved during a slow cooling from the melt by use of the droplet technique. While both techniques enable large undercoolings to be attained before solidification, the droplet technique permits accurate measurement of the actual crystallization temperature. During slow cooling, uniform dispersions (5 to 10 μ droplets) of Sn, Pb, Bi, In and Sn-Bi and Sn-Pb alloys in a carrier fluid were observed to undercool up to 160°C before solidification. The undercooling in the Sn-rich alloy systems was found to remain constant at 117°C up to 20% by wt Pb or Bi. These results provide confidence that the new developments in droplet preparation permit the attainment of the maximum undercooling before solidification.

The undercoolings of metallic systems are compared with droplet studies on water and aqueous solutions where slow cooling to glassy state in otherwise crystallizable binary systems is possible.

INTRODUCTION

The interest in the solidification of liquid metals and alloys at high undercoolings has been quite active, especially since the development of rapid quenching from the melt techniques generally termed "splat cooling". These techniques have been used to demonstrate the existence of numerous metastable alloy structures[1] and in some cases noncrystalline solids.[2] However, the "splat cooling" technique is rather qualitative in that the actual temperature at which solidification commences as well as the cooling rate involved may not be determined precisely.[3]

Solidification at large undercoolings may also be achieved during a slow cooling from the melt by the droplet technique.[4] In order to obtain large undercoolings such as those associated with homogeneous nucleation of crystals from the melt, the liquid must not contain catalytic impurities, nor can it be allowed to solidify in a container whose walls may catalyze nucleation. As a consequence, large samples of undercooled liquid are unsuitable for obtaining large undercoolings because they inevitably contain nucleation catalysts of variable potency. If the liquid can be dispersed into a large number of small drops, only a few of the drops will contain nucleation catalysts. By maintaining droplet isolation by means of surface films which are not nucleation catalysts, the effects of any impurities or inclusions can be restricted to the few drops in which they are located. When a sample of liquid metal in the form of a dispersion is undercooled, the majority of the drops will not begin to solidify until a temperature is reached at which homogeneous nucleation occurs at a high rate. Since growth of crystals from the liquid at the large undercoolings expected for homogeneous nucleation is very rapid, the formation of the first nucleus in any one drop will lead to the rapid solification of the entire drop. Therefore, once the undercooling necessary for nucleation is achieved, there is a close analogy in terms of the rapidity of the solidification process between "splat cooling" and the droplet technique. While both techniques enable large undercoolings to be obtained before solidification, only the droplet technique permits an accurate measurement of the actual crystallization temperature.

DETAILS OF THE EXPERIMENTAL METHOD

Emulsification Procedure

To create emulsions of tin, alloys of tin, and the other metals studied, droplets of the liquid metal were stabilized by a chemical reaction that deposits a surface layer on the droplet. The procedure that was followed was derived from the earlier work of Turnbull[4] for the preparation of the emulsions of mercury droplets. Although the standard Waring blender is suitable for emulsions of mercury, since the metal is liquid at

52

room temperature, it was necessary to adapt the blender procedure for metals having melting points as high as about 350°C.

The requirements that have been established for the formation of a stable metal emulsion are: 1) a suitable carrier fluid that remains liquid above the melting point of the metal to be emulsified to below the lowest temperature of nucleation expected; 2) an oxidant which will raise the metal droplet to a charged state and perhaps supply a suitable anion to coat the droplet surface in preference to an oxidant that reacts to form a gaseous or soluble product; 3) an acid catalyst to promote the oxidation process and interact with or remain involved with the determination of the surface coating; 4) a method by which the emulsification can be carried out above the melting point of the metal within the time limits set by the half-life of the peroxide if a peroxide is used as oxidant.

The primary requirement of the carrier oil was that it have low volatility and tolerate an oxidizing atmosphere at temperatures up to about 350°C. Suitable stability and inactivity was obtained from a polyphenyl ether with a boiling point above 500°C and stability to air oxidation to well over 400°C.

The surfactant reagents included both an oxidizing agent which reacted directly with the metal to leave a product which was surface active in promoting droplet stability and an acid catalyst which promoted a particular reaction path for the oxidant. Three types of oxidants were used in the preparation of the emulsions: chemicals such as sulfuric acid, oxygen and sulfur which reacted directly with the metal and formed metallic compounds; sulfates, oxides and sulfides; salts of metals much lower in the electrochemical series such as copper or silver isophtalate which reacted either by substitutional or electrochemical oxidation; and peroxides such as benzoyl peroxide or di-tertiary-butyl peroxide which formed the salt of the corresponding organic acid or alcohol by decomposition and reaction with the metal. Without the use of a modifier or acid catalyst, the peroxidation reactions took several pathways leading to a variety of undercoolings for any one metal or sample.[5] The prime requirements for the acid catalyst were: no reaction with the metal itself, stability in solution at the temperatures of the reaction, no reaction with the oxidant to form a surface inactive product and catalysis of the oxidation of the metal with more uniform undercooling results. Isophthalic acid proved to be the most functional acid catalyst and was adopted for use throughout the studies.

Measurement of Undercooling

After preparation, the emulsion was transferred to a sample tube of a differential thermal analysis apparatus. A standard heating and cooling rate of approximately 10°C/min was employed throughout the study. Experiments determined, however, that varying the heating or cooling rate by a factor of two did not shift the heat evolution peaks characteristic of a particular emulsion by more than 1°C or about the accuracy of the sample temperature measurement. The melting temperature of each metal during heating served as a calibration point and the undercooling to nucleation was determined by the difference in temperature between the melting point and the peak of the exothermic reaction during cooling. A typical thermogram for an Sn-10% wt Pb alloy is presented in Figure 1.

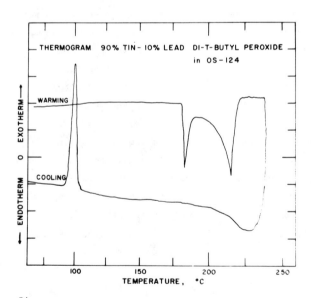

Figure 1. Thermogram for an emulsion of Sn-10%Pb. Warming curve indicates the temperatures of eutectic melting and final melting at 183°C and 216°C, respectively. Cooling curve indicates uniform undercooling of 117°C to a crystallization temperature of 99°C. [5]

Following from these procedures and methods the maximum undercoolings before solidification were determined for pure tin, lead, bismuth, indium and alloys of Sn-Pb and Sn-Bi. In all cases, the measured undercoolings were equal to or greater than the values reported in the literature.[6] Maximum undercoolings for the Sn-Pb system[7] are presented in Figure 2. Alloys of Sn-Pb or Sn-Bi with less than 20% Pb or Bi undercooled to the same extent as pure tin, 117°C. The Sn-Pb alloys undercooled with respect to the liquidus. The Sn-Bi alloys undercooled with respect to the solidus in the primary solid solution range up to 7% Bi and from 10 to 20% Bi nucleated at a constant temperature of 83°C. A detailed discussion of some of these results and the interesting aspects of alloy solidification have been reported elsewhere.[8]

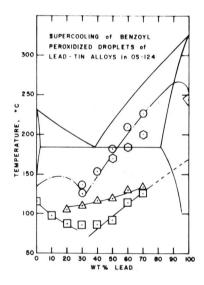

Figure 2. Phase diagram for Sn-Pb indicating undercoolings corresponding to various peroxidation reactions at the droplet surfaces. Squares represent maximum supercooling for benzoyl peroxidized emulsions. The maximum undercooling observed by Hollomon and Turnbull[7] is indicated by the •—•—•— curve.

DISCUSSION

The new developments in emulsification procedures combined with differential thermal analysis have provided the techniques necessary to conduct a wide range of research projects into the solidification behavior of highly undercooled metals and alloys. The results obtained from these

initial studies have provided confidence in these procedures and measurement techniques so that the maximum undercooling before nucleation
may be achieved.

One area in which the droplet technique can be used to advantage
deals with the solidification of alloys. For example, if alloy droplets
solidify as a single phase solid solution in the composition range
where the equilibrium structure is a two-phase mixture, then the critical
nucleus may not be the equilibrium structure, but rather a metastable
structure with a lower nucleation barrier. From a knowledge of the composition and solidification temperature of such droplets it is possible
to ascertain whether they represent metastable extensions of primary
solid solutions and if local interfacial equilibrium existed during
solidification. With the 'splat cooling' technique it has been possible to test for local interfacial equilibrium only in the relatively
few alloy systems which exhibit retrograde solubility.[9] Since the droplet technique can be used to measure the solidification temperature in
a variety of alloy systems, it can provide information on the thermodynamic and kinetic conditions at the liquid-solid interface during rapid
solidification.

The capability of accurately measuring the solidification temperature can also be employed to study the formation of metastable phases
during solidification at high undercooling. By the incorporation of
quenching techniques the metastable phases that may form at high undercooling can be retained for structural study at room temperature or below. Information about undercooling for nucleation of metastable phases
could then be related to the stability of these structures and to their
subsequent reversion behavior upon annealing.

The procedures involved in the droplet technique also have been applied to form emulsions of water and aqueous solutions which undercool
to homogeneous nucleation temperatures that are concentration dependen-
dent.[10] For many aqueous solutions a rapid quenching procedure (less
drastic than 'splat cooling') has been used to determine the concentration dependence of the glass transition temperature.[11] The combined results for the homogeneous nucleation temperature and the glass transi-

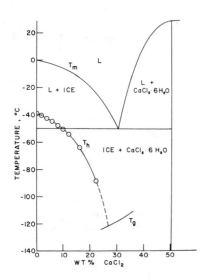

Figure 3. Phase diagram for $CaCl_2$ $-H_2O$ including the temperatures of maximum undercooling indicated by the curve T_h, and the measured glass transition temperatures indicated by the curve marked T_g.

tion temperature in aqueous systems indicates that over a particular concentration range (near the normal eutectic for each system) emulsions from these systems will undercool to the glassy state without an intervening nucleation and crystallization. An example of this behavior, shown in Fig. 3, occurs in the $CaCl_2-H_2O$ system, where the undercooling to nucleation reaches the glass transition curve at 27% by wt $CaCl_2$.

A comparison of the results presented in Figures 2 and 3 indicates that there is one major difference between the effect solutes have on the nucleation temperature of pure water and the effect that lead and bismuth have on the nucleation temperature of Sn. The depression of the nucleation temperature, T_h, for water is always greater than, but proportional to, the depression of the melting point, ΔT_m. In the Sn–Bi and Sn–Pb systems, ΔT_h equals ΔT_m; thus, T_m-T_h remains constant, i.e., constant undercooling. The observation of a more rapid depression of T_h than T_m in the case of the aqueous systems may indicate an enhancement of the tendency to approach the glassy state around eutectic concentrations. Not enough data on metallic systems is available to rule out the possibility that alloys that exhibit very nonideal melting point depression curves or unusually low eutectic temperatures may also display the enhanced depression of T_h versus T_m. In this regard, it is interesting

57

to note that systems that are now readily 'splat cooled' to the glassy
state are exactly those that have such nonideal melting point depression curves and low eutectic temperatures.

ACKNOWLEDGMENT

This work was supported in part by grants from the University of
Wisconsin Graduate School and from the American Foundation for Biologi-
cal Research.

REFERENCES

1. Giessen, B.C. in "Developments in the Structural Chemistry of
   Alloy Phases", (1967), B.C. Giessen, ed., Plenum Press, New York,
   227 (1969).
2. Giessen, B.C. and C.N.J. Wagner in "Liquid Metals, Chemistry and
   Physics", S.Z. Beer, ed., Marcel Dekker, Inc., New York, 633 (1972).
3. Ruhl, R.C., Mater. Sci. and Eng., 1, 313 (1967).
4. Turnbull, D., J. Chem. Phys., 20, 411 (1952).
5. Rasmussen, D.H., Ph.D. Thesis, U. Wisconsin [Madison] (1974).
6. Walton, A.G. in "Nucleation", A.C. Zettlemoyer, ed., Marcel Dekker, New York, 279 (1969).
7. Hollomon, J.H. and D. Turnbull, Trans. AIME, 191, 803 (1951).
8. Rasmussen, D.H. and C.R. Loper, Jr., Acta Met. [in press].
9. Baker, J.C. and J.W. Cahn, Acta Met., 17, 575 (1969).
10. Rasmussen, D.H. and A.P. MacKenzie in "Water Structure at the
    Water Polymer Interface", H.H.G. Jellinek, ed., Plenum Press,
    New York, 126 (1972).
11. Angell, C.A. and E.J. Sare, J. Chem. Phys., 52, 1058 (1970).

METASTABLE ALLOY PHASES BY GETTER-SPUTTERING

B. Cantor and R.W. Cahn

School of Applied Sciences,
University of Sussex, Brighton, U.K.

INTRODUCTION

Metastable phases formed by liquid-quenching have been studied mainly in the metallurgical field[1]; whereas those formed by vapour-quenching have been studied for the purposes of thin film physics.[2] The objective of the present work was to make a detailed comparison of metastable phases formed by the two techniques. A variety of alloys of Al-Ni, Al-Cu, Al-Fe, Ag-Cu, and Cu-Zn were quenched from the vapour by co-sputtering; the resulting structures were investigated by electron microscopy and compared with structures obtained previously by liquid-quenching and thermal evaporation.

EXPERIMENTAL TECHNIQUE

Alloy thin films were co-sputtered from 99.999% pure sectored sheets of the constituent elements in an r.f.getter-sputtering unit[3,4] (Figure 1). The alloys were deposited onto 1 mm thick cleaved slices of NaCl single crystal bonded onto 1.5 mm thick Cu sheet which was cooled directly with liquid N_2 and rotated at 60 rpm to ensure homogeneous co-deposition. The sputtering gas was 99.9% pure Ar which was further purified by a getter in the sputtering chamber. Operating conditions were chosen to minimise substrate heating while maintaining a steady discharge (Figure 2); this required low r.f. power ($\sim$0.5 kw), low axial magnetic field ($\sim$20 Oersted) and high gas pressure ($\sim 10^{-2}$ torr), and produced slow deposition rates of $\sim$2Å thickness per second.

Deposited films, typically 2000Å thick, were removed from the substrate by dissolving the NaCl in water, and were examined in a transmission electron microscope. Annealing experiments were performed with an in-situ heating stage in the electron microscope. Film compositions were determined on specially prepared 1-2µm thick films by energy-

59

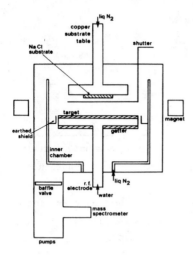

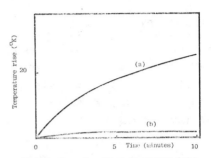

Figure 2   Temperature rise at the substrate during
sputtering (a) high r.f. power
(b) low r.f. power.

Figure 1. Schematic diagram of
r.f. sputtering apparatus

Figure 2. Temperature rise at
the substrate during sputtering
(a) high r.f. power; (b) low
r.f. power.

Figure 3. Typical electron
diffraction pattern from
sputtered alloys.

Figure 4. Precipitation of
$NiAl_3$ particles from sputtered
Al-25.0 at%Ni after 6 minutes
at 300°C.

dispersion x-ray microanalysis, with a low electron accelerating voltage of 5 keV to obtain valid absorption and fluorescence corrections.

AS-DEPOSITED MICROSTRUCTURE

In all cases, the deposited alloy films produced no contrast on bright-field electron micrographs, and the corresponding diffraction patterns consisted of 6-8 diffuse rings (Figure 3). This indicated that the films were polycrystalline with equiaxed grain sizes of 50-100 Å, a microstructure similar to that observed previously in other sputtered alloys[5,6] (Table 1).

TABLE 1: GRAIN SIZES OF SPUTTERED ALLOYS

Alloy	Al-Ni, Al-Cu, Al-Fe Ag-Cu, Cu-Zn (present work)	Fe-C, Ag-Co[5]	20/10 stainless steel, α/β brass[6]
Grain Size	50-100Å	70Å	<100Å

SOLID SOLUTIONS

Table 2 shows maximum solid solubilities which were obtained by co-sputtering, together with the maximum solubility in corresponding equilibrium, liquid-quenched, and thermally evaporated alloys. Both methods of vapour-quenching (sputtering and thermal evaporation) are more effective than liquid-quenching in forming extended solid solutions. Short-range ordering in liquid alloys can limit the as-quenched solubility by promoting the nucleation of equilibrium phases, and a detailed model has been proposed for this effect.[12] For vapour-quenched alloys, however, short-range ordering is not possible and this limitation on maximum solubility is removed. Differences in solubility limits between sputtered and thermally evaporated alloys are almost certainly caused by different substrate conditions (see next two sections).

TABLE 2: MAXIMUM SOLUBILITIES IN QUENCHED ALLOYS

		Maximum Solid Solubility (at%)			
Solvent	Solute	equilibrium	sputtered	liquid-quenched	thermally evaporated
Al	Cu	2.5	28.5	18.0[7]	—
Al	Ni	0.02	20.9	7.7[8]	11.5[10]
Al	Fe	0.026	12.9	4.4[9]	7.5[11]
Ni	Al	21.0	16.2	—	—

INTERMETALLIC COMPOUNDS

Table 3 shows metastable intermetallic compounds which were formed in sputtered alloys, and the corresponding liquid-quenched and thermally evaporated structures. For sputtered Al alloys, a discordered bcc phase replaced a variety of equilibrium ordered bcc phases; for sputtered Cu-Zn, the complex γ-brass phase was replaced by the simpler β-brass structure, an effect which has been observed previously for Ag-Zn and Cu-Al alloys.[17] These results support a previous suggestion that vapour-quenching tends to produce simple, disordered crystal structures.[18]

TABLE 3: METASTABLE INTERMETALLIC COMPOUNDS IN QUENCHED ALLOYS

Alloy	Composition (at%)	Sputtered Phases	Liquid-Quenched Phases	Thermally-Evaporated Phases
Al-Ni	∿40-80Ni	disordered bcc	—	β(NiAl)[15]
Al-Cu	∿45-60Cu	disprdered bcc	ordered bcc[13]	ordered bcc[16]
Cu-Zn	∿15-70Zn	Cu,Zn & β(CuZn)	β(CuZn)[14]	—

Whether an alloy can form ordered or complex phases depends on the extent of surface diffusion during deposition. The mean diffusion distance of a surface atom before being restricted by adjacent depositing atoms is given approximately by:

$$\bar{x} = (2D_s t)^{1/2} = (2\nu t)^{1/2} a \exp(-Q_d/2kT)$$
$$= (2\nu a/R)^{1/2} a \exp(-Q_d/2kT)$$

where D_s is surface diffusivity, t time, ν vibration frequency, a inter-atomic spacing, Q_d surface diffusion activation energy, k Boltzmann's constant, T substrate temperature and R deposition rate.[2,4] Inserting the appropriate data, there is effectively zero diffusion on liquid nitrogen-cooled substrates, and depositing atoms cannot form ordered or complex phases.[4] However, there is extensive diffusion on uncooled substrates and this explains the lower solubility limits and ordered bcc phases in thermally evaporated Al-Ni and Al-Cu (Tables 2 and 3).

AMORPHOUS PHASES

Thermally evaporated Ag-Cu alloys with $\sim$27-73 at%Cu are amorphous when deposited onto amorphous substrates at 77K, and form an amorphous/fcc phase mixture on cleaved NaCl at 77K.[19] The same alloys sputtered on cleaved NaCl at 77K showed a single phase fcc structure (Table 2) similar to that obtained by liquid-quenching.[20] This suggests that for similar conditions of substrate material and substrate cooling, sput-tering is less effective than thermal evaporation as a method of va-pour quenching. Such a hypothesis is supported by evidence that sput-tered atoms are slightly more energetic than thermally evaporated atoms under comparable conditions.[21]

ANNEALING BEHAVIOUR

Table 4 shows the lowest temperature (T_p) for which precipitation was initiated within 10 minutes in sputtered metastable phases. The super-saturated solid solutions were more resistant to heat treatment than comparable conventional age-hardening Al alloys. For liquid-quenched alloys, resistance to heat treatment has been ascribed to a loss of vacancies to foil surfaces.[22] In sputtered alloys, the as-deposited grain size is less than the width of precipitate free zones, and grain boundaries can act as vacancy sinks to produce low solute mobility and high resistance to thermal treatment. The precipitation sequence in sputtered Al solid solutions was:

supersaturated fcc $\sim$50Å grain size	$\longrightarrow$	grain growth to $\sim$200Å grain size	$\longrightarrow$	Precipitation of $NiAl_3$, $CuAl_2$ or $FeAl_3$

TABLE 4: ANNEALING BEHAVIOUR IN SPUTTERED ALLOYS

Alloy	as-deposited structure	annealed structure	$T_p(°C)$
Al-Ni	supersaturated fcc(Al)	Al+NiAl$_3$	300-400
Al-Cu	supersaturated fcc(Al)	Al+CuAl$_2$	350
Al-Fe	supersaturated fcc(Al)	Al+FeAl$_3$	500
Al-Ni	disordered bcc	NiAl+{Ni$_2$Al$_3$} or {Ni$_3$Al}	300

The low as-deposited grain size promotes rapid grain growth, and also provides sufficient heterogeneous nucleation sites to favour direct precipitation of equilibrium phases rather than the formation of intermediate phases.

The precipitation sequence for sputtered Al-Ni alloys with the disordered bcc structure was:

$$\text{disordered bcc} \longrightarrow \begin{array}{c}\text{ordering to single} \\ \text{phase NiAl}\end{array} \longrightarrow \begin{array}{c}\text{precipitation of} \\ \text{Ni}_2\text{Al}_3 \text{ or Ni}_3\text{Al}\end{array}$$

For the alloys which contained Ni$_2$Al$_3$ at equilibrium, this was effectively a two-stage ordering process. Initially, the atoms ordered on the bcc lattice to form non-stoichiometric NiAl; subsequently, the vacancies ordered on the Ni sub-lattice to form equilibrium Ni$_2$Al$_3$.

In sputtered Al-25.0 at%Ni, the precipitation behaviour was markedly different from that of the other alloys investigated. Slow nucleation and rapid growth produced large particles of NiAl$_3$ similar to those which have been observed previously for crystallization of amorphous alloys[23,24] (Figure 4). One essential criterion for such rapid growth is that the transformation involves no change of composition and is not limited by solute diffusion. However, precipitation of other Al-Ni intermetallic compounds from sputtered alloys takes place by copious nucleation and relatively slow growth. Clearly, some subsidiary requirement must be fulfilled for the type of precipitation shown in Figure 4, but it is not yet clear what this requirement might be.

REFERENCES

1. Anantharaman, T.R. and C.R. Suryanarayana, J. Mat. Sci. 6 (1971) 1111.
2. Chopra, K.L., "Thin Film Phenomena" (McGraw-Hill: New York) (1969).
3. Holland, L. and C.R.D. Priestland, Vacuum 22 (1973) 133.
4. Cantor, B. and R.W. Cahn [to be published].

5.   Dahlgren, S.D. and M.D. Merz, Met. Trans. 2, 1753 (1971).
6.   Navinsek, B., Thin Solid Films, 13, 367 (1972).
7.   Ramachandrarao, P., M. Laridjani, and R.W. Cahn, Z. Metallk, 63, 43 (1972).
8.   Tonejc, A., D. Rocak, and A. Bonefacic, Acta Met. 19, 311 (1971).
9.   Tonejc, A. and A. Bonefacic, J. Appl. Phys. 40, 419 (1969).
10. Dwzevic, D., A. Bonefacic, and D. Kunstelj, Scripta Met. 7, 833 (1973).
11. Collet, O., A. Fontaine, and A. Guinier, Mat. Sci. Eng. 19, 277 (1975).
12. Bose, S.K. and R. Kumar, J. Mat. Sci. 8, 317 and 1795 (1973).
13. Ramachandrarao, P. and M. Laridjani, J. Mat. Sci. 9, 434 (1974).
14. Greenholz, M. and R.W. Cahn [to be published].
15. Dwzevic, D., Scripta Met. 9, 543 (1975).
16. Takahishi, N. and K. Mihama, Acta Met. 5, 159 (1957).
17. Michel, P., Ann. Phys. [Paris], 1, 719 (1956).
18. Anseau, M.R., J. Mat. Sci. 9, 1189 (1974).
19. Mader, S., A.S. Nowick, and H. Widmer, Acta Met. 15, 203 (1967).
20. Duwez, P. R.H. Willens, and W. Klement, Jr., J. Appl. Phys, 31, 1136,(1960).
21. Chopra, K.L., J. Appl. Phys., 37, 3405 (1966).
22. Thomas, G. and R.H. Willens, Acta Met. 12, 191 (1964); 13, 139 (1965); and 14, 1385 (1966).
23. Lucas, J.M., D. Phil. thesis, Sussex (1969).
24. Köster, U. and P. Weiss, J. Non-Cryst. Sol. 17, 359 (1975).

PROPERTIES OF PLASMA-SPRAYED ALUMINIUM-COPPER ALLOYS

K.D. Krishnanand and R.W. Cahn

Materials Science Laboratories
School of Applied Sciences
University of Sussex, Brighton, UK

INTRODUCTION

An awkward paradox stands in the way of those who wish to use the high strength of splat-quenched alloys for industrial purposes. The molten alloy must be atomised into small drops to ensure rapid quenching, but the standard extrusion methods of compacting the quenched flakes into artifacts of useful size requires the application of heat, and this often reduces the strength which it is hoped to achieve. Even when strength is not reduced, a multistage process inevitably entails high costs. One attractive solution to this paradox is to use plasma-spraying because this technique combines melting, quenching, and compacting in a _single_ operation. The experiments reported here represent a preliminary attempt, using Al-Cu alloys as test materials, to assess the potential and limitations of this intrinsically promising method. In the process, one radical improvement in experimental technique was established.

Moss first demonstrated solid solubility extension and a metastable phase in two silver alloys[1]; later, another team extended terminal solid solubility and achieved enhanced strength in Al-V alloys[2,3]. Thin foils only 100 μm thick were sprayed to achieve very fast quenching. Superalloys have also been obtained with enhanced mechanical properties.[4] A metastable form of Al_2O_3 has been produced by plasma-spraying[5] and supersaturation in the TiO_2-Cr_2O_3[6] and Al_2O_3-Cr_2O_3[7] mixtures was achieved by this technique. The observation that such refractory powders can be obtained in supersaturated form highlights the fact that the plasma technique entails no container problems. In the above-quoted and some other papers (e.g., Ref. 8) the influence of a range of variables such as powder size, substrate conductivity and preparation, gas flow rate, arc power, on the quality and adhesion(and to a lesser extent the quenching rate)of the deposit have been analysed.[*]

67

Experiments were done with pre-alloyed powders of Al-6at.% and Al-12at% Cu, prepared by argon atomisation of the alloy.[9] All size fractions were supersaturated and, to judge from dendrite dimensions, had cooled at $\sim 10^4$K/sec. These were melted in and sprayed from a 15 RW, PLA 7000 DC plasma-gun made by Associated Engineering Developments, Ltd., Rugby. The 45-75 μm (nearly spherical) size fraction of the powder ($\sim$20% of the total) was used for best results.

In the arc-plasma technique, a stream of powder in a carrier gas is fed into a highly ionised, rapidly flowing argon plasma which is at a very high temperature. The plasma and powder particles together are projected at the substrate, which in our experiments was a water-cooled, grit-blasted copper plate. (The grit-blasting is an essential procedure before each run). The residence time of each particle in the plasma is only $\sim$0.1 msec, and so the flow rate, particle size, powder feed rate, distance from substrate and plasma arc power must be delicately balanced to be sure to melt all the particles without heating the substrate too much. For instance, heating of the substrate by the hot plasma plume could be reduced by cutting down the arc power, but then the particles are apt to freeze in flight, short though their trajectory is. The problem cannot normally be overcome by using very fine particles, because these are hard to inject properly into the gas stream. Increasing the substrate distance to reduce substrate heating again is apt to lead to freezing in flight. Substrate heating, which is plainly detrimental if very rapid quenching is desired, is also sensitive to the traverse rate of the gun, which is oscillated parallel to the substrate, and to the thickness of the deposit which has already been built upon. The substrate metal is also important: fortunately it has been found[10] that substrates of good conductivity give poor adhesion, so deposits are readily stripped from a copper substrate.

It appeared that factors making for good melting and deposition in general also result in unacceptable substrate heating. Under the usual operating conditions, the substrate heated rapidly to $\sim$970K, which suffices to quickly anneal out any metastable phase. Experiments were

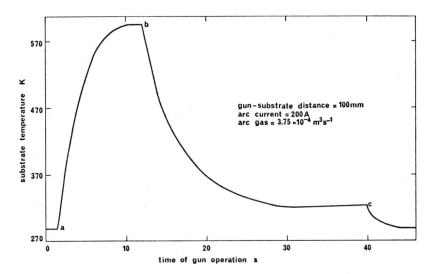

Figure 1. Time-temperature profile of surface of water-cooled copper substrate. (a): Gun switched on. (b): N_2 cross-blast switched on at 5 psi pressure. (c): Gun switched off.

therefore done with a cross-blast of cold nitrogen from a flattened nozzle, placed near the substrate. The cold gas blows away the hot plasma plume without appreciably diverting the molten alloy droplets: the droplets are not cooled enough to freeze them in flight, but substrate heating (which is largely due to the plasma plume) is drastically reduced. Figure 1 (supra), prepared from an oscilloscope record of a thermocouple embedded in the substrate surface, demonstrates this.

Numerous experiments were done, with and without cross-blast at different nitrogen pressures, and using spherical copper and aluminium powders combined with SEM examination of the deposits, to ensure conditions combining effective melting with effective quenching from the melt. The state of the droplets at impact can be judged from the SEM image of the deposit.

The following conditions were chosen as giving optimum performance for both alloy compositions:

69

Spraying done in air. Substrate: water-cooled and grit-blasted copper
plate. Arc gas (Ar) flow: 3.75 x $10^{-4}m^3$/sec. Powder feed gas flow:
7.5 x $10^{-5}m^3$/sec. Arc current: 500A at 20V. Gun-substrate distance:
100 mm. Gun traverse rate (in oscillation): 150 mm/sec. Powder feed
rate: 250 mg/sec. Nitrogen cross-blast pressure (when used):
7 x 10^4Pa (∿10psi). (This pressure was double that used to obtain
Fig. 1; the faster cross-blast neutralised the higher arc current of
500A, as compared with 200A in Fig. 1). Spraying time: variable, depending upon deposit thickness required; it takes ∿25 sec to deposit
300 μm. Deposits all had a density greater than 90% of theoretical;
the average was 93%. Some deposits were reduced 30% by rolling to consolidate them further. Microhardness tests were made on a Leitz tester
at 0.025 Kg load as stripped, polished deposits. The error bars shown
in Figures 2-4 indicate ± standard deviation from the mean.

Lattice parameters were measured by diffractometry, using (331)
and (420) fcc reflection, typically to ±0.0001 μm.

Solute contents were deduced on the presumption that Vegard's Law
applies.[11] Peaks were clear and symmetrical. Guinier transmission diffraction patterns taken through the thickness of the stripped deposits
also showed symmetrical but fairly wide peaks, indicating a variance
of composition of typically ±3at.% on an alloy containing average of
∿8at.% Cu in supersaturated solution. The deposits were thus more homogeneous than those produced by rotary splat-quenching, but less homogeneous than those produced by a two-piston apparatus.[11]

RESULTS

Figure 2 shows hardness measurements of the Al-6at.% Cu alloy, in various conditions, both as deposited (and rolled), and after aging at
423K. Figure 3 shows corresponding measurements for the 12at.% Cu alloy.
(The results for pure Al in Fig. 2 are included to show that hardening
cannot be attributed in any substantial measure to the 0.2at.% alumina found to be trapped in all deposits.)

The initial hardness values obtained with cross-blast (that is,
with efficient quenching) were ∿165 DPN for the more dilute alloy and

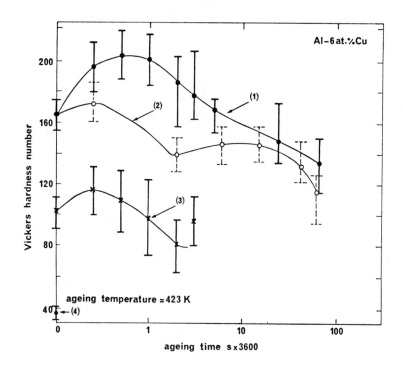

Figure 2. Hardness of plasma-sprayed Al-6at.% Cu alloys, ∿220 μm thick as sprayed. (1) With cross-blast, 30% rolling reduction. (2) No cross-blast, rolled. (3) As-sprayed, no cross-blast or rolling. (4) Pure Al, as sprayed. Aging at 423K.

∿300 DPN for the more concentrated alloy. The latter is the highest hardness yet recorded for an aluminium alloy, so far as we are aware — though it is true that this value includes a contribution from the cold-rolling. It compares with a peak hardness of 135 DPN for water-quenched and optimally aged Al-2at.% Cu (dural)[12], and with ∿270 DPN in the outer, harder zone of splat-quenched Al-4at.% Fe flakes.[13] When these flakes were compacted and hot-extruded to form a useful section, the hardness dropped to ∿170 DPN.[13]

Since the purpose of these experiments was primarily to assess the usefulness of plasma-spraying as a single-stage process for splat-

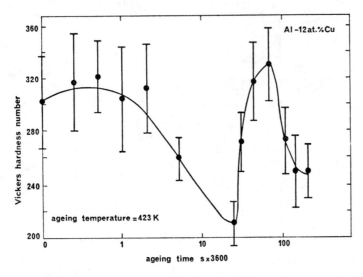

Figure 3. Hardness of plasma-sprayed Al-12at.% Cu alloys, ⌐220 μm
thick as sprayed, then reduced 30% by rolling. Aging at 423K.

quenching and compacting, no detailed x-ray or TEM investigation was
mounted to interpret our findings. However, some diffractometer measurements were made to assess mean supersaturated solute contents in
various deposits. Thus, for the 12% alloy, we found that a 220 μm
deposit made with cross-blast contained 6.3at.% Cu in solution; one
120 μm thick, without cross blast, 7.7at.%Cu; one 120 μm thick with
cross-blast, 9.5at.%. (These findings show the sharp deterioration
with increasing thickness.) Typically, a 220 μm deposit of the 6% alloy, made with cross-blast, contained 3.7at.% Cu in solution. As a further aid to interpretation, isochronal annealing measurements as sum-
marised in Fig. 4 were made.

It is plain that in Fig. 2, curve (1) represents an effectively
quenched alloy with remaining potential for aging, whereas (2), without cross-blast, is already fully aged in the as-deposited condition
and thereafter overages. There is an indication of secondary aging.

72

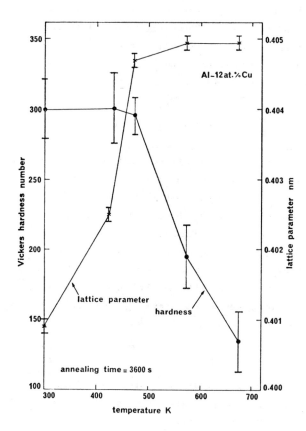

Figure 4. Changes in lattice parameter and microhardness on iso-chronal annealing for 1 hour of an Al-12at% Cu alloy, plasma-sprayed with cross-blast to a thickness of 120 μm, and roll-ed 30% before aging.

Figure 3 shows a large scatter on hardness values in the 12% alloy, which is consistent with the approx. ±3% variance in solute content revealed by Guinier diffraction. There is slight, barely significant primary aging. The large hardness dip on continued aging, followed by very large secondary aging, is a novel, as yet unexplained phenomenon. It may well be due to an unknown phase, but to test this requires further examination. Figure 4 suggests, but does not firmly prove, that this secondary aging is absent at much higher aging temperatures, otherwise one would expect an upturn of hardness at the highest aging temperature in that Figure.

73

Figure 4 also indicates that over a range of aging temperatures (300-470K), reduction of solid-solution hardening is neatly balanced by another factor, which can only be increased precipitation-hardening. Certainly this graph makes it clear that the major part of the high initial hardness of 300 DPN is due to supersaturated solid-solution hardening.

CONCLUSIONS

Subject to the careful optimisation of operating conditions, and using the new technique of cross-blasting to divert the hot plasma plume, it is possible to achieve very substantial hardness-levels in as-sprayed Al-Cu coatings. Layers thicker than 200-300 μm begin to fall substantially in supersaturation and hardness. It may be possible to improve this by direct cold-gas quenching of the deposit as it grows, or by further improvement of spraying conditions, to the extent that metastable layers of several millimetres can effectively be built up. It is already clear, however, that very hard surface coatings can be applied by this technique.

ACKNOWLEDGMENT

We are grateful to Dr. M. Laridjani for his help with the Guinier camera, and to Mr. R. Hill for experimental assistance, and to the Science Research Council for financial support.

REFERENCES

1. Moss, A.R. and W.I. Young, Powder Met., 7, 261 (1964).
2. Moss, M., Acta Met., 16, 321 (1968).
3. Moss, M. and D.M. Schuster, Trans. ASM, 62, 201 (1969).
4. Moskowitz, L.N., R.M. Pelloux, & N.J. Grant, Proc. 2nd Int'l Conf. on Superalloys, Seven Springs, New York (1972).
5. McPherson, R., J. Mater. Sci., 8, 851 (1973).
6. Barry, T.I., R.K. Bayliss, and L.A. Lay, J. Mater. Sci. 3, 229 (1968).
7. McPherson, R.M., J. Mater. Sci., 8, 859 (1973).
8. Marynowski, C.W., F.A. Holden, and E.P. Farley, Electrochem. Tech., 3-4, 109 (1965).
9. Lebo, M. and N.J. Grant, Met. Trans., 5, 1547 (1974).

10. Grisaffe, S.J. and W.A. Spitzig, NASA Tech. Note D-1705 (1963).
11. Cahn, R.W., K.D. Krishnanand, M. Laridjani, M. Greenholz, and R. Hill, Proc. Second International Conference on Rapidly Quenched Metals, §2, N.J. Grant and B.C. Giessen, Eds., Mat. Sci. Eng. 23 (1976).
12. Silcock, J.M., T.J. Heal, and H.K. Hardy, J. Inst. Met. 82, 239 (1953).
13. Thursfield, G. and M.J. Stowell, J. Mater. Sci. 9, 1644 (1974).
*   The use of plasma spraying onto a rotating copper disk in an inert gas filled chamber to produce bulk quantities of amorphous Cu-Zr has recently been reported [B.C. Giessen, N.M. Madhava, R.J. Murphy, R. Ray, and J. Surrette, Met. Trans. 7A (1976) (Acc.)]. (Eds.)

APPLICATION OF THE SPLAT-COOLING METHOD TO NON-METALLIC SYSTEMS

A. Revcolevschi and R. Collongues

Laboratoire de Chimie Appliquée de l'Etat Solide
Vitry, France

1.0 INTRODUCTION

The technique of ultra-fast quenching of materials from the melt (splat-cooling) developed about 15 years ago by Duwez and his collaborators[1] has since been extensively applied, as evidenced by a large volume of publications quoted in two very complete reviews.[2,3]

·It is surprising to notice in these reviews that the application of splat-cooling has been almost exclusively limited to metals and alloys. Ionic or covalent type compounds like oxides have been totally neglected, except for some experiments by Sarjeant and Roy.[4-6]

It seems however that quite recently ceramists in different countries have expressed interest in the techniques.

We shall review here briefly the work carried out so far in the field of oxides.

2.0 EXPERIMENTAL DEVICES

Splat-cooling of oxides requires essentially two sets of conditions:

- those generally considered in ultra-fast quenching from the melt,[7] the main one being the necessity to spread the liquid on a substrate into an extremely thin layer.
- those associated with the manipulation of molten oxides (availability of high temperatures and necessity to operate in oxidizing conditions, generally in air, in order to avoid non-stoichiometry by oxygen deficiency).

Different devices have been constructed. We shall consider them according to the method used to spread the liquid on the substrate.

2.1 Squeezing Methods

The molten sample is squeezed between two metallic bodies into a thin foil.

2.1.1 Rollers

The end of a rod-shaped sample is heated in a plasma or an image furnace: the resulting liquid globules are dropped and rolled between two rollers rotating at 1200 rpm.[8,9] Films a few tens of μm thick are collected; their surface can reach a few square centimeters. Experiments are carried out in air. Melting points well above 2000°C can be reached.

2.1.2 Hammer and Anvil Technique

Ultra fast quenching by spreading a molten sample between a metallic body and a high velocity piston has been suggested by Pietrokowsky[10]; the method has been, in the case of metals, generally associated with r.f. induction heating in a levitation type device.

This method has been adapted to oxides, in devices involving appropriate heating methods.

In the setup of Yajima, et al., the oxide to be quenched, in powder form, is injected into the flame of a plasma torch. The resulting molten droplets are crushed between two metallic bodies propelled one against the other.[11]

In the device constructed in the C.N.R.S. — Laboratory at Odeillo in France, the sample is melted into a globule at the focus of a 3kW solar furnace (Figure 1). The spring loaded hammer sliding on a rail

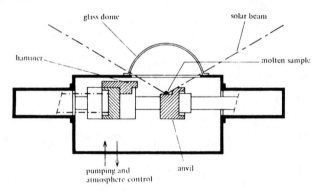

Figure 1: Hammer and anvil type device associated with a solar furnace.[9]

is released magnetically. The moving parts are devised so that the solar
radiation is not masked before quenching. Experiments may be carried
out in air at temperatures above 3000°C. For experimentation under controlled atmosphere, the device is placed into a chamber with the radiation penetrating through a glass dome.[9,12] The quenched samples are
made up of two parts: one actually squeezed between the two metallic
plates, which has an average thickness of about 100–300 μm varying
with the type of material, and another part resulting from the bursting
of the sample upon squeezing and consisting of lamellæ 10 to 50 μm
thick which have undergone the best quenching conditions. A modification of this device is presented in Reference 13.

Laser melting has recently been used and temperatures above 2000°C
have been reached.[14]

2.2 Shock Wave Quenching (Gun Method)

A molten oxide sample is blown onto a metallic substrate by a shock
wave generally originated in a "gun" by the bursting of a thin polyethylene diaphragm. The guns which are used are similar in design to
those described by Duwez.[7] The heating methods are of two sorts:

2.2.1 Resistance Heating

In the device of Sarjeant and Roy[4] the sample is melted in a V
notch of a metallic strip connected to the output of a transformer.
The strip made of platinum or iridium is heated in air; the liquid
is blown on a glass or metallic substrate. An alternative to such a

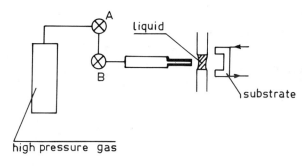

Figure 2. Splat-
cooling adapted to
an image furnace
floating zone.[9]

device consists in melting an oxide sample inside a r.f. heated platinium DTA type crucible. The shock wave forces the liquid through a hole made at the bottom of the crucible and propels it on a copper substrate placed a few centimeters below. [15-17]

The temperatures available in such devices depend on the metal which is used: generally 1600°C with platinum and 2100°C for iridium.

2.2.2 Concentration of Light

The device presented in Figure 2 is a combination of a "gun" similar to those generally used in splat-cooling units and a floating zone. High-pressure argon is admitted into a copper tube reservoir placed between two electrovalves, A and B. Opening of valve B bursts a polyethylene diaphragm separating into two parts a cylindrical chamber. The shock wave is led through an alumina tube onto the floating zone. The droplets impinge on a water-cooled copper substrate. The floating zone is established between two cylindrical rods prepared by compression and sintering, placed at the focus of a bielliptical image furnace. Temperatures up to 2300°C can be attained when the light source is a 6.5kW xenon lamp. Experiments are carried out in air without any risk of contamination. [9]

An exploding wire technique has also been considered in association with an image furnace. [9]

In all the methods involving shock wave quenching, the quenched samples are in the form of very thin lamellæ welded one to the other into a lace type design.

2.3 Flame Spraying Technique

Oxide powder is introduced and melted in the flame of an oxyhydrogen burner or a plasma torch. The molten droplets are sprayed on a metallic rotating disk[9,18] or a cold substrate. [19]

3.0 COOLING RATE DETERMINATION

Cooling rates have been theoretically estimated for oxide splat-cooling. [20] They have also been experimentally evaluated[21] by a method similar

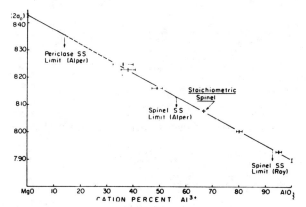

Figure 3. Plot of
solid solution lat-
tice parameter a_0
versus aluminium cat-
ion percent in the
system MgO-Al$_2$O$_3$, indicating extension of
solid solubility.[5]

to that developed by Burden and Jones[22] for metallic alloys, consisting
in correlating the solidification rate to the interlamellar spacing of
an eutectic microstructure. Such eutectic interlamellar spacing has
been measured on electron microscopy replicas of NiO-CaO splat-cooled
samples. Cooling rates of 10^5°C sec^{-1} have been found for a hammer and
anvil type technique.[22]

4.0 NEW METASTABLE PHASES

Ultra-fast quenching from the melt is particularly interesting because it makes possible the retention at room temperature of metastable
phases impossible to retain by conventional quenching methods.

The application of splat-cooling to oxides has led to some new metastable phases which we shall consider here according to the generally
accepted classification[2,3,7]: supersaturated solid solutions, non-equilibrium crystalline phases and amorphous structures.

4.1 Supersaturated Solid Solutions

The only case to have been reported so far seems to be that of
solid solubility extension in the MgO-Al$_2$O$_3$ system where liquidus
temperatures are above 1900°C.[5] The only compound in this system,
spinel MgAl$_2$O$_4$, dissolves at high temperatures large amounts of
alumina yielding a cation deficient solid solution, the composition

81

TABLE 1[10]

Oxide	Stable modification	modification obtained by splat-cooling	previous information on the metastable modification	m.p. °C
Al_2O_3	α (corindon)	γ , δ (spinel)	prepared by plasma spray quenching (25)	2050
Bi_2O_3	α (monoclinic)	β (tetragonal)	β stable above 710°C	825
WO_3	α (monoclinic)	β (tetragonal)	β stable above 770°C	1470
Nb_2O_5	α (H form)	δ (hexagonal)	already prepared by hydro-xide gel decomposition	1490
Ta_2O_5	α (H form)	δ (hexagonal)	previously unknown	1870

TABLE 2: Binary Oxide Systems in which Glasses have been Prepared.

Glass former	Modifier yielding glass structures by :	
	Rapid cooling	Splat-cooling
Nb_2O_5 and Ta_2O_5	K_2O, BaO, PbO (28)	same + Ln_2O_3 (8)
TiO_2	K_2O, Rb_2O, Cs_2O (29) (30)	CaO, SrO, BaO (6)
	BaO (31)	Ln_2O_3 (8)
Al_2O_3	CaO (31), SrO, BaO (19)	PbO (much larger concentration range) (15)
	PbO (32)	Ln_2O_3 (11) (35)
Ga_2O_3	CaO (31)	PbO (15), Ln_2O_3 (33)
Fe_2O_3		CaO, SrO, BaO, PbO (15), Ln_2O_3 (8)

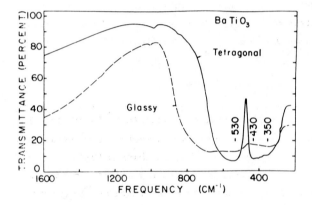

Figure 4. Infra-red spectra for the splat-cooled glass and crystallized form of BaTiO3 (bands from 500 to 650 cm^{-1} and 690 to 850 cm^{-1} are attributed to respectively sixfold and fourfold coordination of Ti^{4+}.[6]

limits of which have been determined by Roy, et al.[23] $MgAl_2O_4$ also dissolves periclase, MgO, in the amounts defined by Alper, et al.[24]

Splat-cooling has considerably extended the range of solid solubility on both sides of the stoichiometric composition. Figure 3 indicates the extension of the solubility. The lattice parameter of pure alumina is that of γ alumina (spinel structure), easily accessible by splat-cooling. A continuous spinel structure solid solution can then be found from pure alumina up to 80 mol.% MgO.

4.2 Non-Equilibrium Crystalline Phases

A few metastable modifications of pure oxides have been prepared by splat-cooling and are presented in Table 1.[4]

4.3 Amorphous Structures (Glasses)

One of the most attractive features of splat-cooling is that in some cases it is possible by this method to retain in the solid state a random arrangement of atoms derived from that of the liquid state. In the case of metals more than thirty amorphous metallic alloys have been reported.[2]

Similar results have been obtained with pure or mixed oxides. We shall not consider here the oxide systems based on well-known glass formers such as SiO_2, GeO_2, B_2O_3, P_2O_5, etc., in which vitreous phases are prepared by normal cooling. It should be pointed out, however, that in these systems splat-cooling extends the composition range in which glasses can be prepared.

4.3.1 Pure Oxide Glasses

Splat-cooling has led to the following pure oxide glasses: V_2O_5 (m.p. 675°C), TeO_2 (m.p. 733°C), MoO_3 (m.p. 795°C) and WO_3 (m.p. 1470°C).[4] In the case of the last two oxides the glass phase coexists with crystalline modifications.

In the particular case of vanadium pentoxide glasses prepared by splat-cooling a detailed study of the structure and semi-conducting properties has been carried out;[16, 17, 26] short range order seems to be similar to that found in orthorhombic V_2O_5. An ESR study of the glasses has shown that the mobility of the 3d electron of the paramagnetic ion V^{4+} is smaller in the glass than in the crystal. Long range order is reestablished above 200°C.

4.3.2 Binary Oxide Systems

Splat-cooling has shown the "potential glass forming" ability of some oxides which by themselves do not yield glasses even when splat-cooled, but which associated to a second oxide of very different cationic field (charge/radius) may form vitreous structures in large composition ranges by fast quenching from the melt: these oxides are Nb_2O_5, Ta_2O_5, TiO_2, Al_2O_3, Ga_2O_3 and Fe_2O_3.

Table 2 summarizes the results which have been obtained.

The properties of some of these glasses have been studied: Ln_2O_3-Al_2O_3[11,27], TiO_2-BaO.[6] In this last system infra-red spectrometry measurements have indicated a six-fold coordination for Ti^{4+} in both glass and crystalline structure. However, in the glass a shift towards a four-fold coordination appears (Figure 4).

The glass forming ability of the glass formers mentioned in Table 2 has been evaluated by Stanworth in terms of atomic radius and electronegativity of the corresponding elements.[34]

5.0 CONCLUSION

This very brief review indicates an increasing interest of ceramists in splat cooling. It seems, however, that the work carried out so far in the field of oxides has consisted mainly of the development of experimental devices and exploring the possibilities of the method, rather than applying it to specific problems.

Since splat quenching of oxides yields cooling rates of the same order of magnitude as those obtained for metals, one might predict that a wider application of the method to ceramic systems will lead to results as spectacular as those reported for metals in Refs. 2 and 3.

It is difficult to foresee possible applications of splat-cooled oxides. It should be pointed out that their utilisation in the form of foils is impossible due to sample brittleness. The magnetic properties of the new iron oxide glasses will be worth studying. It will also be interesting to study the chemical properties of splat-cooled amorphous materials and compare them to those of the equivalent crystallized material. In this respect we note the important difference of the solubility in water of amorphous and crystallized vanadium pentoxide (see Ref. 37).

It is also expected that the next few years will see the application of splat-cooling to non-metals other than oxides.

6.0 BIBLIOGRAPHY

1. Duwez, P., R.H. Willens, and W. Klement: J. Appl. Phys., 31, 1136 (1960).
2. Anantharaman, T.R. and C. Suryanarayana: J. of Mater. Sci., 6, 1111 (1971).
3. Jones, H. and C. Suryanarayana: J. of Mater. Sci. 8, 705 (1973).
4. Sarjeant, P.T. and R. Roy: J. Amer. Cer. Soc. 50, 500 (1967).
5. Sarjeant, P.T. and R. Roy: J. Appl. Phys. 38, 4540 (1967).
6. Sarjeant, P.T. and R. Roy: J. Amer. Cer. Soc. 52, 57 (1969).
7. Duwez, P.: Trans. ASM 60, 607 (1967).
8. Suzuki, T. and A.M. Anthony: Mater. Res. Bull. 9, 746 (1974).
9. Revcolevschi, A., A. Rouanet, F. Sibieude, and T. Suzuki: High Temp. High Press. 7 (1975) [in press]
10. Pietrokowsky, P.: Rev. Sci. Instr. 34, 445 (1963).
11. Yajima, S., K. Okamura, and T. Shishido: Chemistry Letters Chemical Society of Japan — 741 (1973).
12. Rouanet, A., F. Sibieude, and M. Faure: J. Phys. E 8, 389 (1975).
13. Foex, M., F. Sibieude, A. Rouanet, and D. Hernandez: J. Mater. Sci. 10, 1255 (1975).
14. Krepski, R., K. Swyler, H.R. Carleton, and H. Herman: J. Mater. Sci. 10, 1454 (1975).
15. Kantor, P., A. Revcolevschi, and R. Collongues: J. Mater. Sci. 8 1359 (1973).
16. Rivoalen, L.: Thesis — Paris University (1974).
17. Rivoalen, L, A. Revcolevschi, J. Livage, and R. Collongues, "Amorphous vanadium pentoxide", J. Non Cryst. Solids [in press].
18. Poudrai, J.: Thesis, University of Orleans, France (1974).
19. Frank, B. and J. Liebertz: Glastechn. Ber. 41, 6, 253 (1974).
20. Sarjeant, P.T. and R. Roy: Mater. Res. Bull. 3, 265 (1968).
21. Revcolevschi, A.: "Cooling rate determination in splat cooling of oxides", J. Mat. Sci. [in press].
22. Burden, M.H. and H. Jones: J. of the Inst. of Metals 98, 249 (1970).
23. Roy, D.M., R. Roy, and E.F. Osborn: Am. J. Sci. 251, 337 (1953).
24. Alper, A.M., R.N. McNally, P.H. Ribbe, and R.C. Doman: J. Amer. Cer. Soc. 45, 263 (1962).
25. Lejus, A.M.: Thesis — Paris University (1964).
26. Kahn, A., J. Livage, and R. Collongues: Phys. Stat. Sol. (a) 26, 175 (1974).
27. Coutures, J.P., A. Rouanet, G. Benezech, E. Antic, and P. Caro: C.R. Acad. Sci. 280, c, 693 (1975).
28. Baynton, P.L., H. Rawson, and J.E. Stanworth: Proceedings of the IVth Int. Glass. Congress, 52 — Paris (1956).
29. Rao, B.V.J.: Physics Chem. Glasses 4, 22 (1963).
30. Rao, B.V.J.: J. Amer. Cer. Soc. 47, 455 (1964).
31. Rawson, H.: Inorganic Glass Forming Systems, Academic Press, 199 New York (1967).
32. Harari, A. and J. Théry: C.R. Acad. Sci. 264, 84 (1967).
33. Coutures, J., F. Sibieude, A. Rouanet, M. Foex, A. Revcolevschi, and R. Collongues: Rev. Int. Hautes Temp. et Réf. 4, 263 (1974).
34. Stanworth, J.E.: J. Amer. Cer. Soc. 54, 61 (1971).

35. Yajima, S., K. Okamura, T. Shishido: <u>Chemistry Letters — Chemical Society of Japan</u> 1327 (1973).
36. Tarte, P.: <u>Physics of Non-Crystalline Solids</u> , Inter-Science Publishers, 549, New York (1965).
37. Livage, J. and R. Collongues, Proc. 2nd Int'l Conf. on Rapidly Quenched Metals, M.I.T., Cambridge, Massachusetts, 1975, 82, N.J. Grant and B.C. Giessen, Eds., <u>Mat. Sci. Eng.</u> <u>23</u> (1976).

SPLAT QUENCHING OF IRON-BASE ALLOYS
IN A CONTROLLED ATMOSPHERE

J.V. Wood   I.R. Sare

University of Cambridge, U.K.   CSIRO Division of Tribophysics,
                                   Australia

INTRODUCTION

Although many devices have been developed to rapidly quench alloys
from the liquid state, the most popular and most effective with respect to achieving high cooling rates is still the basic 'gun' technique. Since its inception, several modifications have been developed to permit the quenching of a wider range of materials and the attainment of higher cooling rates. Early versions operated with an
inert gas cover to prevent oxidation and now a family of devices designed to operate in completely controlled atmospheres has evolved.
Higher cooling rates are claimed for these more advanced apparati
due to: (1) bad thermal conductivity of inert gas atmosphere, (2) enhanced thermal contact between molten metal and substrate, and (3)
more efficient liquid spreading.

An apparatus for rapid quenching of iron-base alloys in an inert
atmosphere has previously been described by Ruhl and Cohen (1969)
using a complex and expensive resistance furnace. It is necessary to
reduce the thermal mass of any crucible assembly to a minimum when
dealing with high melting point alloys in a closed system. The present work describes an apparatus utilising a radio frequency heating
source capable of being pumped down to 1.5mN m^{-2} (10^{-5}torr) prior
to the admission of an inert atmosphere.

APPARATUS

A schematic diagram of the high temperature controlled atmosphere
'gun' is shown in Figure 1. Alumina and silicon nitride were employed as crucible materials and both react with dissolved carbon in
molten steels under partial pressures of argon. The reaction with
alumina is well known and can be prevented by passing a flow of pure

87

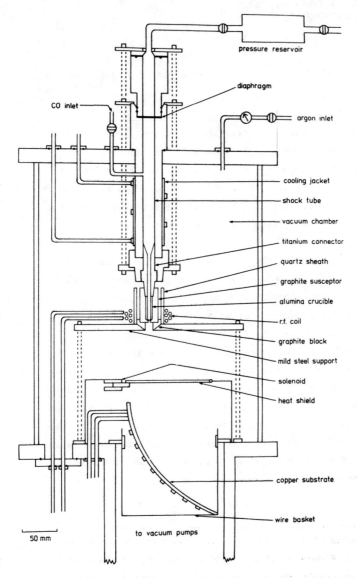

pressure reservoir

diaphragm

CO inlet

argon inlet

cooling jacket

shock tube

vacuum chamber

titanium connector

quartz sheath

graphite susceptor

alumina crucible

r.f. coil

graphite block

mild steel support

solenoid

heat shield

copper substrate

wire basket

50 mm

to vacuum pumps

Figure 1. Schematic diagram of high temperature, controlled atmosphere, 'gun' splat quencher.

88

CO over the melt at a rate of $15cm^3.s^{-1}$. The reaction with Si_3N_4 is selective in that the α phase reacts faster than β and crucibles degenerate very rapidly unless a flow of N_2 is maintained over the melt.

To achieve good coupling from the radio frequency source a graphite susceptor surrounded the crucible with a thickness just exceeding the coupling skin depth ($\sim$3mm). The stainless steel shock tube cannot be directly fitted to the graphite susceptor because of its low melting point. The connector is made from pure titanium channelling the shock wave from the shock tube directly into the crucible. A carbide is formed by diffusion of carbon into the titanium thus giving a coherent refractory surface. The focusing arrangement and the relatively long crucible ensure that there is no splash-back of molten metal during quenching.

To achieve the high temperatures needed it was necessary to use a seven-turn closely wound r.f. coil. In a closed system functioning below atmospheric pressure, there was considerable arcing problems both between the individual turns of the coil and between the coil and graphite susceptor. The former was overcome by dipping the coil in alumina cement and subsequently baking the composite. The latter problem was solved by inserting a tight fitting quartz collar on the inside of the coil. Direct temperature measurement was not possible with this assembly, but a series of standard power-time settings were derived by measuring the temperature in the crucible without the shock tube being present. Prior to heating, the chamber was pumped down to $3mN.m^{-2}$ (2×10^{-5}torr) after a preliminary argon flush. Pure argon was admitted to a pressure of $60kN.m^{-2}$ (500torr) and heating took four minutes to achieve at least 100K superheat above the specimen melting point.

Quenched foils generally adhered firmly to the roughened water-cooled substrate, though on occasions, large thick coherent splats were produced which pulled away from the substrate during cooling. Such specimens were discarded.

Values for the surface tension of the transition metals used in the present work are at least twice those of metals and alloys commonly investigated by rapid quenchers (e.g., Fe = 1872 dyn.cm^{-1}, and Al = = 914 dyn.cm^{-1} [Allen, 1972]). Hence, a molten iron stream is relatively difficult to break up into small droplets. To achieve thin splats it was necessary to have an orifice size of 0.4mm and an orifice to substrate distance of 100mm. Carbon does not significantly affect the surface tension of iron, but dissolved oxygen has a dramatic effect (Allen, 1972). This is confirmed by introducing oxygen into the closed system which enables large thin coherent foils to be produced at a shorter orifice — substrate distance. Thus for iron base alloys, the mean thickness of the splat is reduced which compensates for the less efficient heat transfer due to the oxide layer.

Oxygen traces also have an effect on microstructure. Figure 2a is a scanning electron micrograph from the top surface (the surface which is last to solidify) of an austenitic stainless steel quenched in the above apparatus. An equiaxed grain structure is observed which contrasts markedly with that of Figure 2b which was obtained from the same alloy after 10 torr of oxygen had been introduced into the atmosphere prior to splatting. Transmission electron microscopy gives evidence of a significant surface oxide layer in addition to the pores within grains and at grain boundaries. Furrer (1973) suggests that oxide particles like these are beneficial to sintered compacts made from splat quenched Al-Fe alloys, but for steels the large amount of oxide and excessive porosity will hinder compaction and significantly increase sintering times.

Nickel was used as a test material for the apparatus. Davies, et al., (1973) report that very small regions of nickel can be rendered non-crystalline after quenching in an argon blanket. Under normal conditions splat quenched nickel remained crystalline in our apparatus. However, on increasing the partial pressure of air to about 0.1 torr, small non-crystalline regions could be found by transmission electron microscopy (Wood and Akhurst, 1975).

Figure 2. (a) Top surface of austenitic stainless steel foil quenched
in pure argon. (s.e.m.)
(b) As in (a), but with 10torr O_2 present in atmosphere.

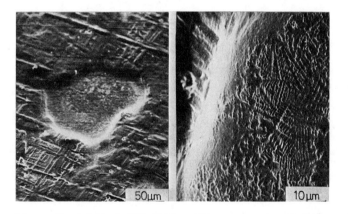

Figure 3. (a) Fe-1.2%C 'lift off' area surrounded by regions which
replicate the substrate.
(b) Detail of (a) showing dendrite growth in 'lift off' area.
(s.e.m.).

The most common method of estimating cooling rates in splat cooling studies utilises the relationship between dendrite arm spacing (d) and cooling rate (r) proposed by Matyja, et al. (1969), and expressed by the general equation: $d = c \cdot r^{-a}$, where a and c are constants. A plot of this function using a value of $a = 0.32$ has been used as a 'master diagram' for several aluminium base alloys, ignoring the effects of chemical composition. Brower, et al. (1970) have measured cooling rate and dendrite arm spacing for a number of iron-base alloys. Whilst the relationship is linear on a log-log plot, the value of a depends on the alloy composition (between 0.32 for an Fe-25%Ni alloy and 0.41 for an Fe-17%Cr-0.7%Mo-1.1%C steel). The values for d achieved in the present apparatus for transformable and non-transformable steels varies between $0.1-0.5\mu m$ in foils examined by transmission electron microscopy. Using the two values for a above this leads to a difference of two orders of magnitude in r for a $d = 0.1\mu m$. Thus unless data are available for a specific alloy composition very little can be deduced and the technique of inter-lamellar spacing of a eutectic alloy must be considered as an alternative (Burden and Jones, 1970).

Using a different crucible arrangement on the present apparatus, Scott (1973) estimated a mean cooling rate for eutectic $Al-CuAl_2$ of 3×10^4 $K.s^{-1}$ for foils of thickness $65\mu m$. However, as much as 60% of the eutectic had become degenerate which is predicted to occur at rates of 10^6 $K.s^{-1}$ (Miroshnichenko, 1966). Also regions of α-Al solid solution were found, which Davies and Hull (1972) claim to occur at rates of the order of 10^9 $K.s^{-1}$. It would thus appear that no one technique can unambiguously be used for the determination of cooling rate, but the general consensus of results indicates a cooling rate region between 10^6-10^8 $K.s^{-1}$ for the present apparatus.

HEAT FLOW IN SPLAT COOLING

A number of anomalous structures are seen in very thin regions of some rapidly quenched alloys produced by the gun technique. In general, it is assumed that such regions have cooled extremely fast compared with

92

the bulk of the foil. Since such areas are often transparent to electrons accelerated at 100kV, the myth that all such electron transparent areas are the fastest cooled has been propagated. In steels, electron transparent regions are characterised by two main structures: (1) featureless grains, and (2) branched dendrites. In both cases, growth occurs parallel to the substrate-foil interface. Indeed, for primary dendrites to be observed in the foil plaine, (as needed to accurately measure secondary dendrite arm spacing) this situation must occur. However, the bulk of any splat is made from grains growing perpendicular to the plane of the foil (Figure 2a). If extensive electrochemical or ion-beam thinning is performed on splat foils, these typical structures are revealed (Wood and Sare, 1975).

It has been observed that not all areas of a splat solidify in direct contact with the substrate. Such an area is shown in Figure 3a, which is a scanning electron micrograph of a bottom surface of a foil, and have been termed 'lift off' areas. Wood and Honeycombe (1974) have correlated such areas with electron transparent regions and Figure 3b shows that within these regions dendritic growth occurs in the plane of the foil. It has been postulated that the primary direction of heat flow occurs to the nearest contact point on the substrate (Wood and Sare, 1975) and that there is, in general, a critical diameter of 'lift off' region whereby the actual cooling rate in such thin areas is less than the mean cooling rate of the remaining splat. This can be observed in austenitic steels where in small thin regions featureless grains are observed, but at a diameter of about 20μm there has been complete breakdown to dendritic structures. In both cases the thickness of the foil is comparable and illustrates the care that must be exercised in applying one-dimensional heat flow theories to electron transparent regions.

REFERENCES

1. Allen, B.C.: 'Liquid Metals', Marcel Decker, New York, 16 (1972).
2. Brower, W.E., R. Stachan, and M.C. Flemings, Cast Met. Res. J. 6, 176 (1972).
3. Burden, M.H. and H. Jones, J. Inst. Metals 98, 249 (1970).

4. Davies, H. and J.B. Hull, Scripta Metall. 6, 241 (1972).
5. Davies, H., J. Aucote, and J.B. Hull, Nature (Physical Sci.) 246, 13 (1973).
6. Furrer, P., Proc. 3rd Int'l Conf. on the Strength of Metals and Alloys [Cambridge], p. 46 (1972).
7. Matyja, H., B.C. Giessen, and N.J. Grant: J. Inst. Metals 96, 30 (1969).
8. Miroshnichenko, I.S.: Crystallisation Processes, Consultants Bureau, New York, p. 61 (1966).
9. Ruhl, R.C. and M. Cohen, Trans AIME 245, 241 (1969).
10. Scott, M.G., PhD Dissertation, U. Cambridge (1973).
11. Wood, J.V. and R.W.K. Honeycombe, J. Mater. Sci. 9, 1183 (1974).
12. Wood, J.V. and I.R. Sare, Met Trans (1975) [in press].
13. Wood, J.V. and J. Akhurst, U. Cambridge (1975) [unpublished work].

METASTABLE PHASE RELATIONSHIPS PRODUCED BY
RAPID FREEZING OF ALLOYS OF Zn AND Cd WITH Cu AND Ag

T.B. Massalski   Y. Bienvenu

Metal Physics Laboratory,   I.R.S.I.D.
Mellon Institute of Science   Metz, France
Carnegie-Mellon University,
Pittsburgh

INTRODUCTION

As Pol Duwez pointed out in his Campbell lecture[1], from the
point of view of metastable alloys which still retain a crystalline
structure after rapid freezing from the liquid state, the main
theoretical motivation for achieving the highest rates of cooling
has been an interest in testing the Hume-Rothery rules. According
to the generally accepted view, two metals should form a complete
series of solid solutions if their structures are the same, if there
is only a small difference in valence and electronegativity, and
if the size difference between the respective atomic radii does
not exceed about 13-15%. There are several binary systems for
which the above conditions appear to be well satisfied and yet a
misability gap is found in the solid state. The well known examples
are provided, among others, by the group of four elements adjoining
in the Periodic Table (Fig. 1).

When mixed vertically, the resulting Cu-Ag and Zn-Cd phase
diagrams are simple eutectics in the solid state despite the fact
that the starting structures are the same; fcc in Cu-Ag and hcp
in Zn-Cd. Since the respective metals belong to the same columns
in the Periodic Table the nominal valence is the same, and the
electronegativity difference is negligible. Some other parameters
of interest are listed below in Table I.

As is well known, early experiments in rapid liquid quenching
have shown that a complete series of solid solutions can be obtained
in the Cu-Ag system.[1] Since the size difference between Cu and Ag
is substantial but not excessive, it seems that this result agrees
with the Hume-Rothery postulates. If the solute partitioning can
be prevented by rapid cooling, and if the free energy of the liquid
is sufficiently raised because of undercooling, there is enough
free energy driving force available to crystalize the solid of the
same composition as the liquid, and one of the Hume-Rothery factors,
namely the effect of the atomic size difference, can be suppressed.
In other words, the size problem plays a lesser role during solidification of a supercooled liquid.

The Zn-Cd system presents a puzzling picture because, up to
the present time, liquid quenching experiments have been unsuccessful
in overcoming the eutectic separation, and a two-phase mixture always
results in the middle of the phase diagram.[1,2] This behavior can
be rationalized in terms of a schematical quenching diagram showing
the relative positions of the respective free energy curves at different rates of cooling.[3,4] The situation for the Cu-Ag and Zn-Cd

95

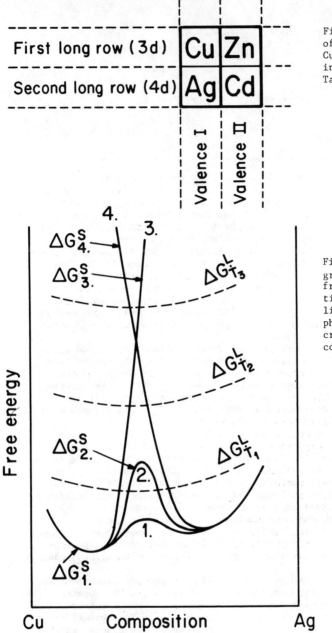

Figure 1. Position of the four elements Cu, Ag, Zn, and Cd in the Periodic Table.

Figure 2. The diagrammatic relative free energy relationships between liquid and solid phases with increasing undercooling.

TABLE

Solvent	Structure	a/c	d_{HR}	at. vol. Ω_o	Hume-Rothery Size Factor				Volume Size Factor				Debye temp. θ_D
					Cu	Ag	Zn	Cd	Cu	Ag	Zn	Cd	
Cu	fcc		2.557	11.8		12.9	4.2	16.5		43.5	17.1	67.4	343
Ag	fcc		2.889	17.1	-11.5		-8.0	3.1	-27.8		-13.7	14.8	223
Zn	hcp	1.85	2.664	15.2	-3.8	+8.2		-10.5	-50.0	-18.4		-18.5	322
Cd	hcp	1.88	2.979	21.6							+42.7		210

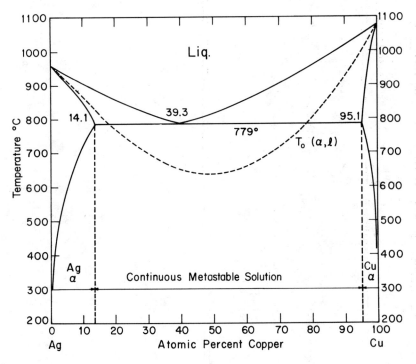

Figure 3a. A diagrammatic trend of the T_0 temperature in silver–copper system.

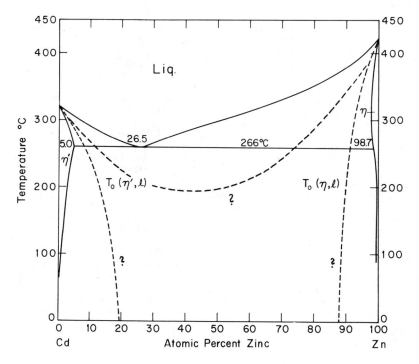

Figure 3b. Two possible diagrammatic trends of the T_o temperature in cadmium–zinc system.

99

systems is illustrated in Fig. 2. It may be argued that for the Cu-Ag system, at some sufficiently rapid cooling rate $\mathring{T}_1$, the free energy of the liquid phase $\Delta G^{Liq}_{T_1}$ at the solidification temperature is situated above the hump in the free energy curve ΔG^{Solid}_1. Hence partitioning is prevented and the T_0 temperature* can become a continuous function of composition across the corresponding phase diagram (Fig. 3a). It appears that in the Zn-Cd system the hump is much larger (i.e., as in curve ΔG^{Solid}_2 in Fig. 2), and the curve ΔG^{Liq}_T then falls below it, preventing the T_0 temperature from becoming continuous. With this interpretation, an eventual success in obtaining a complete metastable range of solid solubility in the Zn-Cd system would depend on our ability to raise the $\Delta G^{Liq}_{T_1}$ curve to the position $\Delta G^{Liq}_{T_2}$, i.e., above the large hump. However, it is also possible that the free energy situation for the Zn-Cd system is like that shown in Fig. 2 by the two curves ΔG^{Solid}_3 and ΔG^{Solid}_4, i.e., that each η phase in Fig. 3b is represented by a separate free energy curve, and that they intersect in the middle of the diagram. Under these conditions, complete misability in a true sense becomes impossible, no matter how fast the cooling rate.

In addition to the above considerations, the phase relations formed by the same four adjoining elements when they are mixed horizontally, or diagonally, present another set of puzzles in the Zn-, and Cd-rich regions. The phase diagrams for the four representative systems are shown in Fig. 4. In each binary system, the primary solid solution, close packed hexagonal η phase, based on Cd or Zn, is followed by an intermediate close packed hexagonal ϵ phase (or ϵ' in Cd-Cu). Again, there is a misability gap between the two hcp phases. Ever since the late 1920's, when the phase fields of the Cu-Zn system were established in detail by metallurgical techniques, this intriguing situation has been of considerable interest. Our recent work has shed some light on these problems.[5]

The two-phase field in each system is limited in temperature by a peritectic equilibrium reaction $\text{Liq} + \epsilon \rightleftharpoons \eta$ which takes place less than $20\,^\circ\text{C}$ above the melting temperature of zinc, or cadmium. The hcp structures of ϵ and η alloys differ mainly in two respects: in their content of the noble metal constituent, and in their axial ratio, c/a. The η phase alloys, which are solvent-rich, have the highly prolate structure with an axial ratio well above the ideal value of 1.633 for the closest packing of hard spheres, as do Zn and Cd, while the ϵ phase alloys have an axial ratio below the ideal value. A shallow minimum occurs in the axial ratio of the ϵ phases at an electron concentration near 1.86.[6]

---

*The temperature T_0 below which a composition-invariant solidification of a supercooled liquid becomes possible is the temperature at which the free energy curves for the liquid and the solid intersect. The actual solidification temperature for the metastable phase is probably somewhat lower than T_0 in each case, in order to overcome nucleation and growth difficulties. Thus, at T_0, composition invariant solidification is possible but no driving force is available for the kinetic process of solidification.

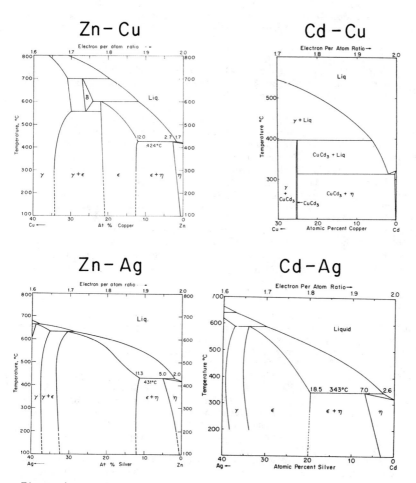

Figure 4. The Zn- or Cd-rich portions of the phase diagrams Zn-Cu, Cd-Cu, Zn-Ag, and Cd-Ag.

101

By contrast, in the η phases, the axial ratio is a rapidly increasing function of composition until the values characteristic of zinc or cadmium are reached.[6]

The rapid freezing technique has been a most useful tool in bringing into focus some of the interesting inter-relationships between the various combinations of the ε-η, η-η, and ε-ε phases.

II.   THE SYSTEMS Zn-Ag AND Zn-Cu

Rapid freezing from the melt of the Zn-Cu and Zn-Ag alloys in the composition range corresponding to the equilibrium two-phase mixture results in metastable extensions of both the ε and η phases for nearly all compositions in the two-phase region.[5] Nevertheless, a discontinuity in structure apparently occurs between the metastable extensions of the stable phases, at nearly the same electron concentration in each system; 1.912 in Zn-Cu and 1.916 in Zn-Ag, i.e., when about 9 out of 100 atoms are no longer Zn or Cd. The occurrence of this discontinuity may be judged from the trends in the respective lattice parameters and the axial ratio, and also from the change in the atomic volume with composition. The results for the Zn-Ag and Zn-Cu systems are illustrated by data in Figs. 5 and 6. It is of interest that the structural discontinuity in both systems appears to be very slight. The observed behavior of the lattice parameters and the atomic volume is consistent with an interpretation that the free energy curves which relate to the ε and η phases are separate free energy curves for each alloy phase and that they intersect at a very steep angle.

III.   THE SYSTEM Cd-Ag

The cadmium-rich end of the cadmium-silver equilibrium phase diagram (Fig. 4d) also conforms to the general alloying pattern described above for the binary systems Zn-Ag and Zn-Cu. Samples of composition falling in the ε, ε + η, and η phase fields of the equilibrium phase diagram were quenched rapidly from the melt.** Variations with composition of the lattice spacings and the axial ratio, shown in Figs. 7, 8, 9, and 10 demonstrate that the ε phase stability range can be extended from 18.5 at.% Ag to 14.5 at.% Ag, and that of the η phase from 5 at.% Ag to about 9 at.% Ag. The shallow minimum observed in the trend of the axial ratio in the ε phases of the Zn-Ag and Zn-Cu systems, close to the ε/(ε + η) phase boundary,[6] also occurs in this system in the extended ε phase, but is barely discernible.

The gun technique used in the present investigation was unable to produce metastable extensions of the ε or η phases over the composition range between 9 and 14 at.% silver. Nevertheless, the trends of the lattice parameters, the axial ratio, the atomic volume (Fig. 10), and particularly their most likely extrapolations, strongly suggest that a continuous transition between the ε and η phases is extremely unlikely.

**The details of the experimental work reported here for the systems Cd-Ag, Cd-Cu and Zn-Cd-Ag were the same as in the earlier publication.[5]

Again, a discontinuous transition between the two hcp structures ϵ and η is indicated at an electron concentration near 1.91 (when about 9 atoms out of each 100 are silver atoms).

IV. THE SYSTEM Cd-Cu

An exception to the alloying pattern described above occurs in the cadmium-copper system (Fig. 4c). The equilibrium stability range of the η phase based on cadmium is extremely small in this system, with a solubility limit of copper in cadmium amounting to less than 0.2 at.% Cu. Moreover, the intermediate phase which is in equilibrium with the η solid solution is an intermetallic compound which is usually designated as Cd_3Cu (since it corresponds to an ordered structure at 25 at.% Cu), whose crystal structure is not isomorphous with that of the disordered hcp ϵ phases mentioned above. The physical reason for the different form of the Cd-Cu phase diagram may be traced to the large difference in size between cadium and copper atoms, ($\sim$ 16%),*** which by far exceeds the difference in size between zinc and copper($\sim$4%), or zinc and silver ($\sim$8%), or cadmium and silver ($\sim$3%).

The experimentally determined variation of the lattice parameters in the η phase structure of the Cd-Cu system, in the rapidly quenched samples (Fig. 11) demonstrates that the range of the η phase can be metastably extended from the equilibrium value of about 0.2 at.% to a metastable value of about 2.5 at.% Cu. Samples with a higher copper content contain a significant amount of the intermetallic compound Cd_3Cu. The diffraction lines from an annealed powder sample with composition Cd- 25 at.% Cu may be adequately interpreted by a complex hexagonal unit cell of dimensions a = 8.124 $\pm$ 002 A and c = 8.755 $\pm$ 002 A, values which are somewhat higher than those indicated by Dey and Quader.[7] Moreover, the indexing of the diffraction lines by these authors proved incorrect, which was also pointed out by Trottier et al.[8,9] in an independent study. The lattice parameters of the splat-cooled Cd_3Cu phase were the same as those given above.

V. THE SYSTEM Zn-Cd

Baker and Cahn[2] investigated the possibility of extending the stability range of the η phase based on zinc by rapidly quenching molten Zn-Cd alloy samples on a copper substrate held at liquid nitrogen temperature. They concluded on the basis of the variation of the atomic volume with composition that the zinc-rich η phase could be extended beyond the equilibrium solubility limit of cadmium in zinc ($\sim$ 1.5 at.%) to about 3 at.% Cd in zinc. Since the liquidus boundary on the Zn-rich side has a retrograde form, this result suggested that the Cd is trapped in the solidification interface, contrary to the principle of local equilibrium.[2] Our experiments, using a copper substrate at room temperature, have shown practically no extension of the zinc-rich η phase boundary. Similar ex-

---

*** These are Hume-Rothery size-factors based on the closest distance of approach of atoms in the structures or the pure elements (see Table 1).

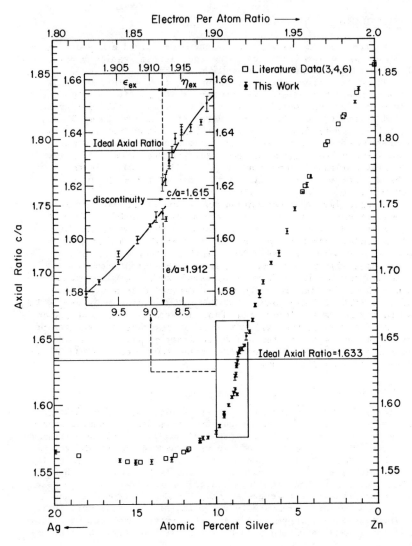

Figure

5. Variation of the axial ratio <u>c/a</u> with silver content in
    rapidly quenched Zn-Ag alloys. (Reproduced from reference 5.)

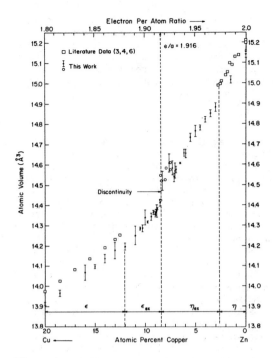

Figure
6a. Variation on the atomic volume Ω with copper content in
rapidly quenched Zn-Cu alloys (open dots refer to lattice
spacings measurements based on (00.2) and (10.2) reflections
only).

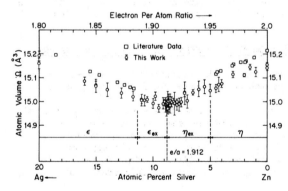

Figure
6b. Variation of the atomic volume Ω with silver content in rapidly
quenched Zn-Ag alloys. (Reproduced from reference 5)

periments, with respect to the solubility limit of Zn in Cd also appear
to have failed in extending the Cd-rich η phase boundary beyond the
equilibrium value (at 266°C) of about 5 at.% Zn. The data for the first
four at.% is shown in Fig. 12. All samples with 5 to 6 at.% Cd contained
a certain amount of the zinc-rich terminal solid solution. Thus, as
already pointed out by the earlier work, the misability gap still persists
in the Zn-Cd system following the rapid freezing.

VI.   THE SYSTEM Zn-Cd-Ag

The composition regions near the Zn-Cd binary wall (e/a > 1.7
electron/atom) of the equilibrium isothermal (200°C) section of the Zn-
Cd-Ag and Zn-Cd-Cu phase diagrams are shown in Fig. 13. The η phase fields,
based on zinc or cadmium, extend little into the binary, or the ternary
sections (less than 3 at.% away from the Zn or Cd corners, in Zn-Cd-Ag),
and are separated along the Zn-Cd binary side by a narrow η-η two-phase
field. The size complication introduced by the system Cd-Cu is responsible
for the complex nature of the relationships in the ternary section Zn-Cd-Cu
and will not be considered further. In the Zn-Cd-Ag section, the equilibrium
ε phase fields based on the Zn-Ag and Cd-Ag binary sides of the phase
diagram, extend deeper than the corresponding η phase fields, and are
separated by a wide two-phase field (ε + ε'), centered along the line of
constant electron concentration of e/a = 1.80, or 20 at.% Ag.

The application of the rapid quenching technique has shown that
the variation of the lattice spacings of the ε hcp phases in the samples
along the composition line $Zn_{(80 - x)}Cd_xAg_{20}$, is consistent with a continuous series of metastable hcp phases, from the stable Cd-Ag ε' phase
to the stable Zn-Ag ε phase. The results for the c parameter and the
axial ratio are shown in Fig. 14. The metastable, supersaturated ε
phases obtained in the middle of the two-phase field (for 30 < x < 50)
were highly unstable at room temperature and transformed within minutes
into a mixture of the equilibrium phases.[****] The lattice parameters
and atomic volumes show practically no departure from a straight line
variation between the values characteristic of the binary ε phases. Thus,
the "Vegards' Law" is obeyed in this continuous metastable series of the
ε phases.

VII.   METASTABLE PHASE RELATIONSHIPS BETWEEN THE VARIOUS hcp PHASES

The equilibrium and metastable relationships between the various
ε and η phases related to Zn and Cd indicate the following general
pattern:
(a)   The ε-η two-phase fields are likely to be a result of
two separate free energy curves in each system, one for the ε and one
for the η phase, which intersect at a steep angle in a narrow region
of composition. This has already been suggested by the earlier
work on Zn-Ag and Zn-Cu systems, and is illustrated in Fig. 15. The
corresponding T_0 curves would then intersect, as illustrated in Fig. 16.

[****] For further details regarding this work, please see Ref. 16.

106

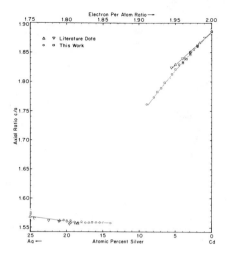

Figure

7. Variation of the axial ratio c/a with silver content in rapidly quenched Cd-Ag alloys.

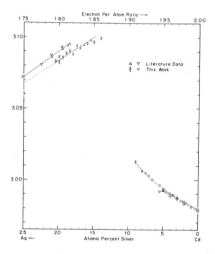

Figure

8. Variation of the a lattice parameter with silver content in rapidly quenched Cd-Ag alloys.

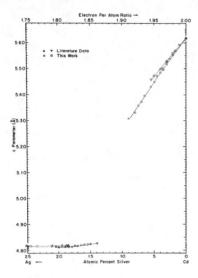

Figure

9. Variation of the $\underline{c}$ lattice parameter with silver content in rapidly quenched Cd-Ag alloys.

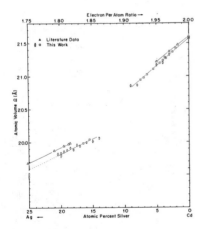

Figure

10. Variation of the atomic volume Ω with silver content in rapidly quenched Cd-Ag alloys.

108

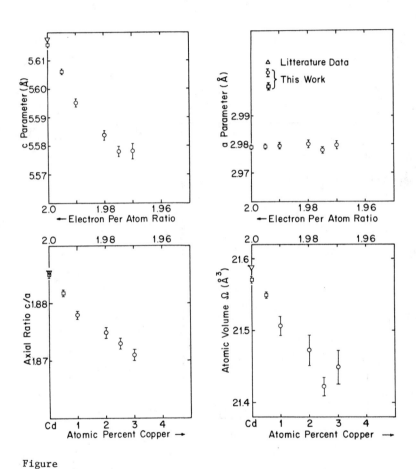

Figure
11. Variation of the lattice parameters with copper content in
    rapidly quenched η phase alloys of the Cd-Cu system.

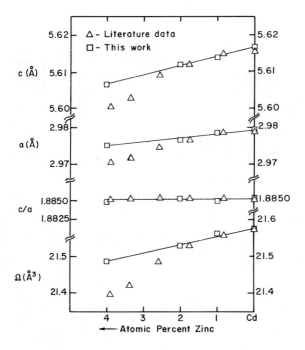

Figure

12. Variation of the lattice parameters, axial ratio, and atomic
     volume in rapidly quenched η phase alloys of the Cd-Zn system.

110

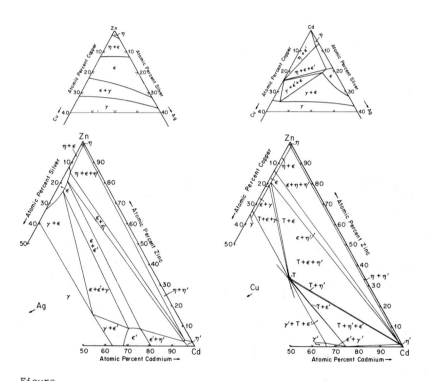

Figure
13. Equilibrium isothermal sections at 200° C in the ternary systems
Cu-Ag-Zn, Cu-Cd-Zn, Ag-Zn-Cd and Cu-Zn-Cd.

111

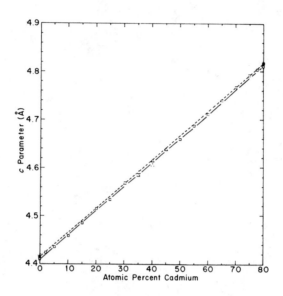

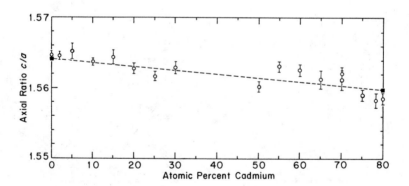

Figure
14. Variation of the c̲ parameter and the axial ratio along the line
of constant electron concentration at 20 at.% silver in the
ternary system Ag-Cd-Zn.

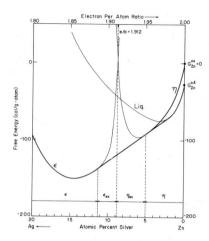

Figure

15. Free energy configuration for the solid ε and η phases and the liquid phase at the peritectic temperature (430°C) for zinc-silver. The free energy of each phase in its stability range has been calculated from thermodynamic data. The arbitrary energy references are $G_{Zn}^{os} = 0$ $G_{Ag}^{os} = +8000$ cal/g-atom. (Reproduced from reference 5.)

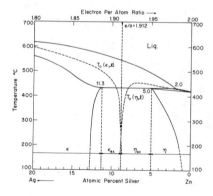

Figure

16. Possible trends of the T_o temperatures in the zinc-silver system. (Reproduced from reference 5.)

113

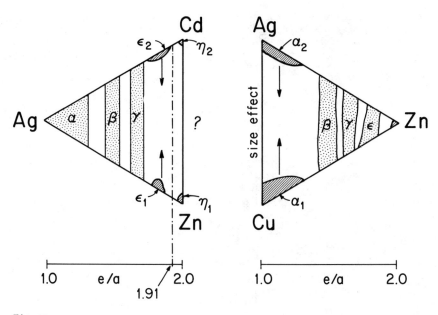

Figure
17. Schematic representation of the phase relationships in the
    ternary systems Ag–Cd–Zn and Zn–Cd–Ag.

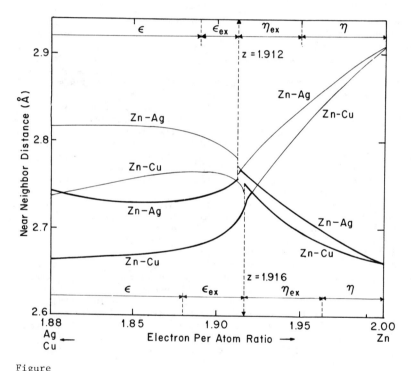

Figure
18. The nearest neighbor distances in the ε and η phases, and
their metastable extensions in the Zn-Ag and Zn-Cu systems.

The present data on the Cd-Ag system reinforces this conclusion, although the intersection point between the corresponding ΔG^{solid} curves in the Cd-Ag system is probably at a much higher temperature, compared with ΔG^{Liq}. This would account for the lack of metastable solid solubility in the region near 1.91 electrons per atom. If this interpretation is correct, the relationship between ε and η phases cannot be represented by a continuous free energy curve with a hump in the middle.

(b) A similar situation to the above may also exist in the Zn-Cd system, between the η and η' solid solutions based on these metals, but this is not indicated by calculations based on the available thermodynamical data[17,18]. Computer calculations based on reasonable models can reproduce the phase diagrams quite accurately, and they indicate a fairly shallow and continuous T_0 trend. This seems to suggest that, if the liquid could be cooled violently enough, it should be possible to obtain a continuous range of metastable η phases. The only way to obtain a separation of the T_0 curves would be to assume two different solids, or have an exceedingly large excess entropy contribution[17]. It may be, therefore, that the cooling rates related to the gun technique are not fast enough because there is a very large ΔG available to drive the solute partitioning. Perhaps deposition from vapor on a cold substrate will resolve this point in the future. At the moment the experimentally observed misability gap remains.

(c) When zinc is replaced by Cd, in the presence of 20 at.% of Ag, i.e. in the ternary ε phase range of the Zn-Cd-Ag system, a complete metastable range of solid solutions is obtained, suggesting that the free energy curve of the ternary ε phase is a continuous function of composition and merely has a hump in the middle. Thus, the situation with respect to the ε phases is like that in Cu-Ag system between α' and α. The equilibrium solid solubility is represented by a misability gap due to the size effect, but this can be overcome by rapid crystallization from the liquid. In both cases the electron concentration representing the conduction electrons is constant.

From the point of view of the general alloying behavior the above results may be illustrated by a schematic representation of the phase relations in the ternary systems Ag-Cu-Zn and Zn-Cd-Ag, as shown in Fig. 17. The size difference aspect present along the Ag-Cu binary wall is apparently reduced on alloying with Zn, and hence the equilibrium miscibility gap between the α' and α phases is followed by continuous ranges of mutual solid solubility between the β, γ and ε phases, as the nominal electron concentration increases. By comparison, the lack of miscibility between Zn and Cd appears to be associated with a problem which cannot be overcome by rapid freezing. This, however, changes rapidly on alloying with silver so that by the time some 9 at.% of silver are added the lack of solubility is eliminated in the metastable range of the ternary ε phases. Further alloying with silver represents the same situation as in the corresponding Cu-Ag-Zn system.

116

It is not possible at this time to postulate the alloying parameter which prevents solid state solubility between the η-η phases and the ε-η phases. The only obvious difference appears to be related to the d band electrons. Both zinc and copper are in the first row while silver and cadmium are in the second row. At first, therefore, it would seem that the same change in the d band structure should occur while passing from zinc to cadmium as on passing from copper to silver. However, the d band is present only a few electron volts below the Fermi level in the noble metals, as compared with zinc and cadmium where the d band is much further below. In case of an hcp structure, the first coordination shell of 12 atoms consists of two sets of six near neighbors, one set at a distance $\underline{a}$, and the other at a distance of $(a^2/3 + c^2/4)^{1/2}$, from the central atom. If the atomic volume is kept constant, or if it changes uniformly, it is evident that the nearest neighbor distance reaches the largest separation at the ideal close packing, and is then smaller than that distance for any other packing. The nearest neighbor distances in the ε and η phases, and their extensions in the zinc-silver and zinc-copper systems, are illustrated in Fig. 18. In both alloy systems the atomic volume is not constant over the electron concentration range 1.8-2.0 and it may be seen that the nearest neighbor distance is largest in the region of the ideal axial ratio. If the trend of interaction energy between d electron shells should follow more or less the same trend as the nearest neighbor distance, a maximum in this energy may occur in a vicinity of the ideal axial ratio. This would explain why the ε and η phases may become increasingly unstable in the region of closest packing, and why the rate of change of the axial ratio with electron concentration is largest in this region. The above discussion does not apply to the case of zinc and cadmium atoms when they are mixed because the axial ratio of the two elements concerned is more or less the same. Thus, although the η-η situation may be in some way related to the d band electrons, further speculation appears to be premature until more detailed data regarding the electronic structure, and its change on alloying is established.

ACKNOWLEDGMENTS

We wish to acknowledge many interesting discussions with a number of colleagues, Dr. U. Mizutani, Dr. L. F. Vassamillet, Dr. J. Bevk, Dr. Karl Gschneidner, Dr. L. Kaufman and Dr. J. C. Warner. Drs. Kaufman and Warner have independently suggested that the T_0 temperature in the Zn-Cd system should be continuous.

This work was supported in part by a Grant from the National Science Foundation, Washington, D. C., which is gratefully acknowledged.

REFERENCES

1. Duwez, P., Campbell Memorial Lecture, Trans ASM 60, 607 (1967).

2. Baker, J. C. and J. W. Cahn, Acta Met. 17, 575 (1969).

3   See for example the review by Sinha, Giessen and Polk, A. K. Sinha, B. Giessen and D. Polk, Chapter 1, 3, "Crystalline and non-Crystalline Solids, Treatise of Solid State Chemistry" ed. N. Bruce Hanney, Plenum Press, New York (1976).

4. Jones, H., Rep. Prog. Phys. 36, 1425 (1973).

5. Massalski, T.B., L. F. Vassamillet and Y. Bienvenu, Acta Met. 21, 649 (1973).

6. Massalski, T.B. and H. W. King, Acta Met. 10, 1171 (1962).

7. Dey, B.N., and M. A. Quader, Ind. J. Phys. 37, 283 (1963).

8. Trottier, J.P., R. Prudhomme, C. Diot and A. Grenaut, La Recherche Aerospatiale 5, 233 (1970).

9. Trottier, J.P., private communication.

10. Hansen, M., and K. Anderko, "Constitution of Binary Alloys", McGraw Hill, New York (1965).

11. Gebhardt, E., G. Petzow and W. Krauss, Z. Metallkunde 53, 372 (1962).

12. Gebhardt, E.,and G. Petzow, Z. Metallkunde 47, 401 and 751 (1956).

13. Petzow, G. and E. Wagner, Z. Metallkunde 52, 736 (1961); 53, 189 (1962).

14. Petzow, G., H. L. Lukas and F. Aldinger, Z. Metallkunde 58, 175 (1967).

15. Petzow, G., private communication (1972).

16. Bienvenu, Y., Ph.D. Thesis. (1973).

17. Warner, J.C., private communication (1974).

18. Kaufman, L., private communication (1975).

CLASSIFICATION AND CRYSTAL CHEMISTRY
OF ORDERED METASTABLE ALLOY PHASES

Bill C. Giessen

Departments of Chemistry and Mechanical Engineering
Northeastern University, Boston, Massachusetts

INTRODUCTION

A number of metastable ordered phases (MOPs) have been prepared in the past 15 years using various quenching techniques. About fourteen of these phases are considered in the following, and they are classified primarily according to the type of their relation to corresponding equilibrium phases.

In this paper, a crystal chemical approach to the classification of metastable alloy phases is introduced, using disordered metastable phases as examples; the presently recognized categories of MOPs are then listed and described; last, some criteria of MOP formation are discussed.

General Considerations on Metastable Phase Formation: Strictly speaking, except at the temperature $T = 0 K$ and at the exact composition of their stoichiometry, intermediate phases cannot be completely ordered. With this proviso, one may classify intermediate alloy phases on the basis of their structures as ordered and disordered; the ordered ones, in turn, may be stoichiometric, having only a narrow homogeneity range of, say, <1%, or non-stoichiometric, i.e., based on a certain ordered structure type, but with a broad homogeneity range and thus partially disordered.

In the rich and complex field of equilibrium alloy phases, ordered phases (as defined here) vastly outnumber the disordered ones, consequently, the major crystal chemical compilations[1,2] of alloy phases are almost totally devoted to stoichiometric ordered phases. For metastable metallic phases, however, the frequency of occurrence so far has been the reverse and more metastable disordered phases than MOPs have been described.

Since the advent of rapid quenching from the liquid and vapor, a sizeable number of metastable metallic intermediate phases have been produced,[3-5] besides a large number of extended terminal solid solutions and non-crystalline solids. The crystalline intermediate phases are mostly disordered and those known at present generally belong to certain families such as the following:

[a] Metastable Hume-Rothery phases based on noble metals, e.g., ζ-(Au-Sb)[6] and γ-(Au-Sn);[7]

[b] Inter-B metal phases, which form phase fields[1,5,8] together with related stable phases, as discussed below;

[c] Inter-transition metal phases such as ζ-(Nb-Ni),[9] which have a crystal structure identical with that of the transition metal or alloy with the same average group number (AGN).

By contrast, relatively few MOPs have been formed so far by rapid quenching methods. One reason for this may be simply that not enough systems fertile for MOP formation have been studied, as shown by the following consideration: It is reasonable to assume that MOPs will tend to form in systems already containing several ordered equilibrium phases. As a rule of thumb, ordered binary phases occur most commonly in systems whose components are neither too close to nor too far removed from each other in the Periodic Table, i.e., systems that have inter-mediate differences in valence and electronegativity. (If the component elements are far removed from each other, e.g., if they cross the Zintl line,[2] they have fixed valences and large electronegativity differences; this results in the formation of few, highly stable ordered phases. If, on the other hand, they are too close, stoichiometric compounds will again not be favored as a result of the occurrence of extended terminal solid solutions.) Amorphous phase formation may also intervene to prevent retention of MOPs. Most of the rapid quenching work done to date, however, has been concerned with extensions of terminal solid solutions or with systems where the tendency for amorphous phase formation is high; this has reduced the likelihood of MOPs being formed.

Metastable Disordered Phases: We review here a well documented, simple alloy chemical relationship found to be valid for a wide range of

120

disordered stable and metastable solid solutions, including both terminal and intermediate ones, all of which are composed of B metal elements of groups I–V, in the second short period and the three long periods.[1,5,8] For these elements and the alloy systems formed by them it was
observed that alloy phases tend to be isostructural if they fall within
certain composition ranges (phase fields) characterized formally by
their location in the periodic table, using the standard long periodic
representation; the location of any intermediate alloy composition is
obtained for binary alloys by connecting the constituent elements and
plotting the composition on this line. A similar procedure may be used
for ternary and higher alloys. Phase fields are thus characterized by
ranges of favorable combinations of two coordinates. The horizontal and
vertical positions of a given alloy composition represent the average
valence electron concentration and the average pseudopotential (e.g.,
the position of q_o, the zero of the pseudopotential in the Fourier
representation), respectively.[8]

In this first-order two-parameter description,[8] the specific nature
of the pair or triplet of elements forming a given phase is not considered. This disregard of the chemical factor, the electrochemical factor
(electronegativity) and the geometrical factor (size) in favor of the
energy band factor (structural energy based on band structure considerations)[1] obviously limits the applicability of this empirical approach
to sets of elements where the former are less effective than the energy
band factor, i.e., elements for which the differences of the former
factors are small or at least similar in all systems used in deriving
the correlation.

Area-of-Stability Plots: For the present discussion, we will also adopt
the concept of the area-of-stability plot,[10] in which the structure
type or types of binary or ternary intermetallic phases with a given
stoichiometry and one common alloying element is presented in terms of
the periodic table position of the other partner(s). Thus, the transition or noble metal elements T combining with lead to form equilibrium
phases TPb with the NiAs type structure are Pd, Pt, and Ir, as treated
below, No. 1b. (Here, as in the following, crystallographic data are
generally taken from Reference 11.)

Figure 1. Stable (below) and Metastable (above) Phases in Pseudobinary AB_3 Systems with $B = Ni$. L is the Number of Close-Packed Hexagonal AB_3 Layers in the Repeat Unit along the c-axis.

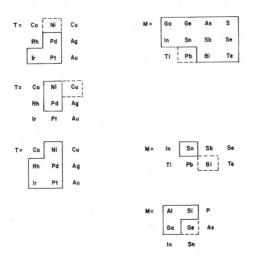

Figure 2: Area-of-Stability Plots for Several Phases of Interest. a) TPb phases (left) and NiM phases (right) with the NiAs-B8 type structure, b) T_2Al_3 phases with the Ni_2Al_3 type structure, c) TBi phases (left) and AuM phases (right) with the NiAs type structure, d) FeM phases with the CsCl structure type.

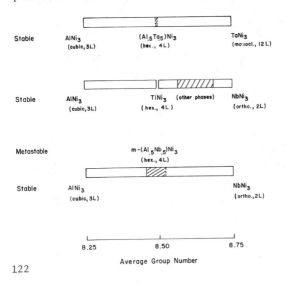

Using crystal chemical arguments, especially structural analogies, one may group the presently known metastable ordered phases prepared by rapid quenching according to the classification scheme in Table I. A list of MOPs is given in Table II; examples taken from this list will be discussed as illustrations, usually with reference to their specific crystal chemistry. Some MOPs have characteristics that lead to their listing in more than one category.

Type No. 1. Analogy to Equilibrium Phase: This is a straightforward case in which one or more of the component elements of an MOP are chemically closely related to (and often isovalent with) the corresponding component element or elements of an isostructural equilibrium phase. Examples are:

[a] m-$(Al_{.5}Nb_{.5})Ni_3$,[12] which has the hexagonal $TiNi_3$ type structure consisting of four ordered, close-packed AB_3 layers stacked in the sequence abac, abac. It is isostructural with the equilibrium phase $(Al_{.5}Ta_{.5})Ni_3$[13] which occurs as an intermediate phase in the pseudo-binary system $AlNi_3$-$TaNi_3$ (see Figure 1a) and the isoelectronic equilibrium phases $TiNi_3$ (Figure 1b), $ZrPt_3$ and others. The quenched-phase plot for the $AlNi_3$-$NbNi_3$ pseudobinary is given in Figure 1c.

The occurrence of this metastable phase strengthens the concept[14] that the stacking sequence adopted by ordered, close-packed transition metal phases with AB_3 stoichiometry depends on the valence electron concentration (or the average group number) which can be changed continuously in pseudobinary systems while maintaining the stoichiometry. It would be interesting to see whether additional ordered AB_3 phases with different stacking sequences could be produced at appropriate intermediate valence electron concentration values in such pseudobinary systems; a preliminary search at other compositions in the $AlNi_3$-$NbNi_3$ system has not been successful.[15]

[b] Metastable NiPb with the NiAs type structure.[16] Because of the immiscibility of the components in the liquid state, this phase could only be prepared via the vapor phase, specifically, by vapor deposition. The area-of-stability plots for the NiAs type, Figure 2a, suggest by

123

TABLE I. Types of Metastable Ordered Intermediate Alloy Phases Classified by Crystal Chemical Relationships.

No.	Type	Description	Examples*
1	Analogy to Equilibrium	Components of phase closely related to those of isostructural equilibrium phase	$(Al_{.5}Nb_{.5})Ni_3$, $NiPb$, Ta_3Ge, $(m\text{-}TiCu_3)$
2	Metastable Extension of Area-of-Stability	Extension of an area-of-stability of equilibrium phases occurring in systems formed by elements in adjacent groups in periodic table.	Cu_2Al_3
2a	Extension with Distortion	As in case 1 or 2, but with a substantial change of the axial ratios, or distortion to lower symmetry	$ThFe$, $m\text{-}NpGa_2$, $\pi''\text{-}GeSb$
2b	Extension with Changes in Stoichiometry	As in case 1 or 2, but with a change in stoichiometry and altered atomic positions	$Au_{6-7}Bi_8$
2c	Extension to Non-Stoichiometric Phase	As in case 1 or 2, but with change to partially disordered, non-stoichiometric phase	$m\text{-}GeFe_{1-x}$
2d	Extension with Change in Structure Type	As in case 1 or 2, but with change of structure to crystal chemically related structure	$\beta\text{-}Zr(Zr_xPt_{1-x})_3$
3	Metastable Phase with Transition Type Structure	Structure of metastable phase forming transition between structures of adjacent binary equilibrium phases	$m\text{-}TiCu_3$, $(\pi''\text{-}GeSb)$
4	Metastable Phase with Unique Structure	Phase without structural relation to any known adjacent phases as defined in 1-3	$m\text{-}ZrPt_2(?)$, $Y(Au\text{-}Ge)(?)$

* double-listed compounds given in parentheses

analogy that a Ni–Pb phase of this structure type might form as a metastable phase, considering the presence of the analogous Pb compound
PdPb on the one hand and the Ni compounds NiSn and NiBi on the other,
as shown; the finding of metastable NiPb thus confirms the usefulness
of area-of-stability plots as a guide in locating MOPs. It is interesting to note that NiPb molecules appear to form in the vapor phase
prior to crystallization.[16]

[c] T_3M phases with the Cr_3Si – A15 type structure which are of great
importance as high-T_c superconductors; here T = V, Nb, Ta and M = Si, Ge,
Sn. Besides the metastable change of the composition of the off-stoichiometric phase "Nb_3Ge" discussed below under another heading, a new
phase Ta_3Ge has been prepared by sputtering;[17] again, analogy with isovalent, isostructural phases containing either Ta or Ge such as Ta_3Sn
and V_3Ge suggested the possibility of forming Ta_3Ge.

Type No. 2: Metastable Extension of Area-of-Stability: For phases be-
longing to this type, the strict crystal chemical analogy between MOPs
and equilibrium phases existing in the previous case (type No. 1) no
longer exists; however, the component elements of an MOP of this type
are still chemically related to those of isostructural equilibrium
phases. An example is metastable Cu_2Al_3 with the Ni_2Al_3 type structure.[18]
The area-of-stability plot, Figure 2b, shows the occurrence of several
T_2Al_3 phases; here T is a Ni-group element. In Cu_2Al_3, Ni in Ni_2Al_3 is
replaced by the neighboring element Cu; the corresponding change of
the average valence electron concentration appears to make the Al-Cu
phase energetically less favorable than Ni_2Al_3 relative to the respective equilibrium phases but does not prevent its formation as a metastable phase.

In the following, four variations of this fundamental type of MOP
are listed and discussed in order of increasing degree of departure
from type No. 2.

Type No. 2a: Metastable extension of Area-of-Stability with Distortion:
In non-equilibrium intermetallic phases with unit cell parameters that
are variable at constant volume, such as the axial ratios of noncubic
phases or internal positional parameters, the second-order structural

125

TABLE II. Metastable Ordered Alloy Phases Prepared by Rapid Quenching.

Phase	Composition or Comp. Range*	Crystal System and Space Group	Structure Type	Mode of Preparation	Type of Metastable Phase**	Ref.
$(Al_{.5}Nb_{.5})Ni_3$	not det.	Hex. -P6/mmc	$TiNi_3$-DO_{24}	Gun	1	12
NiPb	50% Pb	Hex. -P6/mmc	NiAs-B8	Vapor Depos.	1	16
Ta_3Ge	~75% Ta	Cub. -Pm3n	Cr_3Si-A15	Rf-Sputtering	1	17
Cu_2Al_3	~60% Al	Hex. -C3m	Ni_2Al_3-$D5_{13}$	Splat Cooling	2	18
ThFe	~50% Fe	Orth. -Cmcm	CrB-B33	Gun	2a	20
ThCu	~50% Cu	Orth. -Cmcm	CrB-B33	Gun	2a	20
$m-NpGa_2$	~63-67% Ga	Hex. -P6/mmm	AlB_2-C16	Gun	2a	21
τ''-GeSb	~50% Sb	Tetr. -I4/mmm	NaCl-B1-distort.	Gun	2a (3)	22
Al_3Ge_2	~60% Al	Monoclinic	Mg_3Bi_2 -related	H_2O-Quench	2a	15,33
$Au_{6-7}Bi_8$	~55% Bi	Hexagonal	NiAs-B8-related	Gun + Anneal.	2b	23,24
$m-GeFe_{1-x}$	55-90% Ge	Cub. -Cm3m	CsCl-B2	Gun	2c	20,25
$\beta-Zr(Zr_xPt_{1-x})_3$	~70-74% Pt	Hex. -P6/mmc	Ni_3Sn-DO_{19}	Arc Splat	2d	26
$m-TiCu_3$	~75% Cu	Orth. -P6/mnm	Cu_3Sb-DO_a	Gun	3 (1)	30
$m-ZrPt_2$	~63-66% Pt	Hex. -P6/mmm	AlB_2-C16	Arc Splat	4?	26

* Compositions in atomic percent;
** Alternate classification in parentheses.

energy[19] may be reduced by changes of these parameters to the point where the phase becomes energetically at least competitive with other phases in the system and may form upon rapid quenching. This phenomenon is relatively common; examples exist in several systems, among them Th-Fe,[20] Th-Cu,[20] Np-Ga,[21] and Ge-Sb.[22]

    i. The two metastable phases ThFe and ThCu prepared by splat cooling[20] have CrB type structures; they represent large metastable extensions of the area-of-stability of this structure type in the ThX area-of-stability plot,[10] both to the left and the right of ThCo which is the only CrB type equilibrium phase of Th with a first-period transition metal.[10] The axial ratios of the metastable phases depart substantially from those of ThCo, indicating that a strong distortion of the CrB type structure is required to lower its energy enough to form the metastable ThX phases.

ii. m-NpGa$_2$ is another example for the formation of an MOP with an axial ratio much different from the axial ratios of the actinide phases UGa$_2$,[10] PuGa$_2$,[10] and even α-NpGa$_2$,[21] a competing phase of the same type at the same composition. The crystal chemistry of NpGa$_2$ is discussed in more detail in another paper presented at this conference;[21] the feature of interest in the present context is that MN$_2$ equilibrium phases (where M = actinide and N = B-metal of group I-IV) with the AlB$_2$ type tend to have axial ratios c/a > 0.94 if N = Ga;[10] the MOP m-NpGa$_2$ with c/a = 0.825[21] represents a metastable extension of a second, low-axial-ratio branch of the NpN$_2$ phases with the AlB$_2$ type structure. m-NpGa$_2$ is thus metastable with respect to α-NpGa$_2$.

iii. The ordered, tetragonally distorted NaCl-type structure (c/a = 1.05$_6$) of metastable π''-GeSb has been discussed[22] in terms of the valence electron concentration of this phase. The structure of π''-GeSb can be regarded as representing a transition between the tetrahedral fourfold atomic coordination in elemental Ge and the approximately octahedral sixfold atomic coordination in elemental Sb via a square planar fourfold coordination (with two additional, close second nearest neighbors) in π''-GeSb. It was assumed[22] that the non-integral valence electron concentration of 4.5 causes this distortion of the NaCl type in π''-GeSb; the NaCl type is also found in SnSb and other phases with higher valence electron concentration.[23] π''-GeSb is thus an example of the loss in symmetry and distortion of a structure type that may occur in metastable phases located outside of its area-of-stability.

Type No. 2b: Metastable Extension of Area-of-Stability with Change in Stoichiometry: For many ordered or disordered equilibrium phases such as NiAs type phases or Hume-Rothery phases[2] some representatives falling within their areas-of-stability show variable stoichiometries; this behavior generally results from the tendency to maintain an energetically favorable valence electron concentration value by a shift of

127

composition. A stoichiometry change of this kind may also be found in a metastable phase lying <u>outside</u> of the area-of-stability of the structure type associated with the parent phase.

An example is metastable $Au_{6-7}Bi_8$.[23,24] This phase is an ordered, off-stoichiometric representative of the NiAs type family. The area-of-stability plots for TM phases of the NiAs type with M = Bi and T = Au are shown in Figure 2c; it can be seen that Bi forms NiAs type phases prolifically but that the metal partner in these phases must be a transition metal; Au, on the other hand, forms a NiAs type phase only with Sn. The metastable extension of the TM area-of-stability into the Au-Bi system appears to require a departure from the TM composition; the exact stoichiometry and supercell of $Au_{6-7}Bi_8$ are still under investigation.

Type No. 2c: Metastable Extension to Non-Stoichiometric Phase: In this interesting variant of Type No. 2, above, the metastable phase is off-stoichiometric (and hence partially disordered), with a considerable homogeneity range. A specific example is m-$GeFe_{1-x}$ with $0.2 < x < 0.8$; this phase is an off-stoichiometric solid solution based on an unobserved GeFe phase with the CsCl-type[20,25] (there are several phases at or near GeFe with other structure types[11]). Equilibrium stoichiometric and non-stoichiometric FeM phases with the CsCl type are listed in the area-of-stability plot, Figure 2d. The phase m-$GeFe_{1-x}$ is of special crystal chemical interest because the off-stoichiometry results from the formation of vacancies in the Fe partial structure; at Ge-rich compositions (high vacancy concentrations) the structure of m-$GeFe_{1-x}$ can therefore be characterized alternatively as a primitive cubic Ge structure with its cubic (ditetrahedral) holes partially filled by iron atoms, i.e., as a metastable solid solution of Fe in a primitive cubic modification of elemental Ge not observed in the unalloyed state.

Type No. 2d: Metastable Extension with Change in Crystal Structure: A stable ordered phase may be extended to an adjacent composition as a metastable phase with a concomitant change of its crystal structure to a crystal chemically related type. An example is found in the Zr-Pt system where a metastable, off-stoichiometric phase β-$Zr(Zr_xPt_{1-x})_3$

128

($x \sim 0.05$) forms upon replacement of $\sim 5\%$ of the Pt atoms in the equilibrium phase $ZrPt_3$ by Zr.[26] While $ZrPt_3$ has the close-packed ordered
$TiNi_3$ type structure with a four-layer stacking sequence as described above
for $(Al_{.5}Nb_{.5})Ni_3$,[12] the stacking sequence has changed in β-$Zr(Zr_xPt_{1-x})_3$
to the two-layer sequence of the hcp $MgCd_3$ type structure; the change
in composition is thus associated with the formation of a metastable
phase of closely related structure.

Type No. 3: Metastable Phase with Transition Type Structure: A metastable phase that is structurally related to an adjacent equilibrium
phase (as in the previous Type No. 2d) may be related at the same time
to another adjacent equilibrium phase with different composition and
structure and may form a transition between these phases. This situation also occurs in equilibrium phases, as observed already in 1931 by
Goldschmidt.[27] In general, however, a phase with a transition type
structure will be higher in energy than the weighted average of the
terminal phases; a transition phase is therefore not likely to form
in equilibrium, as shown by a consideration of the free energy diagram.[28,29] Some transition phases might therefore be expected to be
retained in the metastable state.

An example is π''-GeSb, treated above (Type No. 2a) in connection
with the NaCl type phase field; as shown there, its structure can be
interpreted as forming a transition between the structures of Ge and
Sb.[22] Another example is metastable m-$TiCu_3$,[30] which is located between $TiCu_4$ (orthorhombic, ordered structure based on the hcp Mg-A3
type atomic arrangement) and TiCu (tetragonal, TiCu type ordered
structure based on the bcc W-A2 type atomic arrangement). (There may
be one or more additional stable phases between m-$TiCu_3$ and TiCu.[22])
As shown in Reference 30, the fundamental atomic arrangement in m-
$TiCu_3$ is of the hcp Mg-A3 type but the pseudohexagonal basal plane
net has a strongly distorted axial ratio ($b/a \sim 1.624$ instead of 1.732
for the orthohexagonal basal plane net of the Mg type), relating m-
$TiCu_3$ also to the bcc W-A2 type (with a corresponding axial ratio
"b/a" = 1.414). This shows the metastable Ti-Cu phase to be an ordered transition phase connecting the atomic arrangements of the
structures of two equilibrium phases.

129

It may be noted that m-TiCu$_3$ is also an example for the MOP type No. 1 (analogy to equilibrium phases) because it is isostructural with ZrAu$_3$ (including details such as the characteristic axial ratio value discussed above); ZrAu$_3$ is an equilibrium phase containing two elements from the same groups as Ti and Cu, respectively.[2]

Type No. 4: Metastable Phase with Unrelated Structure: If the structure of a new MOP is not crystal chemically related to other phases in any of the ways specified as types No. 1-3, above, or the miscellaneous cases, below, it is classified as unrelated, even if the structure type is known and occurs in other portions of the periodic table. So far, no MOP has been shown to belong to this category, although it is quite possible that some of the as yet unidentified metastable phases with more complex diffraction patterns, such as γ(Au-Ge),[31] X(Pb-Bi),[32] or one of the metastable Al-Ge phases[33] γ_2 or γ_3 belong in this category.

m-ZrPt$_2$ which has the Ni$_2$In or AlB$_2$ type structure[26] is tentatively listed here because it is not yet clear how closely it is related to other transition metal phases.

Miscellaneous Cases: Without classifying these special cases further, we list here some additional phase relationships:

i. In the types Nos. 1-3 (supra), the crystal chemistry of a new MOP was considered with respect to known equilibrium phases. However, a new MOP that is unrelated to any equilibrium phase may instead be related to another MOP found earlier and classified under Type No. 4 (supra). The new MOP will then belong to one of types Nos. 1-3 with respect to the phase found earlier, with the modification that the prototype itself is metastable.

ii. An MOP may be found at the composition of an equilibrium phase with the same structure but different lattice parameters or with a different structure. An example of the former case has been treated above in m-NpGa$_2$[21]; examples of the latter case are m-ThNi which replaces an equilibrium phase of different structure,[34] and m-GeFe$_{1-x}$ discussed above which replaces the equilibrium phase FeGe$_2$.[20]

iii. Two further cases to be mentioned are the common one of a
simple, stable extension of an ordered equilibrium phase from the
stoichiometric composition by rapid quenching and its corrolary, the
metastable extension of an ordered phase from an off-stoichiometric
equilibrium composition to the stoichiometric composition. A composition change of the latter type has been achieved in the important case
of the Cr_3Si-A15 type superconducting alloys in the Nb-Ge system; while
this phase is stable only with an excess of Nb, metastable alloys have
been prepared at the Nb_3Ge composition by sputtering, with a concomitant rise in T_c from 6.9K at the equilibrium composition with 22at.%Ge
to 17K at 25at.%Ge.[17]

OCCURRENCE OF METASTABLE ORDERED PHASES

The classification scheme given above (Table I) provides a framework
for presenting and correlating the empirical evidence contained in the
examples listed in Table II. Several questions arise in a further discussion of the observed MOPs:

What are the energetics of the new, metastable phases?

What are the criteria for their formation and do they pose major new
questions to be answered by alloy chemical considerations?

How common is their occurrence?

These questions will be treated briefly.

Energetics: Calorimetric measurements on MOPs are required to decide
whether a given MOP is metastable only with respect to the equilibrium
phase or phases but stable with respect to the constituent elements,
i.e., whether it has a negative free energy of formation, or whether it
is metastable in the absolute sense. (In a related study of the disordered metastable phases ζ-(Au-Sb) and π-(Au-Sb),[35] both phases were
found to have positive heats of formation, making them metastable in
the absolute sense; metastable γ-(Au-Sn), on the other hand, is metastable only with respect to the equilibrium phases.)[36]

Such measurements also enable one to make a direct comparison of the
energies of formation of competing structure types at the same composition; this is of importance as a test for cohesive energy calculations.

131

Criteria for the Occurrence of MOPs: The foregoing discussion of the
MOPs listed in Table II shows that all known MOPs are chemically related to one or more equilibrium alloy phases; hence, they are all
subject to the same considerations based on crystal chemical criteria
such as component size, valence, electronegativity and band structure
effects[1] as these phases. Beyond this, the study of MOPs may be of
theoretical value by indicating the changes, e.g., distortions, required for a phase to occur at compositions lying outside of its area-
of-stability. Considerations of this kind may yield significant crystal chemical relationships.

Frequency of MOP formation: Considering the large number of alloy systems studied in non-equilibrium, and the number of equilibrium phases
existing in these systems, the number of observed MOPs has been relatively small. As Table II shows, nearly all listed MOPs contain at
least one d-electron metal, i.e., a transition metal or noble metal.
This is probably due to the facts that more systems containing d-
electron metals have been studied in non-equilibrium than other types
of systems and that the d-electron metal systems studied generally contain several stable ordered phases. While many B-B metal systems have
also been studied, there are substantially fewer ordered equilibrium
phases in these systems, and hence MOPs were rarely found. Only few
A-B metal systems (where A is an electropositive and B an electronega-
tive s/p-electron metal) have been studied to date, but such systems
generally contain several intermediate phases; it will be interesting
to see whether a corresponding number of MOPs will be found in these
systems.

The relatively small number of MOPs may also be connected with the
high glass-forming ability of many binary alloys.[5,37,38] It is quite
possible that for many binary compositions the amorphous state consti-
tutes the ground state of the alloy as a single phase, i.e., all other
single phases with the same composition have a higher energy (under the
constraint that transformation into more than one phase is excluded[39]).
This concept agrees with the frequency of formation of a metallic glass
instead of an MOP, especially upon vapor deposition; it must be confirmed by cohesive energy calculations.

132

ACKNOWLEDGEMENTS

Financial support of research at Northeastern University under ONR Contract N14-75-C-713 is gratefully acknowledged. The author thanks for partial personal support under NSF-MRL Grant DMR 72-03027-A05 at the Materials Research Center, Massachusetts Institute of Technology, during the preparation of this paper.

REFERENCES

1. Pearson, W.B., _The Crystal Chemistry and Physics of Metals and Alloys_, John Wiley and Sons, New York (1972).
2. Schubert, K., _Kristallstrukturen Zweikomponentiger Phasen_, Springer-Verlag, Berlin (1964).
3. Giessen, B.C., in _Developments in the Structural Chemistry of Alloy Phases_, B.C. Giessen, Ed., Plenum Press, New York, 227 (1969).
4. Jones, H. and C. Suryanarayana, _J. Mater. Sci._ 8, 705 (1973).
5. Sinha, A.K., B.C. Giessen, and D.E. Polk, in _Treatise on Solid State Chemistry_, III, N.B. Hannay, Ed., Plenum Press, New York, 1 (1976).
6. Predecki, P., B.C. Giessen, and N.J. Grant, _Trans. Met. Soc. AIME_ 233, 1438 (1965).
7. Giessen, B.C., _Z. Met._ 59, 805 (1968).
8. Giessen, B.C., _Adv. in X-Ray Analysis_ 12, C.S. Barrett, G.R. Mallett, and J.B. Newkirk, Eds., Plenum Press, New York, 23 (1969).
9. Ruhl, R.C., B.C. Giessen, M. Cohen, and N.J. Grant, _J. Less Comm. Met._ 13, 611 (1967).
10. Dwight, A.E., in _Developments in The Structural Chemistry of Alloy Phases_, B.C. Giessen, Ed., Plenum Press, New York, 181 (1969).
11. Pearson, W.B., _A Handbook of Lattice Spacings and Structures of Metals and Alloys_ 2, Pergamon Press, Oxford (1967).
12. Giessen, B.C. and R. Ray, _J. Less Comm. Metals_ 23, 95 (1971).
13. Giessen, B.C. and N.J. Grant, _Acta Cryst._ 18, 1080 (1965).
14. Havinga, E.E., J.H.N. Van Vucht, and K.H.J. Buschow, in _Ordered Alloys_, Proc. Third Bolton Landing Conference, 1969, Claitor's Publ., Baton Rouge, 111 (1970).
15. Giessen, B.C. [unpublished results].
16. Ricci-Bitti, R., J. Dixmier, and A. Guinier, _Compt. rend._ 266(B), 565 (1968).
17. Hanak, J.J., J.I. Gittleman, J.P. Pellicane, and S. Bozowski, _J. Appl. Phys._ 41, 4958 (1970).
18. Ramachandrarao, P. and M. Laridjani, _J. Mater. Sci._ 9, 434 (1974).
19. Heine, V. and D. Weaire, _Solid State Physics_ 24, H. Ehrenreich, F. Seitz, and D. Turnbull, eds, Academic Press, New York, 249 (1970).
20. Segnini, M., Ph.D. Thesis, Northeastern University, Boston (1972).
21. Giessen, B.C. and R.O. Elliott, in _Proc. Second Int'l Conf. on Rapidly Quenched Metals_ (1975) 8II, N.J. Grant and B.C. Giessen, Eds., _Mater. Sci. and Eng'g_ (1976).
22. Giessen, B.C. and C. Borromee-Gautier, _J. Sol. St. Chem._ 4, 447 (1972).

133

23. Giessen, B.C., U. Wolff, and N.J. Grant, <u>Trans. Met. Soc. AIME</u> <u>242</u>, 597 (1968).
24. Wang, R., T.X. Mahy, and B.C. Giessen [to be published].
25. Segnini, M., R. Ray, and B.C. Giessen [to be published].
26. Raman, R.V. and B.C. Giessen [to be published].
27. Goldschmidt, V.M., <u>Kristallchemie und Roentgenforschung</u> in: <u>Ergeb-</u> <u>nisse der Technischen Roentgenkunde</u>, Akad. Verlagsges., Leipzig (1931).
28. Giessen, B.C. and R.H. Willens in <u>Phase Diagrams; Materials Sci.</u> <u>and Tech. III</u>, A.M. Alper, Ed., Academic Press, NY, 103 (1970).
29. Jones, H., <u>Rep. Prog. Phys. 36</u>, 1425 (1973).
30. Giessen, B.C. and D. Szymanski, <u>J. Appl. Cryst. 4</u>, 257 (1971).
31. Anantharaman, T.R., H.L. Luo, and W. Klement, <u>Trans. Met. Soc.</u> <u>AIME 233</u>, 2014 (1965).
32. Borromee-Gautier, C., B.C. Giessen, and N.J. Grant, <u>J. Chem. Phys.</u> <u>48</u>, 1905 (1968).
33. Koester, U., <u>Z. Metallk. 63</u>, 472 (1972).
34. Giessen, B.C. and R. Ray [unpublished results].
35. Jena, A.K., B.C. Giessen, M.B. Bever, and N.J. Grant, <u>Acta Met. 16</u>, 1047 (1968).
36. Jena, A.K., B.C. Giessen and M.B. Bever, <u>Met. Trans. 4</u>, 279 (1973).
37. Giessen, B.C. and C.N.J. Wagner, in <u>Physics and Chem. of Liquid</u> <u>Metals</u>, S.Z. Beer, Ed., Marcel Dekker, New York, 633 (1972).
38. Turnbull, D., in <u>Solidification</u>, Am. Soc. of Metals, Metals Park, Ohio, 1 (1971).
39. Baker, J.C. and J.W. Cahn, ibid., 23.

HIGH RESOLUTION MICROANALYSIS AND MICROSTRUCTURAL CHARACTERISTICS OF SPLAT QUENCHED ALUMINIUM-COPPER ALLOYS

D.B. Williams and J.W. Edington
Department of Metallurgy and Materials Science
University of Cambridge, England

INTRODUCTION

Transmission electron microscope studies of splat quenched alloys show that, in general, many important microstructural features are in the dimensional range 50-100nm. This scale however is below the ultimate spatial resolution of conventional microanalytical techniques which is between 100 and 200nm. Thus no data are available concerning fine scale composition fluctuations which might be expected to accompany the formation of the microstructure. This paper reports initial micro-analytical studies of splat-quenched Al-Cu alloys, using the ultra high resolution technique of combined electron microscopy and energy analysis (E.M.E.A.). This specialised technique is capable of detecting Cu composition changes of ∿0.5at.%[1] with a spatial resolution <10nm. The aim of the study is to correlate the microstructural features observed by transmission electron microscopy, with the microanalytical data obtained using E.M.E.A. The Al-Cu system was chosen because

[a] it is susceptible to analysis by E.M.E.A.[1] and

[b] basic microstructural information concerning the splat-
quenched condition already exists.[2,3]

EXPERIMENTAL

Seven alloys, up to and including the eutectic composition (17.3at.%Cu) were prepared from 99.999% Al and Cu. Specimens (∿100mg) were splat quenched from 850°C using a modified Duwez gun system, operated under reduced pressure (400torr) of high purity argon. Details of the apparatus are described elsewhere in this conference proceedings.[4] Specimens obtained are extremely porous, adhering closely to the water-cooled copper substrate; large areas are transparent to 100kV electrons.

135

Figure 1. Diffraction pattern from as quenched single phase Al-17.3at.%Cu.

The technique used for micro-analysis, E.M.E.A., studies the energy variation in electrons that have been transmitted by thin foil specimens in the electron microscope. Most transmitted electrons traverse the specimen with no energy loss, but a certain proportion undergo various forms of inelastic interaction. In light elements, these interactions are particularly simple, the predominant type known as plasmon interactions. The energy lost by a high voltage (100kV) electron in initiating a plasmon interaction depends upon the free electron density, which changes as composition is changed. In Al-Cu alloys it is found that the energy loss increases linearly with Cu content. E.M.E.A. uses a simple electron spectrometer beneath a conventional electron microscope, to analyse the energy distribution of electrons forming the bright field image. This energy distribution is simply interpreted in terms of composition change. The spatial resolution of the technique is limited by the localisation of the plasmon interaction, which is ∿10nm. Details of the technique, apparatus and experimental procedure have been described elsewhere.[5]

RESULTS AND DISCUSSION

General microstructural characteristics resembled those previously reported.[2,3] Specimens containing <5at.%Cu were single phase in all electron transparent regions. All alloys containing >5at.%Cu were similarly single phase in their thinner regions. Figure 1, supra, shows a <100> diffraction pattern from a single phase eutectic alloy. There is no indication of solid solution breakdown to G.P zones. Despite obtaining large regions of the eutectic alloy in the single phase condition, often <50nm in thickness, no regions of amorphous structure as reported by other investigators[6] were observed.

136

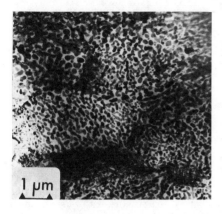

Figure 2. Degenerate and radial eutectic structures in Al–17.3 at%Cu.

With increasing specimen thickness (slower cooling rate) and increasing Cu content, the characteristic precipitation sequence of G.P. zones /θ"/θ'/θ was observed. However, particular features of the precipitation reactions were noted which have not been previously reported. These are described below.

In alloys containing >12at.%Cu, the predominant precipitation mode is that of the degenerate eutectic structure shown in Figure 2. The α and θ phase separation becomes less degenerate as the specimen thickness increases, eventually assuming a linear morphology. However, high voltage (1MeV) electron microscopy reveals that such structures are not true lamellar eutectics, but are colonies of radial lamellae, similar to those reported[7] in a previous study using electron microscope replica techniques. Conventional lamellar eutectic structures (Figure 3) were only observed in specimens ion-beam thinned from the bulk.

Using the average radial lamella spacing, a cooling rate of 10^6 °C sec^{-1} was calculated.[7] However, similar calculations on true lamellar structures from thicker regions gave a value range of 10^7–10^8 °C sec^{-1}. This apparent anomaly probably arises because [a] the calculation is not strictly valid for the radial structure

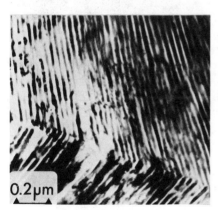

Figure 3. Lamellar eutectic structure in ion-beam thinned Al–17.3 at.%Cu.

137

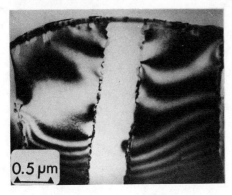

Figure 4. Grain boundary θ precipitation in Al-12at.%Cu.

which appears to be another form of the degenerate eutectic, and [b] substantial (∿10%) errors may occur in the cooling rate calculation due to the difficulties in estimating the original foil thickness of ion-beam thinned material. However, all values are well within the range reported by other workers[3,4] using the same equipment.

In specimens containing between 5 and 12at.%Cu two new microstructural features were often observed. First grain boundary precipitation occurred, as shown in Figure 4. X-ray microanalysis using EMMA 3 and conventional electron diffraction techniques indicate that the precipitates are equilibrium θ phase. The precipitation appears to be characteristic of normal grain boundary precipitation,[1,8] giving rise to distortion of the boundary plane around the particle. Accompanying

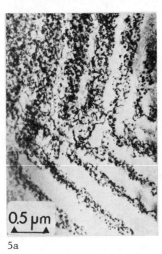

Fig. 5a. Banded precipitation in Al-7at.%Cu. (Bright field)

Fig. 5b. Dislocation substructure associated with with Fig. 5a. (Weak beam)

5a

5b

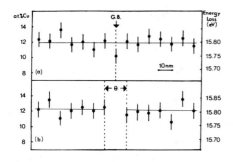

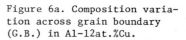

Figure 6a. Composition variation across grain boundary (G.B.) in Al-12at.%Cu.

Figure 6b. Composition variation across grain boundary in Al-12at.%Cu containing θ precipitates.

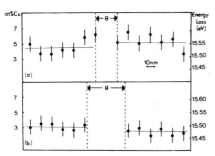

Figure 7a. Composition variation across α lamella in a degenerate eutectic structure.

Figure 7b. Composition variation across α lamella in a radial lamellar structure.

this grain boundary precipitation, matrix particles were commonly observed to assume a banded morphology as shown in Figure 5a. Precipitation is heavy and weak beam electron microscopy (Figure 5b) reveals the particles to be clearly associated with matrix dislocations and loops. Such a precipitation mode could be interpreted in terms of a cellular substructure arising as a result of solute segregation during solidification, wherein the banded regions correspond to cell walls. Alternatively, the phenomenon may arise from bulk slip processes occurring as a result of quenching strains.

Tilting the foil until the precipitate bands are parallel to the electron beam enabled the plane of the band to be identified as {111}, which is the slip plane in f.c.c. materials. Thus it is considered that the latter explanation is more reasonable. In this case precipitation would occur after solidification, on the dislocations formed during the slip process.

139

Using E.M.E.A., we have investigated microsegregation of Cu in the following specific situations.

First, a grain boundary in single phase Al–12at.%Cu showing no signs of boundary precipitation was examined. The resulting composition profile is shown in Figure 6a. There is no observed variation in composition to within $\pm$5nm of the boundary. The high supersaturation is maintained.

This implies that at such boundaries, solute segregation >10nm has not occurred during the solidification process. Thus an independent estimate of the cooling rate may be obtained. Assuming Newtonian cooling conditions an average diffusion coefficient $\bar{D}$ may be estimated from available date.[1] If diffusion is considered to be significant from the melt temperature to $T_{m/2}$ where T_m is the alloy melting temperature $\bar{D}$ $\sim$10^{-10}cm^{2}sec^{-1}. For a diffusion distance $x = \sqrt{(\bar{D}t)}$, of 10nm, the time t allowed for diffusion is $\sim$10^{-2}secs. During this time the total temperature change is $\sim$700°C, thus giving a cooling rate of 1.4×10^5°C sec^{-1}. This simple calculation implies that at cooling rates >10^5°C sec^{-1}, diffusion profiles should not be detectable, even using E.M.E.A. Furthermore conventional interlamellar spacing calculations certainly do not overestimate the actual rate of cooling.

Secondly, grain boundaries containing precipitates were studied. An identical result was obtained (Figure 6b). No solute segregation was detected within $\pm$5nm of the grain boundary θ. Thus it is proposed that solute rejection occurs into the liquid phase during solidification $(D_L \gg \bar{D})$, and that nucleation of θ occurs at the interface between two solidifying grains. No bulk solid segregation is required. This conclusion is supported by the occurrence of precipitation at the foil edge in Figure 4 where the solute-rich liquid finally solidifies. Surface nucleation occurs and the slower cooling rate of the thicker foil perimeter allows limited precipitate growth. The observed volume fraction of θ was too small to lower the bulk matrix supersaturation by an amount which would be detectable using E.M.E.A.

Thirdly, measurements were made across α lamellae in both degenerate and radial lamellar structures. The results are shown in Figures 7a and 7b respectively. Again, there is no evidence for solid state diffusion, although in both situations the α solid solution composition is ∿5at.%Cu. This is ∿2.5at.%Cu above the equilibrium α composition at the eutectic temperature indicating, [a] that equilibrium segregation has not occurred in the liquid, and [b] subsequent solid-state diffusion has not occurred to reduce this remaining supersaturation, either in the degenerate or radial eutectic structure. This may be interpreted in terms of independent α and θ nucleation within the liquid phase since the absence of solid-state diffusion has already been demonstrated in Figures 6a and 6b.

It was noted frequently during analysis of the single phase regions in specimens containing >5at.%Cu that a point-to-point (∿50-100nm) variation in bulk Cu composition of up to 5at.% occurred. This inhomogeneity appeared more significant as the Cu content increased and consequently may arise either from incomplete mixing during melting or liquid phase Cu clustering. This inhomogeneity would appear to be consistent with x-ray peak width measurements [3,9] which have been interpreted in terms of a range of solid solution composition within a particular foil. If such variations also occur on a larger scale, the inhomogeneous microstructural distribution characteristic of splat-quenched thin foil specimens may be explained. However, such factors as local cooling conditions and specimen reheating must also contribute significantly to this phenomenon.

SUMMARY

Transmission electron microscopy has revealed a wide variety of microstructures in splat quenched Al-Cu alloys. All alloys up to 17.3at.%Cu were produced in the single phase condition in their thinnest regions but retained their crystallinity. In thicker areas, grain boundary and matrix θ precipitation occurs. The latter mode is associated with {111} slip bands in the lower Cu content alloys. With increased copper additions, the predominant precipitation mode is that of the degenerate eutectic, with radial lamellar eutectic structures present in the thickest regions of the foil. True lamellar structures are only observed in regions ion-beam thinned from the bulk.

High resolution microanalysis using E.M.E.A. has demonstrated for the first time that bulk solid state diffusion does not occur in the thinnest regions of the splat. This conclusion predicts a cooling rate well within that calculated by conventional interlamellar spacing methods. Even if grain boundary θ precipitation has occurred, the required copper segregation and θ nucleation must have occurred in the liquid state.

Finally, it was observed that splat quenched aluminium-copper alloys are inhomogeneous on a <100nm scale. The degree of inhomogeneity increases with increasing copper content.

REFERENCES

1. Doig, P. and J.W. Edington, Phil. Mag. 28, 961 (1973).
2. Ramachandrarao, P., M. Laridjani, and R.W. Cahn, Z. Metallk., 63, 43 (1972).
3. Scott, M.G. and J.A. Leake, Acta Met. 23, 503 (1975).
4. Wood, J. and I. Sare, Proc. Second International Conference on Rapidly Quenched Metals, MIT, 81, N.J. Grant and B.J. Giessen, editors, MIT Press, Cambridge, Massachusetts (1976).
5. Edington, J.W. and G. Hibbert, J. Microsc. 99, 125 (1973).
6. Davies, H.A. and J.B. Hull, J. Mat. Sci. 9, 707 (1974).
7. Burden, M.H. and H. Jones, J. Inst. Metals 98, 249 (1970).
8. Aaron, H.B. and H.I. Aaronson, Acta Met. 16, 789 (1968).
9. Jansen, C., Ph.D. Thesis, M.I.T. (1971).

ON METASTABLE DILUTE BINARY $\underline{Ni}B_x$ AND $\underline{Fe}C_x$ ALLOYS OBTAINED BY SPLAT QUENCHING

M.C. Cadeville, G. Kaul,   Ph. Maitrepierre
M.F. Lapierre, and C. Lerner
                                            I.R.S.I.D.
Lab. de Structure Electronique des Solides St. Germain en Laye, France
Strasbourg, France

I. INTRODUCTION

The work presented here is part of a more general investigation of the electronic structure of interstitial impurities (H,B,C,N) in transition metals (1,2). The essential role played by such impurities (mostly carbon and boron) in the metallurgy of steel is a strong incentive for such studies. For instance, progress in the understanding of the phenomena (3) controlling the microstructure of boron-containing steels is still dependent on more precise data (4,5) on the nature and extent of Fe_γ-B and Fe_α-B solid solutions.

Our earlier investigations (1,2) were concerned with dilute alloys of boron and carbon in F.C.C. metals (Ni, Co, Pd, Pt) and alloys (Ni-Fe). The results led us to propose models for the electronic structure of carbon in an F.C.C. (ferromagnetic or paramagnetic) metal. Moreover, our studies emphasized the limited solubility of boron in Ni, Co and Pt, which hindered the investigation of the electronic properties.

The present investigation dealt with dilute $\underline{Ni}B_x$ and $\underline{Fe}C_x$ alloys (with $0 < x < 0.04$). The interstitial solubility was extended by splat-quenching of the alloys from the melt, a technique frequently used to extend terminal solubility (6). Splat-quenched Ni-B and Fe-C alloys have been prepared by Ruhl and Cohen (7), but these authors did not study the low concentration range in detail.

The present study consisted in a characterization of the quenched alloys and the determination of their physical properties, leading to a better understanding of the electronic structure. The experimental procedures and results are summarized below.

II - EXPERIMENTAL PROCEDURES

Sample Preparation: The base constituents (metal 4N8, crystalline boron 2N8, reactor graphite) were first melted and homogenized in an induction furnace. Small samples (about 100mg in weight) were splat-quenched in an anvil-piston apparatus under argon atmosphere. The resulting splat-specimens were foils about 30mm in diameter and 5 to 100 μm thick. The thickness depended essentially upon the initial

weight and temperature of the melt before quenching. Parallelepipeds
with constant thickness (about 3mm in width and 10 to 15 mm in length)
were cut from the foils and were used for resistivity measurements,
x-ray diffraction, electron microscopy and Mössbauer spectroscopy.

Resistivity Measurements: At low temperature, the largest contribution
to the resistivity of a dilute ($X \sim 1\%$) alloy is due to isolated impurities which scatter conduction electrons more efficiently than impurities concentrated in clusters or well defined precipitates. Consequently, resistivity measurements of splat-quenched alloys constitute
an easy test of the quenching "efficiency."

The measurements were carried out at 4.2K by using the conventional
4-point potentiometric method. As shown in Fig 1, quenching becomes
efficient when the sample thickness is smaller than about 30μm. However
when the sample thickness is commensurate with the electron mean free
path Λ (about 5 and 15 μm respectively for pure iron and nickel) the
measurement of the resistivity has no particular physical meaning.
This "macroscopic size effect" is observed in pure nickel and iron
(cf. insert, Fig 1) and in very thin alloys. Another disadvantage
of the resistivity as a probe is its non-selective character.

X-ray Diffraction: In contrast to resistivity, X-ray diffraction allows
one to identify the different phases, to estimate their respective pro-
portions and to define the nature and concentration of the solid solution.

X-ray diffraction patterns were obtained at room temperature with
filtered cobalt $K\alpha_1$ radiation in a Seeman-Bohlin camera by transmission,
calibrated with silicon; the films were examined in a microdensitometer
The uncertainties in lattice spacings result essentially from the position and smoothness of the sample inside the camera.

A disadvantage of X-ray diffraction is its lack of sensitivity
to phases which constitute only a small volume fraction. Therefore,
we used preferentially the Mössbauer technique for the FeC_x alloys
to determine the percentage of retained austenitic phase. We do not
give details here about this method of determination which has been widely
used. Other Mössbauer results concerning the martensitic phase will

144

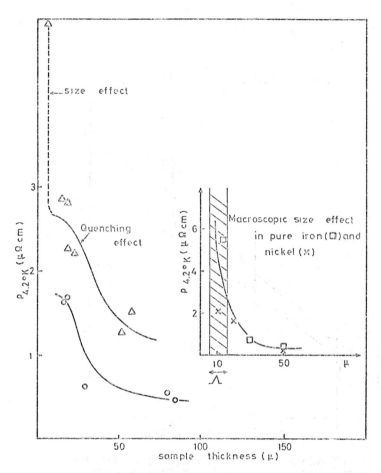

Figure 1. Residual resistivities versus sample thickness in $\underline{Ni}B_x$ (o) (x = 0.01) and $\underline{Fe}C_x$ (△) (x = 0.005) alloys; Insert: macroscopic size effects in pure iron (□) and nickel (X) compared with the order of magnitude of the electronic mean free path (Λ) at 4.2K.

145

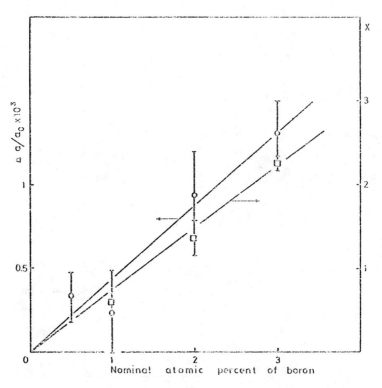

Figure 2. Relative parameter increase ($\Delta a/a_o$) (0) and percentage of boron (X') as Ni_3B boride in splat $\underline{Ni}B_x$ specimens as a function of nominal boron content.

146

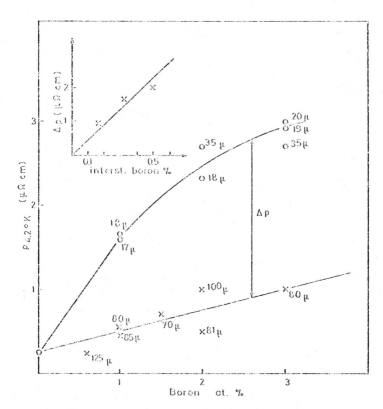

Figure 3. Dependence of the residual resistivity as a function of boron atomic percent in the thickest samples [80 to 100μm (X)] and the thinnest samples [∿20 – 30μm (0)] in NiB$_x$ alloys. Insert: dependence of Δρ ($\rho_{20\mu m}$ – $\rho_{100\mu m}$) on the interstitial boron content deduced from the parameter measurements.

be published elsewhere (7).

Electron Microscopy: Whereas X-ray diffraction and resistivity average
out on a "large" volume of sample (a few mm^3), transmission electron
microscopy (100 kV) allows a truly microscopic characterization of the
samples (nature, size and distribution of phases). Thin foils were
prepared either directly from quenched samples (with edges thin enough
to be transparent for 100 keV electrons) or from electrolytically polished samples (twin jet technique). Selected area diffraction experiments were also conducted for phase identification.

III. - RESULTS

III.1 NiB$_x$ Alloys:

X-ray Diffraction: All splat-cooled specimens contain the F.C.C. phase
plus the orthorhombic phase Ni$_3$B. Only the thinnest ones ($\sim$ 10μm) were
investigated in detail. The lattice parameter of the F.C.C. phase
increases slowly with the total boron content (Fig. 2). Each point in
the Figure is an average value of measurements on several specimens,
which explains the large scatter of the data. The parameter increase
due to 1% interstitial boron has been previously estimated to be about
$(10 \overset{+}{-} 1) \cdot 10^{-3}$ Å in a F.C.C. lattice (2). Thus the estimated interstitial boron content in our alloys is about 20% of the nominal boron
content.

The concentration X' of boron present in the alloy as Fe$_3$B boride
was estimated by comparing the intensities of the most intense line
of the Ni$_3$B spectrum (.121) with the (200) line of nickel, and by cal-
ibrating with an alloy where all the boron had been precipitated as
Ni$_3$B. X' is given by:

$$X' = 0.82 \ \frac{I_{121}}{I_{200}}$$

Our results given in Figure 2 indicate that X' represents about 80%
of the total boron content. Thus our X-ray investigation shows that
in the thinnest splat Ni-B specimens boron is distributed between the
interstitial solid solution ($\sim$20%) and the Ni$_3$B phase ($\sim$80%).

148

Figure 4. Ni-B$_{0.03}$ alloy (80μm splat quenched foil): transmission electron micrograph.

Figure 6. Ni-B$_{0.03}$ alloy (8μm splat quenched foil) – Selected area electron diffraction pattern: (121)* reciprocal lattice section with superlattice spots from the metastable phase.

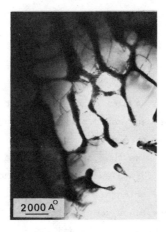

Figure 5. (a, left; b, right) – Ni-B$_{0.03}$ alloy (8μm splat quenched foil): transmission electron micrographs showing the dendritic solidification structure.

Resistivity: As shown in Figures 1 and 3, the values of the residual
resistivity depend on both sample thickness and impurity concentration.
Like the lattice parameters, the resistivities increase with the concentration; hence we believe that both effects are connected. If the
resistivity difference ($\Delta\rho$) between the thick samples (80 to 100μm) and
the thin samples (20 to 30μm) is attributed to the portion of boron
atoms which enter the nickel lattice we obtained:

$$\Delta\rho \simeq 4 \text{ to } 5 \quad \mu\Omega cm/\%B.$$

This value is similar to the residual resistivities associated with
boron or carbon in other transition metals.

Electron Microscopy: We investigated only the most concentrated alloys
(3 at% B). Two kinds of specimens were studies: foils about 80 μm in
thickness and foils with thickness less than 8μm. In the thick foils
the microstructure (Fig. 4) consists of fine grains of nickel (0.5
to 1μm) separated by a phase identified as Ni_3B. One can consider that
such a structure is the normal solidification structure for an alloy
of that composition. The dislocation density inside the nickel grains
is that expected in a rapidly quenched metal.

On the other hand, the thinnest (i.e., the most rapidly quenched)
foils show a fine microstructure with definite dendritic aspect (Fig.
5 a and b). Selected area diffraction on some thin regions yield
characteristic diffraction patterns (cf. Fig. 6). These diffraction
patterns reveal superlattice spots on all reciprocal lattice sections
of nickel. Dark field images demonstrated that these superlattice
spots were associated witht the B-rich phase filling the interdendritic spaces. A careful check revealed that these spots were true superlattice spots and could not be explained by double diffraction effects.
One must therefore assume that a new metastable phase with an F.C.C.
structure such as that of nickel and a lattice parameter three times
as large was formed upon quenching. Moreover, this phase exhibits a
cube $||$ cube orientation with respect to nickel. The metastable
phase appears to be completely isomorphous to the $M_{23}I_6$ compounds
(where M = Fe, Ni, Cr, Mo and I = B, C) frequently found in boron-
containing steels (9). Its composition should therefore correspond

150

to the formula $Ni_{23}B_6$ in the present case.

III.2 $\underline{FeC_x}$ ALLOYS

X-ray Diffraction: The photographs obtained after long exposure times
revealed only martensite peaks for c $\leqslant$ 0.02 and both austenite and
martensite peaks for c $\geqslant$ 0.03. The tetragonal doublet (200,020-002)
is well resolved for c $>$ 0.02, and was used for the determination
of the lattice constants. For each alloy the peak maximum was recorded and calibrated with an appropriate line from the silicon standard.
At high carbon concentrations a low angle asymmetry was observed in
the peak (200, 020) due to the slight orthorhombic character of the
martensite (a $\neq$ b). We have not resolved this doublet (200,020) and
we give the values of an average parameter $\overline{a,b}$. We obtain

$$c = 2.866 + 0.025 \, x$$
$$\overline{a,b} = 2.866 - 0.003 \, x$$

These results are in good agreement with published data.

Resistivity Results: Over the whole range of concentrations investigated the residual resistivity increases linearly with carbon content
(Fig. 7) giving a resistivity increment of 5$\mu\Omega$cm per atomic percent
carbon. This value is in good agreement with that measured by Wagenblast and Arajs[10] in the dilute ferritic phase of 4.9$\mu\Omega$cm/%C. The
absence of a term varying as x^2 in the resistivity –vs–concentration
curve seems to indicate that the number of carbon atoms on nearest
neighbor sites is not enhanced.

Magnetization Results: The magnetization measurements were carried out
at 4.2K in a superconductor coil giving fields up to 150 kOe. The
value of the saturation moment was obtained by extrapolating the magnetization values measured in high fields to H=0. The apparatus was
calibrated with an ellipsoid of pure bulk iron. The magnetization
change with carbon content was compared to that of a fine iron splat
specimen of 2.09μ_B, a value clearly smaller than the magnetic moment
of pure iron of 2.12μ_B. This difference may be due to defects (vacancies, strains, etc.) or impurities retained by quenching. As shown
in Fig. 8, the magnetization of the martensitic phase increases

151

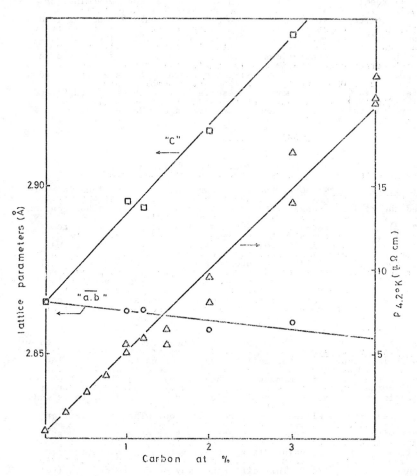

Figure 7. Change of the residual resistivity (△) and lattice parameters a,b (○) and c (□) as functions of nominal carbon concentration in $\underline{Fe}C_x$ alloys.

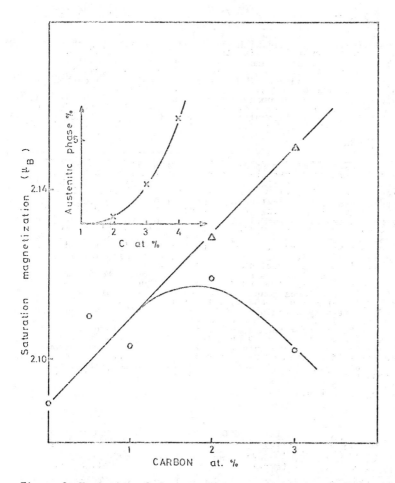

Figure 8. Variation of the saturation magnetization in $\underline{Fe}C_x$ alloys: without correction for the contribution of the austenitic phase (o) and with correction (Δ); Insert: percentage of austenitic phase deduced from Mössbauer spectra.

linearly with carbon content:

$$\Delta\mu = + 2\mu_B \pm 0.5\mu_B/\text{carbon atom}.$$

By assuming that i) the total moment variation is due to the difference between the charges Z_σ displaced by carbon in the $3d_\sigma$ spin sub-bands ($\Delta\mu = Z_\uparrow - Z_\downarrow = +2$) and ii) the total displaced charge Z equals the carbon valence: $Z = Z_\uparrow + Z_\downarrow = 4$, we obtain the following values of Z_σ : $Z_\uparrow = 3$ and $Z_\downarrow = 1$.

The result is quite compatible with the "weak ferromagnetic" character of iron: both 3d sub-bands ($\uparrow$ and $\downarrow$) emerge at the Fermi level with high density of states and contribute to the screening of a carbon atom. By contrast, in a ferromagnet such as nickel, where the $d_\uparrow$ band is totally filled, previous results [1,2] showed that $Z_\uparrow \sim 0$ and $Z_\downarrow = Z$.

IV. CONCLUSIONS

In the main part of dilute splat-quenched Ni-B alloys, boron is distributed between interstitial solid solution ($\sim 20\%$) and Ni_3B boride ($\sim 80\%$), regardless of the total boron content ($0 < x < 0.03$). In some very thin areas of the samples, i.e., the most rapidly cooled regions, a new metastable phase, isomorphous to the $M_{23}I_6$ compounds was identified by electron diffraction. The resistivity associated with interstitial boron in nickel is estimated to be about 4 to $5\mu\Omega\text{cm}/\%B$.

The $\underline{Fe}C_x$ splat specimens consist, for the most part, of martensitic phase. The resistivity associated with carbon in this phase ($5\mu\Omega\text{cm}/\%C$) is slightly different from the value obtained in the dilute ferritic phase. Magnetization measurements ($\Delta\mu = + 2\mu_B \pm 0.5\mu_B/\text{carbon atom}$) indicate that both $3d_\sigma$ ($\sigma = \uparrow$ and $\downarrow$) bands of iron contribute to the screening of carbon.

REFERENCES

1. Cadeville, M.C., F. Gautier, and C. Robert, Proc. 13th Low Temp. Conf., Boulder, 4, 325 (1972).
2. Cadeville, M.C. and C. Lerner, Paper submitted for publication to Phil. Mag., June (1975).
3. Maitrepierre, Ph., D. Thivellier, and R. Tricot, Met. Trans. 6A, 287 (1975).
4. Goldschmidt, H.J., Interstitial Alloys, Butterworth, London (1967).
5. Hayashi, Y. and T. Sugeno, Acta Met. 18, 693 (1970).

6. Duwez, P., Trans. A.S.M., 60, 607 (1967).
7. Ruhl, R.C. and M. Cohen, Trans. AIME 245, 241-251/253-257 (1969).
8. Cadeville, M.C. and J. Friedt [unpublished results] (1975).
9. Henry, G., Ph. Maitrepierre, B. Michaut, and B. Thomas, Proc. 8th International Conference on Electron Microscopy, Canberra, 666 (1974).
10. Wagenblast, H. and S. Arajs, J. Appl. Phys. 39, 5885 (1968).

CRYSTAL STRUCTURE OF A METASTABLE ALUMINIUM-NICKEL PHASE OBTAINED BY SPLAT COOLING

K. Chattopadhyay, P. Ramachandrarao, S. Lele, and T.R. Anantharaman
Banaras Hindu University
Varanasi, India

Under equilibrium conditions the solubility of nickel in aluminium is limited to 0.023at.% at room temperature. Tonejc, Ročak, and Bonefacic have reported some extension of solid solubility as well as formation of a metastable phase in foils of Al-Ni alloys produced by the piston-and-anvil technique of rapid solidification. The present paper deals with the structure of an aluminium 7.15at.% nickel alloy splat-cooled by the "gun" technique that can generate higher cooling rates than the piston-and-anvil technique.

EXPERIMENTAL

The Al-7.15at.%Ni alloy was prepared by vacuum induction melting of the required amounts of aluminium and nickel (99.999+%) in a graph-ite crucible. The alloy was subsequently forged and homogenized by vacuum annealing. Quantitites of the alloy of about 100mg were splat-cooled by the "gun" technique.

X-ray diffraction patterns were obtained with filtered Cu Kα radia-tion using a Philips 114.6mm Debye-Scherrer camera. Some of the flakes were also subjected to similar x-ray examination after mildly deform-ing them in an agate mortar. The flakes were ductile and could not be powdered.

For electron metallography the thinner flakes were utilized. Some areas in these foils were thin enough to transmit electrons and were studied at 100kV in a Philips EM 300 electron microscope.

RESULTS AND DISCUSSION

X-ray Diffraction Analysis: Our x-ray patterns consisted of reflections due to the metastable aluminium solid solution (α-Al) and a set of ex-

tra reflections. For the sake of brevity only the first fourteen of
the over thirty extra reflections are presented in Table I along with
TRB's results. As can be easily seen, there were more extra reflections
in our patterns than reported by TRB. The agreement in $\sin^2\theta$ values
between the common reflections in TRB's and our work is remarkably good
and justifies the conclusion that the same metastable phase was produced in both studies. One significant difference, however, is the complete
absence of the equilibrium Al_3Ni phase in our foils. This may well be
attributed to the higher cooling rates achieved in the "gun" technique.

TRB have tentatively explained the extra reflections on the basis
of an orthorhombic cell of space group Pmna with a=6.40, b=7.56 and
c= 9.56 Å. This unit cell is related to that of equilibrium Al_3Ni in
that the c parameter for the metastable phase is twice that for Al_3Ni,
while the other two parameters are nearly identical. It is relevant to
note that in the range of $\sin^2\theta$ being shown in Table I twenty-three
reflections should appear on the basis of this unit cell and space
group, but only seventeen could be accommodated against the fourteen
reflections observed by us. TRB actually observed only nine of these
reflections. Further, including the extra reflections observed by us,
the agreement between the observed and calculated $\sin^2\theta$ values is rather
poor, resulting in a mean deviation of 0.0009.

We have been successful in explaining our x-ray results on the basis
of a monoclinic cell with a=6.855Å, b=7.418Å, c=4.926Å, and β=102.8°. All the
parameters of this cell are close to those of the equilibrium orthorhombic Al_3Ni phase that has been assigned the space group Pnma. Thus the
space group of the monoclinic cell might be either of $P2_1/m$ or $P2_1$, since
only these space groups have equivalent points in common with the space
group of Al_3Ni. The only forbidden reflections in these space groups
are Ok0 reflections with k=2n. On this basis twenty reflections are
theoretically permitted in the range reported in Table I. Of these,
nineteen were actually observed by us and only the rather low angle
001 could not be seen. Further, the mean deviation is $\sin^2\theta$ in our
case is only 0.0004 which is more satisfactory than in the case of TRB.

The volume of the proposed cell is 244.12 Å^3 and gives an average

158

atomic volume of 15.26 Å3 which is intermediate between that of pure
aluminum (16.60 Å3) and Al$_3$Ni (14.56 Å3).

Effect of Heat and Cold Work: TRB[1] have investigated the thermal stability of the metastable phase and found that it decomposes in 10 minutes at 300°C. Our work confirms this finding.

The metastable phase was found by us also to be extremely unstable
with respect to plastic deformation at room temperature. X-ray powder
patterns of the ground foils were found to have only Al$_3$Ni and α-Al reflections. We take this as indicative of the close structural relationship between the metastable phase and Al$_3$Ni and as lending further support to our interpretation of the x-ray patterns in terms of a monoclinic distortion of the orthorhombic unit cell of Al$_3$Ni.

Electron Metallography: Earlier literature on liquid quenching suggests
that a non-equilibrium phase like the present one could result from
decomposition of a metastable solid solution or through direct solidification from the melt. One of the aims of our electron-microscopic
studies was to ascertain, if possible, the actual mode of formation of
the metastable phase. Our most significant finding in this regard has
been the detection of an amorphous or microcrystalline phase in the
alloy foils. Contrastless areas (see Fig.1a) which give rise to diffraction patterns characteristic of an amorphous phase (Fig. 1b) were
detected in the as-quenched foils. In the course of observation these
areas started to decompose into lamellar products. The nature of the
interface, and the mode of its advancement into the untransformed area
(see Fig.1a) strongly suggest a diffusion-controlled eutectoid-type
(or eutectic-type?) transformation of the amorphous phase.

Electron diffraction patterns obtained from the decomposed regions
consisted of single-crystal patterns from one phase and spots as well
as rings from the other. The interplanar spacings obtained from the
former correspond to the monoclinic metastable phase established by our
X-ray examination and could be accordingly indexed (Fig. 1c). The spot-
cum-ring pattern could be attributed to α-Al.

Further work is in progress to obtain the crystallographic orientation relationships and growth kinetics associated with the formation

of the metastable phase.

We are grateful to Dr. V.V.P. Kutumba Rao and Mr. A.K. Mishra for supplying the alloys. Financial assistance by the U.G.C., New Delhi to one of us (K.C.) is gratefully acknowledged.

REFERENCE

1. A. Tonejc, D. Ročak and A. Bonefacic, <u>Acta Met.</u>, <u>19</u> (1971) 311.

Table I. Observed and calculated values of $\sin^2\theta$ for reflections from the metastable phase in splat-cooled Al-7.15 at % Ni.

Unit cell: TRB - orthorhombic a=6.40 Å, b=7.56 Å, c=9.56 Å;

Ours - monoclinic a=6.86 Å, b=7.42 Å, c=4.93 Å, β= 102.8°.

S.No.	TRB			Ours			
	hkl	$\sin^2\theta_{cal.}$	$\sin^2\theta_{obs.}$	$\sin^2\theta_{obs.}$	$\sin^2\theta_{cal.}$	hkl	$I_{obs.}$
1	101	.0210	.0230	.0238	.0241	1̄10	mw
2	111	.0314	.0315	.0309	.0309	1̄01	vs
3	012	.0364	.0365	.0369	.0365	011	mw
4	020	.0416	.0413	.0424	.0417	020	s-b
					.0432	1̄11	
5	021	.0481	.0470	.0472	.0473	101	mw
6	112	.0509	–	.0532	.0532	200	mw
7	200	.0580	.0560	.0572	.0564	120	w
					.0581	1̄11	
8	121	.0626	.0624	.0631	.0626	2̄01	w
					.0640	210	
9	210	.0684	.0689	.0697	.0690	021	m
	013	.0689					
10	103	.0730	.0732	.0734	.0734	2̄11	vw
					.0741	1̄21	
11	030	.0936	–	.0913	.0905	121	mw
12	212	.0944	–	.0960	.0954	201	vw
					.0964	220	
13	220	.0996	–	.1004	.0999	1̄02	w
	031, 023	.1001					
14	004	.1040	–	.1041	.1030	002	mw

Fig.1. Decomposition of an amorphous region in splat
cooled Al-7.15 at %Ni foil: (a) Electron micrograph of amorphous and decomposed lamellar regions
(b) Electron diffraction pattern from the amorphous region (c) Electron diffraction pattern from
the transformed area. The Miller indices correspond to the metastable phase.

CLUSTERING EFFECTS IN SPLAT-COOLED Al-TRANSITION METAL ALLOYS

A. Fontaine

Laboratoire de Physique des Solides
Université Paris-Sud, Orsay, France

INTRODUCTION

According to the Hume-Rothery rules, solute atoms may have appreciable
solubility in a solid matrix if the difference between the diameters
of the solute atoms and the solvent atoms is less than 15%. In most
cases, this size effect explains the restricted solubility rather sat-
isfactorily; however, it may be necessary to correct this first rule
by what is called the "relative valence effect".

The atomic radii of the 3d-elements Cr, Mn, Fe, Co, Ni, and Cu are
smaller than the radius of the aluminium atom, with a misfit parameter
of less than 15%. However, the equilibrium diagrams of the Al-T binary
alloys exhibit only small α-Al solid solubility ranges, restricted to
$4 \cdot 10^{-2}$at.% for Cu, $2.6 \cdot 10^{-2}$at.% for Mn, and $0.37 \cdot 10^{-2}$at.% for Co,
and only very narrow regions for the three ferromagnetic elements Fe
$(240 \cdot 10^{-6}$at.%), Co $(100 \cdot 10^{-6}$at.%) and Ni $(250 \cdot 10^{-6}$at.%). Quenching
from the liquid state permits the preparation of metastable Al solid
solutions containing up to 3at.% of the solutes Cr, Mn, Fe, Co, Ni,
and Cu.

The difficulty of preparing random solid solutions of the 3d elements in an Al-matrix varies from one solute atom species to another
and is almost independent of the size effect for these six elements.
This is due to the tendency of solvent-solute pairs to associate and
to form intermetallic compounds, where the metallic bonds are no
longer isotropic. The structure of the splat-cooled Al-T alloys is
expected to reflect this chemical effect.

HARDENING IN Al-T ALLOYS

In the Table, following, microhardness data of Al-T alloys and Al-2at.%
Cu are listed, as reported by different authors.[1-3] We point out that

TABLE. Summary of Hardnesses and Related Parameters for Al-T Alloys

SOLUTE ATOMS	Cr	Mn	Fe	Co	Ni	Cu	Zn	Al
r_S (Å)	1.42	1.43	1.40	1.38	1.38	1.41	1.53	1.58
$\|\delta r_S/r_S\| = \eta$	0.101	0.095	0.114	0.127	0.127	0.108	0.032	
Concentration, c (at.%)	0.03	0.03	0.03	0.03	0.03	0.02		
H_v (measured) (Kg/mm^2)	120*	110†	300†	350†	350†	76–80**		23
$\Delta = H_v - H_{vAl}$	97	87	277	327	327	53–57		
$\mu (\eta c)^{4/3} \cdot (10^6)$ (dynes/cm^2)	118	109	139	161	161	75.4		

* Reference 1
** Reference 2
† Reference 3

Following Friedel, we have chosen the "radius of the atomic sphere" which is related to the volume of each atom in the structure of the pure element.

the three 3d-elements with the smallest solubility induce the highest hardening, to values about three times larger than those measured for the other three solute atoms.

In general, the effect of a solute atom is to harden the metal. This effect is characterized by a general shift of the stress-versus-strain curve of the pure metal towards higher stress values. The magnitude of this translation depends on the type of impurity and its concentration. In general, addition atoms have the least effect in the form of coarse, well-separated precipitate particles, and the greatest effect for a finely dispersed precipitate with a spacing of the order of 100Å; when the addition atoms form a solid solution, the effect is intermediate.

This latter effect can be discussed in terms of the "free binding energy" between a dislocation and a solute atom.

[1] Isolated solute atoms are centers of elastic distortion, especially if they are much larger or smaller than the solvent atoms. Consequently, the solute atoms and dislocations will interact elastically and exert forces on each other. (For an

edge dislocation, the energy $W_1 = -\frac{1}{2\pi}\frac{1+\nu}{1-\nu}\mu\,|\Delta v_a|$ is about a few tenths of an eV, where $|\Delta v_a|$ is a measure of the size effect, ν is the Poisson ratio and μ is the shear modulus.)

[2] The difference between the compressibilities of the two atoms ($\kappa-\kappa'$) contributes to the interaction energy. This part is proportional to the difference ($\kappa-\kappa'$) and is larger for a screw dislocation than for an edge dislocation.[4,5]

[3] When the elastic term of W is small (e.g., for Zn or Ag in Al), the electrostatic interaction between the impurity and the dislocation must be considered. The anisotropy around an edge dislocation produces a rearrangement of the conduction electron charge, giving rise to a charge dipole. The field of the dipole can act on the nuclear charge of the impurity atom. This interaction is proportional to the difference of the valences between a substitutional impurity and the matrix (for Cu-Zn, $Z = 1$, $W_3 \underset{\sim}{\cup}$ $-0.005eV$[4]).

When the misfit is large, the size effect predominates, and the interaction energy W can be written $W = \frac{1}{2}\mu b^3 \eta$ where $b = 2.86\text{\AA}$, $\mu = 27 \cdot 10^{10}$ dynes/cm^2 for Al, and $\eta = \left|\frac{\delta R_s}{R_s}\right|$ is the size factor.

Under these conditions, the observed hardening is proportional to $\mu(\eta c)^{4/3}$

$$\Delta = H_v - H_v(Al) = K\,\mu(\eta c)^{4/3} \tag{1}$$

If the microhardness is only due to the size effect, the coefficient K is expected to be independent of the specific type of impurity.

This is the case for the Al-based solid solutions Al-3at.%Cr ($K\cup82$), Al-3at.%Mn ($K\cup79$), and Al-2at.%Cu ($K\cup75$). If the coefficient $K\cup79$, which characterizes these solid solutions hardened only by the size effect, is put into formula (1), one obtains a microhardness $H_v = 133$ Kg/mm^2 for Al-3at.%Fe and $H_v = 154$ Kg/mm^2 for both Al-3at.%Co and Al-3at.%Ni. These values are less than one half of the values measured for these three alloys.

165

An explanation of this discrepancy is attempted in the following, using x-ray small angle scattering data.[3] It has been shown by this technique that classical quenching distributes copper atoms in an Al matrix at random. Dartyge, et al.[6], have shown that this is also the case for Al-4at.%Mn; the small angle diffuse intensity is of the order of magnitude of the Laue intensity calculated for a completely disordered solid solution.

The situation is different for Al-2at.%Ni; analysis of the small angle diffuse intensity shows numerous clusters with sizes of <20Å but also heterogeneities with greater diameters. For Al-3at.%Ni the fraction of small clusters increases. They do not completely disappear after annealing at 300°C.[3]

For Al-Fe solutions with Fe concentrations higher than 0.5at.% up to 3at.%, a significant part of the iron atoms in splat-cooled alloys are isolated, but the other solute atoms are in sites containing iron atoms among their twelve neighbors. Annealing for 5 minutes at 300°C increases this clustering effect.[3]

Such clusters in Al-Fe alloys have also been observed by Janot and Lelay[7] by Mössbauer spectroscopy; for very small solute concentrations ($c < 240 \cdot 10^{-6}$at.%), iron atoms are isolated in the solid solution quenched from 640°C, but after annealing at 300°C clusters involving four iron atoms appear. This first stage of precipitation is enhanced by cold working of the alloy.

Thus, the as-quenched Al-3at.%Fe and Al-3at.%Ni alloys and probably the Al-3at.%Co alloy are not "perfect solid solutions" where solute atoms substitute at random for aluminium atoms (albeit with local distortions) but where solute and solvent atoms are bound by isotropic metallic bonds as in pure Al. Rather, splat cooling of these alloys (Al-Fe, Al-Co, Al-Ni) has produced a very fine dispersion of the solute atoms into small clusters formed with neighboring Al atoms. Stable nuclei with imperfect structure are thus created that may be coherent with the matrix. These clusters are too small to be observed by electron microscopy[3] and are assumed to account for the high observed hardness values.

166

The impossibility of dissolving these solute atoms completely in the aluminium matrix is not surprising considering the very low solid solubility of the 3d- elements Fe, Co, and Ni. Furthermore, one may expect (as in molten Al-Sn)[8] that solute clusters already exist in the liquid. The number of these clusters is enhanced by the large magnitude of the supercooling involved in the rapid quenching process.

CONCLUSIONS

Cr, Mn, and Cu atoms substitute at random for Al atoms at solute contents of up to about 3at.%. The situation is different for Fe, Co, or Ni atoms; these last three solute atom species do not dissolve randomly in the Al-matrix and do not produce solid Al solutions of the type represented by Al-Ag, where solute atoms (Ag) can be dissolved in large concentrations with practically no distortions of the Al-matrix.

This difference in the behavior of Cr, Mn, and Cu atoms on the one hand, and Fe, Co, and Ni on the other is not due only to the magnitude of the misfit of the atomic radii. Rather, it is assumed that the imperfect dispersion of the Fe, Co, and Ni atoms arises from a tendency of solvent and solute atoms to form nuclei of intermetallic compounds where bonds are no longer isotropic metallic ones. These clusters are regarded as extended defects in a single phase structure.

REFERENCES

1. Babić, E., E. Girt, R. Krsnik, B. Leontić, M. Očko, Z. Vućić, and I. Zorić, _Phys. Stat. Sol._ 16 K 21 (1973).
2. Buckle, H., _Pub. Sci. Min. de l'Air_, 147 (1960).
3. Fontaine, A. and A. Guinier [to be published].
4. Friedel, J., _Dislocations_, Pergamon Stud. Eds. (1967).
5. Fleischer, R.L., _Acta Met._ 11, 203 (1963).
6. Dartyge, E., M. Lambert, G. Leroux, and A.M. Levelut, _Acta Met._ 20, 233 (1972).
7. Janot, C. and G. Lelay, _CRAS_ 269 (B), 823 (1969).
8. Hohler, J.C. and S. Steeb, S.A.S. Conf., Graz, Austria, August (1970).

RAPID SOLIDIFICATION OF Al-Mg-Si ALLOYS FROM THE LIQUID STATE

S.K. Bose and Rajendra Kumar

National Metallurgical Laboratory
Jamshedpur, India

INTRODUCTION

The concept of the existence of clusters of either like or unlike atoms
in a matrix of monatomic atoms has found increasingly greater acceptance. By centrifuging of the liquid, Singh and Kumar[1] had earlier provided experimental evidence that clusters of both like and unlike
atoms exist in the Al-Si system depending upon composition, although only
clusters of like atoms were anticipated to be present in the liquid.
Al-Si is a system of the eutectic type; however, Singh and Kumar showed that clusters in the hypo- and hyper-eutectic alloys are significantly and markedly different from one another; whereas in the hypo-eutec-
tic and in eutectic alloys, Si atoms associate with Al atoms as Al-Si
clusters, they are present in hyper-eutectic alloys as Si-Si clusters.
This difference in the nature of the distribution of silicon atoms was
further concluded to exist by Bose and Kumar[2] who confirmed previous
measurements on splat-cooled Al-Si alloys by Itagaki, Giessen, and
Grant[3] and showed that the limit of primary solid solubility as measured by the continuous increase in lattice parameter was extended up
to the eutectic composition by rapidly solidifying the alloys in the form
of thin films (at $\sim 10^4 °C/sec$).

Since the association of Al and Si is energetically weak, it is
likely to be destroyed in the presence of elements forming intermetallic compounds with silicon such as Mg_2Si. Kumar and co-workers[4] studied
the effect of such ternary additions of magnesium on the structure of
liquid Al-Si alloys. Their results unequivocally showed that Al-Si
clusters are destroyed in the presence of magnesium in the melt and
replaced by those of Mg-Si. The present investigation was undertaken
to provide confirmation of the above results by studying the effect of

169

TABLE

ALLOY NUMBER	Al%	Mg%	Si%	EQUILIBRIUM STRUCTURAL ANALYSIS	
				Mg_2Si %	Excess Si%
1	94.10	0.75	5.15	1.18	4.72
2	92.65	1.18	6.17	1.86	5.49
3	89.64	3.94	6.42	6.21	4.15
4	87.32	6.47	6.21	10.2	2.48

magnesium on the rapid solidification of Al-Si alloys in terms of lattice parameter and microhardness changes.

EXPERIMENTAL METHODS

Four aluminium-rich ternary Al-Mg-Si alloys were made and their compositions were selected such as to give an Mg_2Si content in the range of 1-10% with an excess "free" silicon content of up to approximately 4-5% in all cases. Their chemical analysis, along with the equilibrium Mg_2Si content is given in the Table (supra).

Rapidly solidified thin films with thicknesses varying between 0.1-0.15mm were prepared from these alloys by utilizing the experimental setup described previously.[5] The films thus obtained were examined metallographically and also by a Cambridge scanning electron microscope operated at 30KV, with a comparative estimate of the solute concentration using the energy dispersive x-ray analyser, and in the unetched and etched condition, using Keller's reagent. The microhardness data were obtained at room temperature using Leitz equipment with a diamond pyramid indenter subjected to a load of 25gms. The measurements were performed on films with smooth and bright areas and an average of ten readings was taken. Narrow regions of the thin films were selected without applying any mechanical cold work to obtain Debye-Scherrer powder patterns. All the photographs were taken in a camera of 114.7mm diameter using filtered CuKα radiation; Nelson and Riley's extrapolation

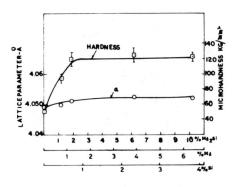

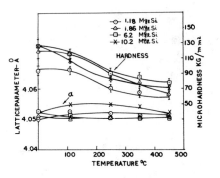

Figure 1. Variation of lattice
parameter and microhardness with
(a) Mg2Si, Mg and Si content;
(b) annealing temperature.

function was used to measure the
lattice parameters accurately.
Changes in the microhardness as
well as lattice parameter were
measured in the "as solidified"
condition and after annealing in
vacuum in the temperature range
of 110-450°C for various times.

RESULTS AND DICUSSION

Figure 1 shows the lattice parameter of the rapidly solidified matrix as a function of the Mg_2Si
content, as calculated on the assumption that Mg associates fully with silicon in the liquid.

The presence of 4-5% free Si
in rapidly solidified supersaturated solid solution would be ex-
pected to decrease the lattice parameter of Al. However, Figure 1 shows
that, despite the presence of free Si, the lattice parameter of the matrix is not less than the value of pure Al. The slight increase in the
lattice parameter with increasing Mg or Si contents indicates that a
small amount of Mg has dissolved in Al. It is suggested that this confirms the assumption that Mg partitions between the Al and Mg_2Si clusters in the liquid state, and, therefore, provides experimental confirmation of the earlier findings of Kumar and co-workers[4] that the
addition of Mg destroys the preferential association of Al-Si atoms
and instead forms Mg_2Si clusters in the liquid state.

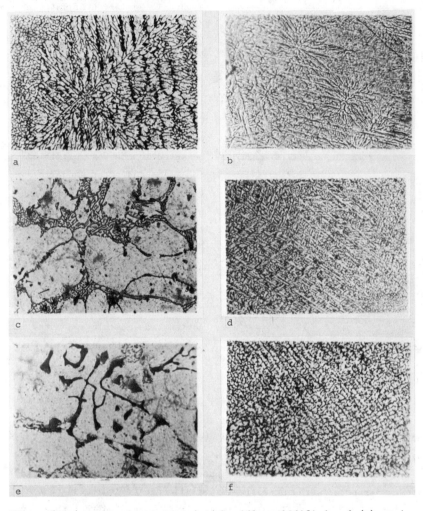

Figure 2. As-cast structures: (a,b) Rapidly solidified and (c) equilibrium cast Al-0.75%Mg-5.15%Si alloy; (d) Rapidly solidified (X350); and (e) Equilibrium cast Al-3.94%Mg-6.42%Si alloy (X1000); (f) Rapidly solidified Al-6.47%Mg-6.21%Si alloy (X350).

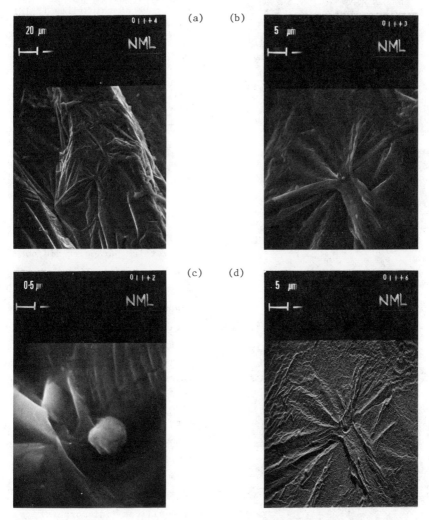

Figure 3. Scanning electron micrographs of rapidly cast thin films of Al – 0.75 Mg – 5.15% Si alloy showing (a) Cast morphology (X 500), (b) Predendritic region (X 2000), (c) Spherical nucleus at the center of the predendritic region (X 20,000), (d) Surface topography under 'y' modulation (X 2000).

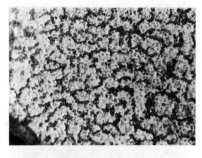

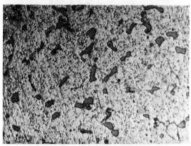

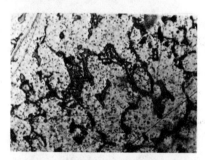

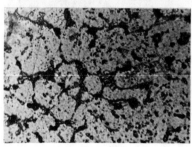

Figure 4. Metallographic structure of thin films (X 1K)
(a) Al - 0.75% Mg - 5.15% Si annealed at 450°C/30 min
(b) Al - 1.18% Mg - 6.17% Si annealed at 45o°C/2 h
(c) Al - 3.94% Mg - 6.42% Si annealed at 250°C/30 min
(d) Al - 6.47% Mg - 6.21% Si annealed at 110°C/30min

The increase in microhardness with increasing Mg_2Si content appears to be purely an effect of dispersion hardening. Upon annealing the thin films at temperatures up to 450°C, the lattice parameter did not undergo any significant change, whereas the microhardness gradually decreased with annealing temperature as shown in Figure 1(b).

Typical metallographic structures of rapidly solidified alloys are shown in Figure 2 which also gives the as-cast structure of an alloy of identical composition solidified under equilibrium conditions by furnace cooling. Although both Mg_2Si (script morphology) and free silicon (idiomorphic) could be detected in equilibrium cast structures, Mg_2Si was masked in rapidly solidified films by the presence of a fine dendritic cellular morphology; the presence of free Si could also not be detected. Solidification characteristics of the alloy containing 0.75% Mg, 5.15% Si were studied by optical metallography, as shown in Figure 2b in which a pre-dendritic starlike structure is obtained in the rapidly solidified thin films before they attain a fully developed dendritic morphology similar to that observed by Kumar and Bose[6] and Biloni and Chalmers[7] in the case of Al–Cu alloys. Scanning electron micrographs of the predendritic region also indicate formation of a predendritic structure as shown in the electron micrographs, Figure 3a.

It was further noticed that the central predendritic region of the electron micrograph, Figure 3 b,c, of the cast thin film consists of a spherical particle which might be acting as a nucleus for the growth of the predendritic structure which is accompanied by normal dendritic solidification. The latter proceeds by a diffusion mechanism which is operative as a result of a solute concentration gradient between the spherical nuclei and the matrix. The surface topography of the cast thin film is also shown in Figure 3d under y–modulation.

The supersaturated solid solution in alloys containing 1.18 and 1.86% Mg_2Si are seen to have undergone decomposition after the specimens were subjected to annealing at 450°C for 30–min and 2–h durations respectively (Figure 4 a,b). Both Si and Mg_2Si have precipitated; the precipitation of Mg_2Si is spread uniformly in the matrix. Particles present at the cell boundaries grow preferentially and coarsen rapidly in alloys containing 6.2% and 10.2% Mg_2Si after annealing at 250° and 110°C for 30 min duration, respectively, as shown in Figure 4 c,d. X–ray diffraction analysis, Figure 5, confirms the precipitation of Si in alloys containing 0.75% Mg, 5.15% Si, and 1.18% Mg, 6.17% Si after annealing at 450°C for 30 min and Mg_2Si in thin films containing 3.94% Mg, 6.42% Si, and 6.47% Mg, 6.21% Si annealed at 250° and 110–450°C for 30–min duration, corroborating the results obtained by metallography.

CONCLUSION

1. The investigation shows that the addition of magnesium to Al–Si alloys reverses the change of the lattice parameter with solute content. This is interpreted as showing that Mg destroys the preferential association of Al–Si clusters in the liquid state.

175

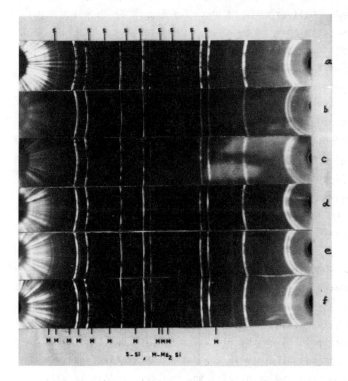

Figure 5. X-ray powder photographs showing the precipitation product in various alloys: (a) Al-0.75% Mg-5.15% Si annealed at 450°C/30 min; (b) Al-1.18% Mg-6.17% Si annealed at 450°C/30 min; (c) Al-3.94% Mg-6.42% Si annealed at 250°C/30 min. (d,e,f) Al-6.47% Mg-6.21% Si annealed at 110°, 350°C, and 450°C for 30 min, respectively.

In contrast to binary Al-Si alloys, rapid cooling does not cause extension of the limit of primary solid solubility.

2. The process of solidification of an aluminium alloy containing 0.75% Mg, 5.15% Si under rapid cooling is characterized by the formation of spherical seed nuclei around which the dendritic structure develops prior to the formation of normal dendritic morphology.

3. The precipitation products were found to be Si in alloys containing 0.75% Mg, 5.15% Si and 1.18% Mg, 6.17% Si and Mg_2Si for the alloys containing 3.94% Mg, 6.42% Si and 6.47% Mg, 6.21% Si after annealing at elevated temperatures.

ACKNOWLEDGEMENT

The authors thank Professor V.A. Altekar, Director of the National Metallurgical Laboratory, Jamshedpur, India, for permission to present this paper.

REFERENCES

1. Singh, M. and R. Kumar, J. Mat. Sci. 8, 317 (1973).
2. Bose, S.K. and R. Kumar, J. Mat. Sci. 8, 1795 (1973).
3. Itagaki, M., B.C. Giessen, and N.J. Grant, Trans. ASM, 61, 330 (1968).
4. Kumar, R., C.S. Sivaramakrishnan, and R.K. Mahanty, Second International Conference on Liquid Metals, Tokyo, September (1972).
5. Kumar, R. and A.N. Sinha, Tr. Ind. Inst. Metals, 22, 9 (1969).
6. Kumar, R. and S.K. Bose, Ind. Met. Soc. Proc., 95 (1970).
7. Biloni, H. and B. Chalmers, TMS-AIME, 233, 373 (1965).

STRUCTURES PRODUCED BY RAPID QUENCHING OF MOLYBDENUM STEELS

I.R. Sare and R.W.K.Honeycombe,

Department of Metallurgy and Materials Science,
University of Cambridge, U.K.

INTRODUCTION

A major limitation of the solid-state solution treatment given to
high alloy steels such as tool steels is that considerable amounts of
the expensive alloying additions are ineffective, because some of the
alloy carbides remain undissolved at the austenitizing temperature.
This means that the microstructure after quenching is very inhomogenous,
and that the full secondary hardening potential of the alloy cannot
be realised. The underlying objective behind this investigation, therefore, was to use the splat-cooling process to dissolve all carbides in
the liquid state to retain a supersaturated solution on quenching,
then to precipitate carbides on subsequent ageing.

Three alloys were chosen for study in the present investigation.
The first has the nominal composition Fe-4%Mo-0.2%C, and was selected
because its behaviour under conventional solid-state quenching and
tempering conditons has been well documented [1]   The second alloy,
Fe-10%Mo-0.5%C, maintained the same stoichiometric ratio of Mo:C as
the first. However, since solid-state solution treatment cannot take
all the carbides into solution prior to quenching it was felt that
this alloy would provide a good indication of the efficacy of splat-
cooling in dissolving and retaining in solid solution an otherwise residual precipitate phase in alloy compositions allied to tool steels. The
study was subsequently extended to a commercial molybdenum-rich high-
speed steel, AISI M1, containing 8.4%Mo-1.5%W-4.1%Cr-1.1%V and 0.77%C.
A controlled atmosphere "gun" splat quenching apparatus was used with
0.25g samples; this is described by Wood and Sare in a separate paper
at this conference.[2]

RESULTS

1. <u>X-ray diffraction analysis</u>: The only phases detected by X-ray diffraction in all three alloys were f.c.c. (face-centered cubic) austenite and a b.c.c. (body-centered cubic) product. The proportion of
austenite varied from run to run with each alloy, but fell within the
approximate limits 0 to 5% for Fe-4%Mo-0.2%C, 0 to 10% for Fe-10%Mo-
0.5%C, and 10 to 70% for M1 high-speed steel. The average lattice
parameters of these phases are presented in Table 1. So few austenite
lines appeared on the diffraction patterns of the two Fe-Mo-C alloys
that no reliable measurement of its lattice parameters could be made.
For comparison Table 1 also lists the lattice parameters of the matrix
phases present in filings of the respective alloys, prepared by solution treatment for 1 hour at 1200°C followed by water quenching.

The lattice parameter determined for the b.c.c. product in the splat-
cooled Fe-4%Mo-0.2%C alloy is not higher than that in the water-quenched filings, within the errors of the measurements, since even solid-
state solution treatment is able to take all the carbide into solution.
In the Fe-10%Mo-0.5%C alloy splat cooling does give a significant increase in the ferrite lattice parameter over that of the water-quenched
filings. Unexpectedly, the martensite in the water-quenched filings
of the high-speed steel M1 has the same degree of solute supersaturation as the ferrite in the splat-cooled material, whilst the austenite
in the filings is considerably less solute rich than its splat-cooled
counterpart. Tetragonality was not detected either in the martensite
of the quenched filings of any alloy, or in the b.c.c. product of the
splat-cooled alloys which could be a result of the limitations of the
X-ray technique employed. The variation in the amount of the austenite
retained to room temperature from run to run was found to be a result
of changes in cooling rate.

2. <u>Electron metallography of the splat-cooled alloys</u>

2.1 <u>Fe-4%Mo-0.2%C</u>: In unthinned regions of the Fe-4%Mo-0.2%C alloy the
ferritic areas invariably consisted of grains elongated in the plane of
the foil, exhibiting no sign of any cellular or dendritic structure

180

(Fig. 1). Areas exhibiting this type of grain morphology have been
shown[3]   to be indicative of heat extraction during solidification
taking place parallel to the plane of the foil. It is apparent that
this phase is not martensitic since it does not contain the laths or
high dislocation density typical of low-carbon martensites. However,
martensite was observed in foils of this alloy thinned by electropolishing. Fig. 2 is a micrograph which shows adjacent areas of ferrite and
finely-twinned martensite. The ledges on the interface between the martensite and ferrite suggest that at some earlier stage one of the phases
was growing into the other. The most likely explanation is that the
material originally solidified as high-temperature δ-ferrite, and that
as cooling proceeded austenite nucleated and started to grow into the
δ-ferrite. However, the cooling rate was so high that this reaction
could not go to completion, and the austenite which had formed subsequently transformed to martensite below its M_s temperature.

2.2 Fe-10%Mo-0.5%C: Several distinct microstructural features were
observed in the Fe-10%Mo-0.5%C alloy. Unthinned foils of ferrite showed a fine dendritic structure (Fig.3), in which dendrite growth had
apparently occurred in the foil plane. A few unthinned areas also revealed a finely-twinned martensitic structure. Second phase particles
were found at the interdendritic ferrite boundaries (Fig.4), and also
as fine precipitate needles within the dendrites. Electron diffraction
and dark field microscopy showed that the coarse interdendritic particles and the fine matrix precipitate were Mo_2C, obeying the usual Pitsch
and Schrader relationship with the ferrite.

2.3 AISI M1 high-speed steel: A number of different microstructures
was observed in the splat-cooled high-speed steel. In the thinnest and
hence fastest cooled flakes the ferrite exhibited two forms. In the
first case the microstructure consisted of elongated grains containing very fine precipitate particles lying in cube directions (Fig. 5).
The corresponding electron diffraction pattern showed pronounced streak-
ing in <100> ferrite directions. Such areas were observed only in-
frequently in the thinnest parts of the electron-transparent areas.
The second ferrite morphology found in unthinned foils was dendritic

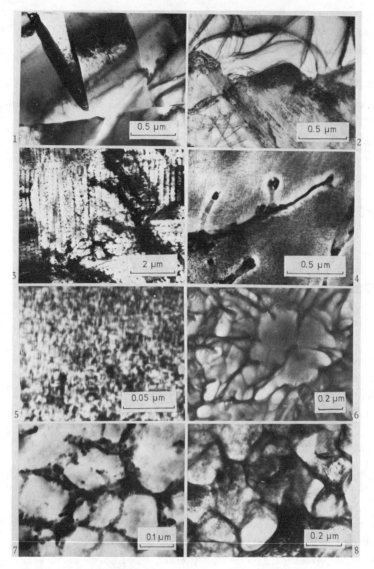

Figures 1-8. Electron Micrographs of Rapidly Quenched Molybdenum Steels.

as shown in Fig. 6. On thinning the thicker splat specimens a dendritic structure was observed in the ferritic matrix (Fig. 7), but with
copious precipitation along the interdendritic boundaries. The precipitate particles produced spots in the diffraction patterns, but
the calculated interplanar spacings could be applied to the three carbides M_2C, M_6C and $M_{23}C_6$, consequently no positive identification could
be made. Austenite was detected only in thinned foils, and as with
the ferrite, two different growth morphologies were observed. First,
small equiaxed grains were evident (Fig. 8) which could be illuminated
individually in the dark field mode. Secondly, long parallel dendrites
of austenite were observed, with precipitation of particles at the
interdendritic boundaries, compatible with the carbides referred to
above.

DISCUSSION

---

1. The nature of the b.c.c. product: Two sets of crystalline reflections were detected by X-ray diffraction analysis of the splat-cooled
alloys. One could be indexed as f.c.c. austenite, and the other as a
b.c.c. product. The metallographic evidence suggests that the latter
may, in fact, consist of two phases, δ-ferrite and martensite. The
morphology of the b.c.c. product in Fig. 1, however, does not resemble
any of the usual microstructures attributed to martensite, and therefore
it can only be δ-ferrite. These elongated grains were observed in the
thinnest, and hence fastest-cooled,[4] parts of the foil, and it is
thought that the cooling has been so rapid that before the δ-ferrite has had
time to transform completely to austenite the temperature has fallen
to a level where diffusion of iron is too slow. Consequently no further
transformation occurs in these δ areas, e.g. to martensite. Similar
considerations apply to the Fe-10%Mo-0.5%C alloy, although the amount
of martensite formed was not as great. Scanning electron microscopy
revealed only regions of a dendritic or cellular solidification morphology, which were found by transmission electron microscopy to be
ferritic (Fig.3). Small regions of martensite were observed, however,
confirming that the b.c.c. product detected by X-ray diffraction was a

mixture of δ-ferrite and martensite.

For the high-speed steel, comparison of the lattice parameters of the austenite in the splat-cooled material and water-quenched filings (Table 1) shows that the former has a much higher solute content than the latter. Since the M_S temperature is depressed by the presence in solid solution of the principal alloying elements in this steel, the austenite in the splat-cooled material would have a much lower M_S temperature, possibly below room temperature. This, coupled with the fact that no martensite was observed in any of the thin foils, leads to the conclusion that the b.c.c. product in the splat-cooled high-speed steel is δ-ferrite only.

2. <u>Solubility enhancement in the high-speed steel</u>: It is evident from the results presented in Table 1 that the lattice parameter of the δ-ferrite in the splat-cooled high-speed steel is the same as that in the Fe-10%Mo-0.5%C alloy. This implies that the former is considerably depleted of solute, because the high-speed steel contains a greater proportion of alloying additions than the Fe-Mo-C alloy, and would therefore normally be expected to have a higher lattice parameter. This solution depletion cannot simply be a result of the copious inter-dendritic precipitation (Fig.7) which accompanied solidification and cooling of the δ-ferrite, for the austenite also exhibited heavy inter-dendritic precipitation.

The reason for the solute depletion of the δ-ferrite is that during solidification the dendrites of this phase reject carbon into the melt, which subsequently solidifies as carbon-enriched austenite dendrites. On further cooling precipitation occurs in both phases, resulting in their depletion of carbon. Hence, by the time carbide precipitation ceases, the ferrite is much leaner in carbon than the average alloy composition, whilst the carbon content of the austenite has fallen to approximately that of the average alloy composition.

That this difference in supersaturation in the two matrix phases of the splat-cooled alloy does exist in high-speed steel M1 is verified by the lattice parameters determined for the quenched filings. Since martensite has the same composition as the austenite from which it forms, the lattice parameters measured for the martensite and austenite in the quenched filings must reflect the same degree of supersaturation, assuming that no autotempering occurred in the martensite phase. However, when the lattice parameters of the phases in the splat-cooled alloy and quenched filings are compared, the δ-ferrite and the martensite in their respective materials have the same lattice parameter, whereas the austenite in the splat-cooled material has a much larger parameter than the austenite in the quenched filings. This can only mean that the austenite in the splat-cooled alloy is more supersaturated than the δ-ferrite.

3. Solidification morphologies of the matrix phases. Apart from the micrographs of twinned martensite, the micrographs of splat-cooled foils from the two Fe-Mo-C alloys show different morphologies of the δ-ferrite. In the Fe-4%Mo-0.2%C alloy the ferritic regions in the thin foils are grains showing no evidence of solute segregation. On the other hand in the Fe-10%Mo-0.5%C alloy the ferritic regions exhibit a cellular or dendritic solidification structure, with the segregation of solute to and formation of carbide at the interdendritic boundaries. The segregation effects associated with cellular and dendritic growth require that solidification takes place between the liquidus and solidus temperatures, for a given alloy composition. If the liquid is undercooled below the solidus, however, solute segregation would not be possible and the entire alloy would solidify with the mean composition of the liquid [5,6] provided that recalescence does not reach the solidus temperature.[7] Since the extent of undercooling is a function of cooling rate, large undercooling will occur in splat-cooling; thus, depending on the range between the liquidus and solidus temperatures, solidification may occur with or without concomitant segregation.

Vertical sections of the ternary equilibrium diagrams of Fe-4%Mo-0.2%C and Fe-10%Mo-0.5%C indicate that the former has a freezing range of approximately 50K whilst the range in the latter is about 150K. Experimental determinations[7,8] of the amount of undercooling which occurs during splat-cooling give values in excess of 100K, so that it is highly likely that the former will solidify as a homogenous solid

185

solution below its solidus temperature, while the latter will exhibit solute segregation as it crystallizes between its liquidus and solidus temperatures.

This same argument can be used to account for the microstructures observed in the splat-cooled high-speed steel. Because of the very large freezing range in high-speed steels the degree of undercooling attainable will, in general, still not be great enough to permit 'diffusionless' solidification below the solidus temperature, and the microstructures will be segregated. In some very thin, localized areas, however, the cooling rate may induce an undercooling sufficient to suppress the commencement of solidification down to a temperature below the solidus, and therefore segregation effects will be absent (for example Fig. 5). The fine carbides present within the grains of this figure are not a result of the solidification process per se, but reflect auto-tempering of the supersaturated solid solution during subsequent solid-state cooling.

4. <u>Applications to high-speed steel technology</u>. It is apparent that the residual carbides normally present in quenched high-speed steels can be eliminated by the splat-cooling process, although further carbides do precipitate during the solidification stage. The latter are, however, present on a much finer scale and in a more uniform distribution than the carbides found in conventional material. Before any judgement can be made about the advantages of splat-cooling as an alternative method of tool steel production, the structures produced by the subsequent consolidation stage would need to be evaluated. In particular one would have to ensure that the high initial solute supersaturations were not destroyed by the subsequent growth of very large carbides in the solid state.

TABLE I. Lattice Parameters of the Phases in Splat-Cooled Material and Water-Quenched Filings.

| Alloy | Treatment | Lattice Parameter ($\mathring{A}$) | |
		Ferrite or Martensite	Austenite
Fe-4%Mo-0.2%C	Splat-cooled	2.875 ± 0.004	(vvw)
	Water-quenched filings	2.874 ± 0.001	(n.d.)
Fe-10%Mo-0.5%C	Splat-cooled	2.890 ± 0.003	(vw)
	Water quenched filings	2.885 ± 0.001	(n.d.)
M1 high-speed steel	Splat-cooled	2.889 ± 0.003	3.637 ± 0.011
	Water-quenched filings	2.889 ± 0.002	3.606 ± 0.009
Calculated		–	3.632

REFERENCES

1. Raynor, D., J.A. Whiteman, and R.W.K. Honeycombe, JISI 204, 349, 1114 (1966).
2. Wood, J. and I. Sare, Proc. Second Internat. Conf. on Rapidly Quenched Metals, Section I, N.J. Grant and B.C. Giessen, Eds. MIT Press, Cambridge MA (1976) [in print].
3. Wood, J. and I. Sare, Met. Trans. [to be published].
4. Ruhl, R.C., Mater. Sci. Eng. 1, 313 (1967).
5. Olsen, W.T. and R. Hultgren, Trans. AIME 188, 1323 (1950).
6. Biloni, H. and B. Chalmers, Trans. Met. Soc. AIME, 233, 373 (1965).
7. Löhberg, K. and H. Müller, Z. Metallkde. 60, 231 (1969).
8. Miroshnichenko, I.S. and G.P. Brekharya, Phys. Met. Metallog. 29, No. 3, 233 (1970).

METASTABLE A-15 PHASE SUPERCONDUCTORS TRANSFORMED FROM METASTABLE BCC Nb₃X SOLID SOLUTIONS MADE BY HIGH-RATE SPUTTER DEPOSITION

R. Wang and S.D. Dahlgren

Battelle, Pacific Northwest Laboratory
Richland, Washington

INTRODUCTION

Recent advances of high-rate sputter deposition have been utilized for fabricating high-field superconductors. A fine-grained $Nb_3(Al_{.75}Ge_{.25})$ superconductor had high critical current densities at 4.2°K of 4.4 x 10^5 A/cm^2 at 100 KOe,[1] 1.3 x 10^5 A/cm^2 at 150 KOe,[2] and 2.6 x 10^4 A/cm^2 at 200 KOe.[2] Critical temperatures were as high as 18.5°K.[1,3] The high critical currents were attributed to the fine grain size of the A-15 phase which formed during heat treatment of the as-deposited metastable bcc phase.[1,3] This paper reports a study of the structural and micro-structural changes accompanying the bcc → A-15 phase transition in sputter-deposited Nb_3X superconductors (where X = Al, Sn, Ge, AlGe, and Si). Phase transformation information is useful to control fabrication procedures and permit tailoring the process to obtain desirable grain sizes and other desirable superconducting properties.

GRAIN SIZES OF BCC AND A-15 PHASES

All the Nb_3X alloys deposited at a high rate of ∿1μm/min and substrate temperature of ∿10°C had the bcc structure. The grain sizes of the bcc phases varied from ∿1000Å for Nb_3Al and Nb_3Sn to ∿50Å for Nb_3Ge, $Nb_3(AlGe)$, and Nb_3Si (Figure 1). After a heat treatment of 24 h at 750°C to form the A-15 phase, the A-15 phase grain sizes were ∿3500Å for Nb_3Al and Nb_3Sn, and 350Å[1,2] for $Nb_3(AlGe)$. All grain sizes were estimated by transmission electron microscopy (TEM).

The grain size of $Nb_3(Al_{.75}Ge_{.25})$ was determined as a function of heat treatment time and temperature with an x-ray line broadening analysis described by Hanak and Enstrom for Nb_3Sn alloys.[4] The relations of the inverse grain size (1/d) and the residual strains of the heat-

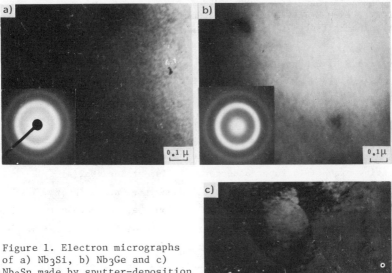

Figure 1. Electron micrographs of a) Nb_3Si, b) Nb_3Ge and c) Nb_3Sn made by sputter-deposition at ∿10°C and a rate of ∿1μm/min.

treated superconductors with the heat treatment conditions are given in Figure 2. The x-ray analysis showed that the grain size of Nb_3 $(Al_{.75}Ge_{.25})$ heat treatment at 750°C for 24 h was 240Å, which is somewhat smaller than the grain size of 350Å[13] previously observed by transmission electron microscopy. The smaller grain size indicated by the x-ray method is probably a more accurate value since it includes small grains that are difficult to count in electron micrographs. A series of 15min to 24h heat treatments at 800°C were made for four deposits. Two free-standing samples and two samples attached to Cu substrates were used for the study. The inverse grain sizes of these four samples had a linear correlation with the logarithm of the heat treatment time. This relationship indicated that the grain size of the A-15 phase was limited between 200Å (extrapolated for 1min when the A-15 grains initially transformed from the bcc phase) and 500Å (for 24h, when the decomposition of the A-15 phase commenced).

190

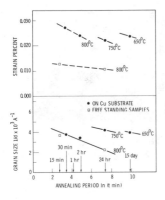

Figure 2. Inverse grain size (1/d) and residual strain (%) obtained from x-ray line broadening analysis for heat treated sputter-deposited $Nb_3(Al_{.75}Ge_{.25})$.

Figure 3. Electron micrographs of $Nb_3(Al_{.75}Ge_{.25})$ showing a) the bcc phase obtained by high-temperature deposition, and b) the A-15 phase resulting from heat treatment at 750°C for 24h.

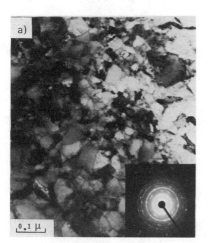

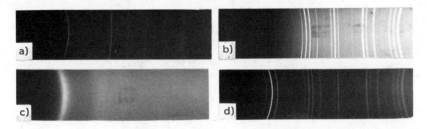

Figure 4. X-ray diffraction patterns from a Unicam camera showing the size difference in $Nb_3(Al_{.75}Ge_{.25})$ superconductors. a) bcc phase made at 200°C, b) A-15 phase formed from a) at 750°C for 24h, c) bcc phase made at 10°C, and d) A-15 phase formed from c) at 750°C for 24h.

From the results on all A-15 phases that were obtained at 750°C for 24h, the grain size of the A-15 phase was approximately 3 to 5 times larger than the grain size of the precursor bcc phase. Thus, when the grain size of the bcc phase increased, the grain size of the A-15 phase was accordingly larger. With this conclusion in mind, a large-grained bcc phase ($\sim$450Å) of $Nb_3(Al_{.75}Ge_{.25})$ was made by depositing at a higher temperature (Figure 3a). After heat treatment at 750°C for 24h the grain size of the A-15 phase was nearly 2000Å (Figure 3b), which was about four times larger than the bcc phase grain size. The different grain sizes of bcc and A-15 phases obtained from both the 10°C and high temperature deposition were also revealed clearly in x-ray diffraction patterns (Figure 4, supra).

TRANSFORMATION KINETICS OF $Nb_3(Al_{.75}Ge_{.25})$ AND Nb_3Si

$Nb_3(Al_{.75}Ge_{.25})$ and Nb_3Si deposits were chosen for a detailed study of transformation kinetics since Nb_3Al, Nb_3Sn, and Nb_3Ge have a transformation behavior similar to that of $Nb_3(Al_{.75}Ge_{.25})$. We also intended to determine whether it is possible to obtain the A-15 phases in Nb_3Si by heat treating the bcc phase. Phase identifications were made by x-ray diffraction with both camera and diffractometer methods. A time-temperature-transformation diagram (Figure 5a) was derived for $Nb_3(Al_{.75}Ge_{.25})$ from 18 heat treatments made between 500 and 900°C for

192

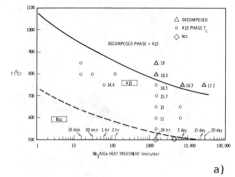

a)

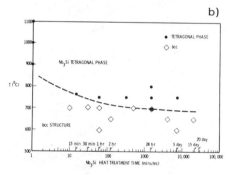

b)

Figure 5. A tentative time-temperature -transformation diagram for sputter-deposited a) $Nb_3(Al_{.75}Ge_{.25})$, and b) Nb_3Si.

15min to 20 days. The highest T_c (18.4°K) was found in alloys heat treated at 750°C for 24h. A short heat treatment for 15min at 800°C also resulted in a well-defined A-15 phase which indicated that the bcc to A-15 transformation occurred quite rapidly. The highest T_c which resulted from optimizing ordering of the A-15 structure was observed in two samples heat treated near the boundary separating the A-15 phase region. This boundary, therefore, may represent the best conditions for obtained a high T_c.

In contrast to $Nb_3(AlGe)$, the bcc phase of the composition Nb_3Si transformed into a tegragonal Ti_3P-type structure for all 19 heat treatment conditions studied (Figure 5b). After annealing at high temperatures (900-1000°C) for short periods (1 min), the diffraction pattern of the Nb_3Si alloy exhibited broad lines with some resemblance to those of the A-15 phase. However, because the diffraction pattern had more features of the Ti_3P type tetragonal phase, we believe it is more properly referred to as the "tetragonal phase". Thus, the A-15 type phase Nb_3Si[5] was not obtained by any of the heat treatments of the bcc phase in this investigation.

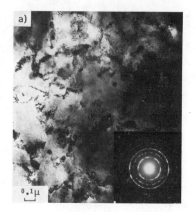

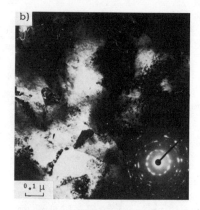

Figure 6. Electron micrographs of Nb$_3$Al. a) Deposited at 10°C, bcc phase only, b) heat treated at 600°C for 24h, both bcc and A-15 phases, and c) heat treated at 750°C for 24h, A-15 phase only.

PHASE TRANSFORMATION OF METASTABLE BCC Nb$_3$Al

The transformation of the metastable bcc phase to the A-15 phase was studied by transmission electron microscopy for Nb$_3$Al since the bcc grains of Nb$_3$Al were large enough for observation of microscopical changes. Nb$_3$Al deposited at 10°C (Figure 6a, supra) shows a large a-mount of coarse incoherent precipitates at the grain boundary and fine coherent precipitates in the grain. Electron and x-ray diffraction patterns showed only the bcc phase before heat treatment. Heat treating at 600°C for 24h resulted in coarsening of both the incoherent and co-

194

herent precipitates (Figure 6b). The diffraction patterns show predominantly bcc phase but also some A-15 phase. Heat treating at 750°C
for 24h resulted in well-defined A-15 grains between 1000 to 5000Å in
size (Figure 6c). The zig-zag grain boundaries indicated that grain
growth was not yet completed. However, both Figures 6a and 6b demonstrate that the bcc → A-15 transformation is a massive transformation,
as previously observed in large grained (>10μm) Nb_3Al alloys made by
arc melting.[6] The coarse incoherent precipitates at the grain boundaries of the bcc phase are attributed to the formation of A-15 phase.

CONCLUSIONS

Our study of the structural and microstructural features for the bcc
→ A-15 phase transition in sputter-deposited Nb_3X superconductors
(where X = Al, Sn, Ge, AlGe, and Si) shows how the grain sizes of both
the bcc and A-15 phases are influenced by the deposition temperature
and heat treatment. The phase transformation bcc → A-15 involves complex solid-state reactions that result from the metastable nature of
these two phases. The grain size of the metastable bcc phase increases
with increasing deposition temperature and the grain size of the A-15
phase was approximately 3 to 5 times larger than the grain size of bcc
phase. This study demonstrates that the grain size of the A-15 phase
can be controlled in the range from 200Å to 2000Å by combinations of
properly selected deposition temperature and heat treatment. The
coarse incoherent precipitates at the grain boundary of as-deposited bcc
Nb_3Al phase may be responsible for the initiation of A-15 phase formation. We believe that the A-15 phases obtained from fine-grained metastable bcc phase by a segregation mechanism would have unique microstructural features which are somewhat different from those in alloys
made by other solid-solution quenching, arc melting, and diffusion
methods. The superconductivity properties resulting from the microstructural aspects in high-rate sputter-deposited superconductors will
be investigated further.

REFERENCES

1. Dahlgren, S.D. and D. Kroeger, J. Appl. Phys. 44, 5595 (1973).
2. Dahlgren, S.D., M. Suenaga, and T.S. Luhman, J. Appl. Phys. 45, 5462, (1974).
3. Dahlgren, S.D., IEEE Transactions on Magnetics, MAG-11, 217 (1975).
4. Hanak, J.J. and R.E. Enstrom, Proceedings of the 10th International Conference on Low-Temperature Physics, Moscow, ed.: M.P. Malkov (1966); (Proizvodstrenno-Izdatelskii Kombinat VINITI, Moscow, S94 (1967).
5. Johnson, G.R. and D.H. Douglass, J. of Low Temp. Phys. 14 575 (1974).
6. Lundin, C.E. and A.S. Yamamoto, Trans TMS-AIME, 236, 863 (1966).

EFFECTS OF RAPID SOLIDIFICATION
IN ARC-SPRAYED ALUMINIUM ALLOYS

H. Warlimont and P. Kunzmann

Swiss Aluminium Ltd.
Neuhausen, Switzerland

INTRODUCTION

The formation of arc-sprayed coatings is associated with rapid solidification of the deposited material. Whereas the spraying parameters
and the technological properties of arc-sprayed coatings have been
studied to some extent,[1] their structural state has not been investigated intensively. In the present work, we report structural observations and property measurements of Al alloy coatings arc-sprayed on a
steel substrate. The results are analysed in particular in terms of
the structural features caused by rapid cooling from the melt.

EXPERIMENTAL PROCEDURE

Specimens were prepared by the arc-spraying technique using a spray
gun with a newly developed closed nozzle and a rectifier with hard
characteristics. Uniformity and reproducibility of the spraying parameters were ensured by using an automatic manipulator for the gun.
The distance from nozzle to substrate was 10cm. The coatings were applied in one or more passes of approximately 1mm thickness each. The
as-rolled surface of the steel substrate was prepared by partial sand
blasting in such a way that both adherent and detachable coatings were
obtained. Six commercial filler wire alloys were used as spraying material. Their analyses along with those of the coatings are listed in
Table 1. The accuracy is $\pm.05\%$ for Mg and $\pm.01\%$ for all other elements.

Cross sections of the coatings were investigated by light microscopy and Vickers microhardness measurements. Specimens for transmission electron microscopy were prepared by electropolishing. The Brin-
nell hardness was measured on the substrate face of detached coatings.
Three point bending tests were carried out with knife edge supports

TABLE 1. CHEMICAL ANALYSES

alloy no.	type AA no.	material	analyses, at.%					
			Si*	Fe*	Cu	Mn	Mg	Zn
1	A199.5 1100	wire	.04	.09	<.01	<.01	<.01	.01
		coating	<.07	<.11	<.01	<.01	<.01	<.01
2	AlSi5 4043	wire	4.5	.08	.02	<.01	<.01	.04
		coating	<5.2	<.09	.02	<.01	<.01	<.01
3	AlMg3.5 5154	wire	.13	.10	.03	.60	2.78	.02
		coating	<.18	<.12	.03	.56	.74	<.01
4	AlMg5 5356	wire	.10	.10	.01	.06	5.73	.02
		coating	<.14	<.13	.01	.05	2.53	.01
5	AlMg4Zn2 5180	wire	.07	.06	<.01	.24	4.17	.80
		coating	<.08	<.08	<.01	.23	2.13	.35
6	AlMg4.5Mn 5183	wire	.11	.07	.03	.32	6.05	.02
		coating	<.13	<.10	.04	.31	3.20	.01

* Si and Fe analyses of coatings include surface contaminates.

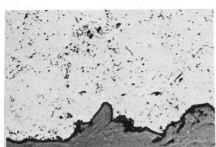

Figure 1a. Alloy 1, 300x, unetched.

Figure 1b. Alloy 1 750x, etch: 1h 10% H_3PO_4

spaced 50mm apart. They were evaluated to obtain the bend elastic limit σ_E^b, the ultimate bend strength σ_E^b, and the elongation to fracture ε^b in the surface fibre. X-ray diffractometer, Guinier and goniometer methods were used to determine the lattice parameters, crystal structures, and texture, respectively.

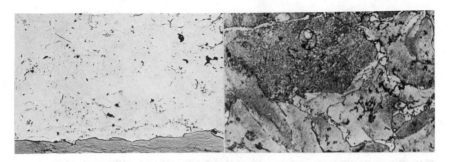

Figure 2a. Alloy 2, 300x,
unetched

Figure 2b. Alloy 2, 750x,
etch: $HF+HCl+HNO_3+H_2O$

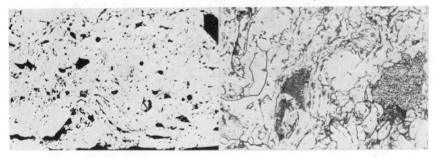

Figure 3a. Alloy 4, 300x,
unetched

Figure 3b. Alloy 3, 750x,
etch: 30 min 10% H_3PO_4

EXPERIMENTAL RESULTS

The light micrographs, Figures 1-3 (supra), show examples of the wide
variation in as-solidified droplet size and shape and in porosity. Alloys numbers 1 and 2 are clearly denser than all other materials. Etching to reveal the cell structure attacked only some individual grains
in each alloy. The cell size varied from 1 to 3μm, independent of composition within the observed range of scatter. This indicates solidification with a cooling rate of about 10^4-10^5 K/s. The electron microstructure varies considerably both as a function of alloy composition
and of location in the coating. Examples of some characteristic microstructural variants are shown in Figures 4-7:

199

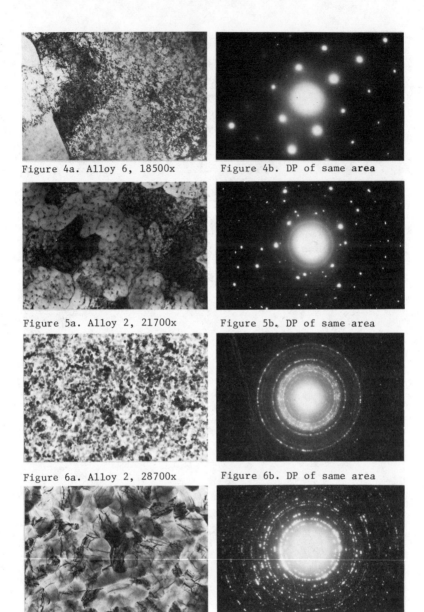

Figure 4a. Alloy 6, 18500x

Figure 4b. DP of same area

Figure 5a. Alloy 2, 21700x

Figure 5b. DP of same area

Figure 6a. Alloy 2, 28700x

Figure 6b. DP of same area

Figure 7a. Alloy 4, 28700x

Figure 7b. DP of same area

TABLE 2. LATTICE PARAMETERS[1]

alloy no.	a, Å	Δa_{exp}, Å	Δa_{calc}, Å	remarks
1[2]	4.048₈	-.0007	-.0006	
1[3]	4.049₇	≈0	≈0	
2	4.043₅	-.0060	-.0087	≈1.5 at.% Si undissolved
3	4.049₄	≈0	-.0014	
4	4.052₉	+.0034	+.0101	≈1.5 at.% Mg undissolved
5	4.055₉	+.0064	+.0073	
6	4.056₅	+.0070	+.0117	≈1.0 at.% Mg undissolved

1) accuracy ±.001Å; 2) as deposited; 3) annealed 1h 400°C.

TABLE 3. LATTICE DISTORTIONS AND INTERNAL STRESSES

alloy no.	$(\overline{\Delta a/a})$ $\cdot 10^3$	$\overline{\sigma}_i$ N/mm^2
1	.5	40
2	3.0	210
3	.3	20
4	3.2	220
5	4.1	290
6	2.5	180

(i) homogeneous solid solution crystallites, Figure 4, (ii) cell structure with precipitates, Figure 5, (iii) fine-grained two-phase structures with varying crystallite sizes, Figures 6 and 7. All structures are plastically deformed.

The lattice parameters of the Al solid solution are listed in Table 2. They vary with alloy content. None of the Guinier patterns exhibited extra reflections, in contrast to the electron diffraction patterns. The volume fraction and particle size of precipitated phases were insufficient for yielding x-ray signals above the background. The x-ray line width was determined and evaluated according to the relation

$$\Delta_{1/2}(2\theta) = 2(\overline{\Delta a/a})\,tg\theta_{hkl}, \tag{1}$$

where $\Delta_{1/2}$ is the excess line width at half peak height, θ_{hkl} is the diffraction angle for the hkl peak, $\overline{\Delta a/a}$ is the average lattice distortion. The measured data along with the computed internal stresses $\overline{\sigma}_i$

201

TABLE 4. MECHANICAL PROPERTIES

alloy no.	t mm	σ_E^{b} [1] N/mm^2	σ_B^{b} [1] N/mm^2	ε^{b} [1] %	H_V [2] kgf/mm^2	H_B [3] kgf/mm^2
1	.6	25.5	43.9	.63	33.8	
	2.1	50.6	91.4	.42		28.6[4], 29.5[5]
2	.6	36.5	62.1	.83	64.5	
	1.9	-	121.4	.54		52.2[4]
3	.7	30.3	54.4	.78	46.0	
	2.0	-	85.2	.37		38.3[4]
4	.7	26.7	51.7	.53	42.8	34.8[5]
5	.4	23.3	50.6	.66	49.1	n.d.
6	n.d.	n.d.	n.d.	n.d.	55.1	n.d.

1) average of 2 measurements; 2) load .1N, 25 measurements per alloy; 3) 10 measurements per alloy; 4) load 25N; 5) load 12.5N; n.d.= not determined since suitable specimens could not be produced.

(using E = 70,000 N/mm^2) are listed in Table 3. As to texture, both the macroscopic samples, as investigated by the x-ray goniometer study, and localized fine grained decomposition structures such as those shown in Figures 6 and 7, exhibited a random crystal orientation distribution.

The stress-deflection curves of all bend tests showed serrated yielding above the elastic limit σ_E^{b}. The stresses σ_E^{b} and σ_B^{b}, listed in Table 4, were determined according to

$$\sigma = 1.5 \, P \cdot l / w \cdot t \qquad (2)$$

where P = measured load, l = distance between the supports (50mm), w = specimen width ($\simeq$10mm), t = specimen thickness. The data are the average of 2 measurements. Hardness values are also shown in Table 4 where the scatter bands were $\Delta H_V \leq \pm 15$ and $\Delta H_B \leq \pm 3$kgf/mm^2.

DISCUSSION

A comparison of the chemical analyses of wire and coating material (Table 1) indicates that $\overset{>}{\sim} 50\%$ of the Mg and Zn content is lost during spraying, whereas the concentration of the other elements remains constant. The comparatively high vapour pressure of Mg and Zn at low temperatures probably results in partial evaporation in the arc and the vapourized fraction is incompletely deposited.

The high cooling rate (10^4–10^5 K/s) indicated by the cell size leads to the supersaturation of the Al solid solution evident from the lattice parameters (Table 2). We calculated the change in lattice parameter assuming that all alloying and major impurity elements were kept in solution according to

$$\Delta a \;=\; \sum_i c_i \, (da_{Al} \,/\, dc_i) \tag{3}$$

where c_i is the concentration of element i (for Si and Fe the wire analyses were used); $(da_{Al} \,/\, dc_i)$ data were taken from the literature. The results are given in Table 2 and should be compared with the difference between $a_{Al} = 4.0495$ and the value obtained experimentally for each coating. The difference between the calculated and the experimental Δa indicates to what extent dissolution is incomplete. The last column in Table 2 gives the probable type and estimated amount of the major element not in solution. The annealing experiment of alloy 1 has proved that Fe is fully dissolved in the as-deposited coating.

The precipitation of Si from alloy 2 indicated by its lattice parameter is also evident from Figure 5. Whereas the precipitate has nucleated on the solidification substructure in Figure 5, the fine two-phase grain structures in Figures 6 and 7 are most probably due to decomposition in the solid state. They are characteristic of rapidly solidified, highly super-saturated alloys such as Al-Fe[2] and Al-Cr.[3]

The variety of the light microstructures, Figures 1-3, is essentially characterized by (i) grains formed from liquid drops flattened during impact on the substrate and subsequently solidified; (ii) drops which have solidified in flight and have been deposited in the partially solidified coating; (iii) pores, arising from solidification of liquid phase drops before their flow had filled all previous surface contours, with incomplete space filling by solid drops. The superior density of alloy 2 appears to be mainly due to its comparatively high fraction of grains solidified after impact. The plastic deformation in the as-solidified structures is considerable as can be seen from the comparatively high dislocation density in Figures 4 and 5. The internal stresses (Table 3) arise probably from (i) deformation by successive impact of drops (characteristic velocity 200 m/s (1)); (ii) localized differential thermal contraction; (iii) long-range thermal stresses due to the temperature gradient between coating and substrate.

The mechanical properties are related to both the solidification structure and the deformation structure. Alloy 1 exhibits hardness values of the half hard temper, part of which may result from the dissolved Fe; the rest stems from work hardening. The strength of alloy 2 is approaching that of comparable cast $AlSi_5Mg$. The mechanical behaviour of alloy 3 is comparatively strongly affected by its manganese content. The strength and hardness values of alloy 4 to 6 are clearly below those of comparable wrought alloys indicating that the high porosity and possible effects of oxide layers at the former drop boundaries reduce the cohesion between the crystallites. The serrated yielding in bending indicates that brittle separation of former drop boundaries occurs at stresses comparable to those required for plastic deformation of the grains. The grains can obviously sustain a higher internal stress than that required to deform the specimens. This is evident from comparing the data in Tables 3 and 4. However, a detailed correlation of structure and properties of the arc-sprayed coatings requires more extensive study which is currently under way.

ACKNOWLEDGEMENTS

We should like to thank Dr. P. Furrer for contributing the electron microscopic results, and Mr. M. Koutny and Miss S. Weber for their meticulous x-ray and metallographic work.

REFERENCES

1. Matting, A. and H.-D. Steffens, Metall 17, 584, 905, 1213 (1963).
2. Furrer, P. and H. Warlimont, Z. Metallkde. 64, 236 (1973).
3. Warlimont, H., W. Zingg, and P. Furrer, Proc. Second International Conference on Rapidly Quenched Metals, M.I.T., 1975, 82, N.J. Grant and B.C. Giessen, Eds., Mat. Sci. Eng. 23 (1976).

FORMATION OF METALLIC GLASSES

Frans Spaepen[51] and David Turnbull
Division of Engineering and Applied Physics
Harvard University, Cambridge, Massachusetts

INTRODUCTION

Metallic glasses formed by melt quenching are alloys which are disordered compositionally as well as topologically. Certain of these alloys, when constrained to remain compositionally disordered, may be more stable in glass than in crystalline form. However, experience indicates that without such constraint, metallic glasses, like non-metallic ones, are thermodynamically less stable than some crystallized state of the system. This means that the temperature, T_g, of the melt↔glass transition has always been observed to lie well below the unconstrained liquidus temperature, T_ℓ. Thus glass formation[1] requires that the melt be quenched through a temperature range, from T_ℓ to T_g, in which it is metastable and where the atomic mobility should remain high enough to sustain very rapid crystal growth. That metal melts can persist for long periods at temperatures deep in their metastable range reflects their very high resistance to the initiation, or homogeneous nucleation, of crystallization. However, the frequency I (volume^{-1} time^{-1}) of homogeneous nucleation increases sharply with undercooling in the fluid regime and if nucleation becomes copious at some undercooling less than $T_\ell - T_g$, the quench rates needed for glass formation may become too high for practical attainment.

According to simple nucleation theory,[2] the large resistance of fluid melts to homogeneous nucleation is due mainly to the relatively large values of the reduced crystal-melt interfacial tension:

$$\alpha = \frac{(N\overline{V}^2)^{\frac{1}{3}}\sigma_{LS}}{\Delta H_f} \tag{1}$$

205

where $\overline{V}$ is the volume/gm. atom of the melt, ΔH_f is the enthalpy of fusion/gm. atom, N is Avogadro's number and σ_{LS} is the crystal-melt interfacial tension/area. Physically, α is the number of monolayers/area of crystal which would be melted by an enthalpy equivalent to σ_{LS}. Experience indicates[2] that α for pure metals is at least $\frac{1}{2}$ and may generally approach the value 0.63 calculated[3] for Hg from measured frequencies of homogeneous nucleation. This means that reduced undercoolings ($\Delta T_r = \frac{T_\ell - T}{T_\ell}$) of $\frac{1}{4}$ to $\frac{1}{3}$, at least, would be needed for the occurrence of copious nucleation in normal sized specimens. However, if the kinetic resistance to atomic transport and crystal growth in the melt remains small to low T_r, it would be practically impossible to quench the melt to a glass.

It is generally supposed that the time constant, τ_D, for atomic transport will scale, at least crudely, with the shear viscosity, η, of the melt. If so, the frequency of nucleation would be negligibly small at and below the glass temperature. If the glass form is to persist, the time constant, τ_u, for crystal growth, must also be extremely large in the regime, $T \overline{<} T_g$. It seems reasonable to assume that τ_u, like τ_D, will scale as η. However, we shall try to show that this scaling, while reasonable for alloys, may lead to gross overestimates of τ_u for pure metals.

From the foregoing analysis, it follows[1] that the likelihood of quenching a melt to a glass, at a given cooling rate, $|\dot{T}|$, and specimen size, will be greater the smaller is the reduced undercooling at T_g or, equivalently, the larger is the reduced glass temperature, $T_{rg} = T_g/T_\ell$.

A rough method, based on simple nucleation and growth theory, for calculating the conditions needed to suppress homogeneous nucleation during melt quenching to T_g was presented elsewhere.[1] Contributions to the nucleation resistance from the transient period required to establish steady-state nucleation were neglected. Applying this method[1] to pure metals with $\alpha = 0.63$ and τ_D scaling as η we have calculated the dependence of I on T_r for $T_{rg} = \frac{2}{3}$ and $T_{rg} = \frac{1}{2}$. From these calculations, we conclude that the minimum quench rates $|\dot{T}|_{min}$ required to effectively suppress homogeneous

nucleation in 0.1 cm^3 specimens are 1°C/sec $\gg$ $|\dot{T}|_{min}$ for $T_{rg} = \frac{2}{3}$ and $\sim 10^{6}$°C for

$T_g = \frac{1}{2}$. If $\alpha \overset{>}{\sim} 0.9$, the frequency of homogeneous nucleation would not reach a detec-

table level even if $T_{rg} = 0$. Davies and coworkers,[4] have used these methods in more

comprehensive calculations of the T_{rg} effects.

Thus, if a glass is to form in melt quenching, the thermodynamically favored

crystallization process must be bypassed. Whether or not this occurs and the glass,

if formed, persists will be determined mainly by the magnitude of the reduced crystal-

melt tension, α, the reduced glass temperature, T_{rg}, and the time constant for crystal

growth τ_u. In the following we shall survey our present understanding of these fac-

tors for metals.

Nucleation and the Crystal Melt Tension

The high resistance of metal melts to homogeneous crystallization, as manifested

by the large undercooling, $\Delta T_r \sim \frac{1}{5}$ to $\frac{1}{3}$, which the melts could sustain, was quite

unexpected when first recognized. It was unexpected because of the tacit assumption,

then prevalent, that the short range order (SRO), as defined, e.g., by a Wigner-

Seitz (W-S) construction, in the melt is only trivially different from the SRO in

the close-packed crystal. A corollary assumption was that the crystal-melt tension

is mostly energetic in origin and, therefore, very small in metal systems because

of the small changes in their volume during melting. It now appears that both of

these assumptions are wrong. The topology of the SRO in the melt can be essentially

different from that in the close-packed crystal and the nature of the melt structure

is such that the energy of creating the crystal-melt interface can be made negligible

only if accompanied by a large entropy loss.

It is generally agreed that the part of the SRO specified by the number, z, of

immediate neighbors to and their separations, r_z, from a reference atom in the

W-S construction is little changed by melting. However, the topology of the SRO, which is defined by the angular distributions of the immediate neighbors, is not determined by z and r_z only. This was demonstrated clearly by the early model studies of Zachariasen[5] on fully connected networks. For example, in 4-connected systems at fixed r_z the topology of the "random" network is distinct from that of the crystal and convertible to it only by a reconstructive process.

Also, it is now realized that even in monatomic systems, where the interatomic forces are undirected, z = 12 coordination at fixed r_z can arise from SRO topologies which are essentially different from that of crystalline close-packing (cc-p). This was made clear by Frank's calculation[6] showing that a 13 atom cluster should have lower energy in a configuration where 12 atoms surround a central atom icosahedrally, i.e. uniformly, than in either of the cc-p configurations. Similar calculations were later made[7,8] for larger clusters with the surprising outcome that the lowest energy configurations are still non-crystalline (though not disordered) at all cluster sizes up to 66 atoms, at least. The reason is that of the tetrahedral and octahedral configurations constituting cc-p, in the ratio of 2 tetrahedra/octahedron, it is the tetrahedral one which has the highest density and lowest energy/atom. When the requirement that the packing pattern persist to infinity is relaxed, a higher proportion of tetrahedra than is in cc-p can be incorporated into small clusters.

Beyond some yet undetermined cluster size, the long range forces presumably dictate that a cc-p configuration will become most favored energetically. However, the foregoing considerations indicate that in bodies of any size formed by the action of extremely short range and undirected interatomic forces, the topology of the SRO should be mostly non-crystalline owing to the prevalence of the tetrahedral configurations. Indeed, model structures generated under these conditions, physically[9,10] or by computer,[11,12] are non-crystalline and exhibit a tetrahedral/octa-

208

hedral (t/o) ratio much higher than that (2/1) in cc-p.

The most dense of these structures is the physically generated[9,10] dense random packing (DRP) of hard spheres. It exhibits a t/o ratio of 15 to 1 and a density about 86% that of cc-p. Space filling is achieved essentially by the incorporation of a small number of large holes all having, in their ideal forms, triangular faces. Apparently crystallization of the DRP structure requires a reconstruction of the SRO which can be effected by nucleation and growth but not by a continuous collapse.

As explained in Cargill's paper,[13] the DRP structure is a model for monatomic glass rather than for the melt. However, the melt and glass are continuously interconvertible with little volume change and the arguments made for the prevalence of tetrahedral elements in the SRO apply to the melt as well as to the glass. Therefore, it is reasonable to suppose that crystallization of the melt will require a reconstruction of the SRO quite similar to that needed for crystallization of the DRP structure. The difficulty of effecting this reconstruction over a sufficient spatial scale is the basis for the microscopic explanation[2,14] for the high resistance of metal melts to nucleation.

These microscopic concepts on the atomic rearrangements in melt crystallization are the basis of a model for the structure and tension of the melt-crystal interface developed recently by Spaepen[15] and Spaepen and Meyer.[16] In this development, the dividing surface, s, is located at a sharply formed junction between the closest packed plane of the cc-p and the DRP hard sphere structure with the provisos:

1. Tetrahedral holes are preferred.

2. Octahedral holes are disallowed.

3. The density is maximum.

There is no volume change, and therefore no change in the strictly volume dependent part of the energy, arising from the formation of this interface. However, the

209

localization of positions in the DRP interfacial layer, which is required for density maximization, results in a large negentropy of interface formation. Spaepen and Meyer's[16] evaluation showed that this negentropy could contribute as much as 0.86 to the value of α at the thermodynamic crystallization temperature, T_m so that $\alpha \sim 0.86$ in this model. Interfacial disorder will, of course, reduce the negentropy but with sizable positive contributions to the interfacial energy. Miller and Chadwick's[17] measurements of σ_{LS} for several metals at T_m indicate α values which actually are near the upper limit given by the model. If α were, indeed, so large, and T independent, there would be little difficulty in achieving quench rates which would suppress homogeneous nucleation in pure metals completely at all T's. However, in its simplest limiting form, the model predicts that α will decrease with T, i.e.:

$$\alpha = \alpha_m (T/T_m) \tag{2}$$

We note that the value $\alpha = 0.63$ calculated from the nucleation rates in Hg melts[3] at $\Delta T_r = \frac{1}{3}$ is in fair agreement with the prediction $\alpha \sim 0.6$ of the model, with the α temperature dependence as in the above equation. Putting the predicted temperature dependence of α into the expression for the homogeneous nucleation frequency $I(T_r, \Delta T_r)$, we obtain:

$$I \simeq \frac{10^{30}}{\eta} \exp \left[- \frac{16\pi}{3} \alpha_m^3 \beta \left(\frac{T_r}{\Delta T_r} \right)^2 \right] \tag{3}$$

In Fig. 1, this nucleation frequency is compared to the one predicted by expression (3), where α is a constant. In the temperature range where homogeneous nucleation can be observed in pure melts ($T_r \sim .7$), the two models give very similar predictions. It is observed that the temperature dependence of the surface tension results in a steeper slope of the $I(T_r)$ curve. The temperature dependence of the viscosity

210

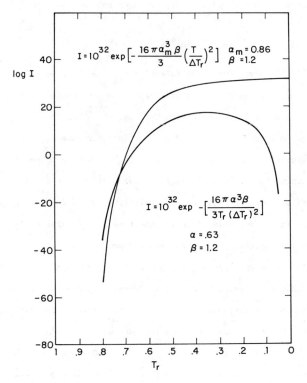

Figure 1.

Comparison between the variation of the nucleation frequency
(I, in cm.$^{-3}$ sec.$^{-1}$) as calculated according to Ref. 1 (constant α),
and as calculated according to formula 3, (temperature dependent α).
The viscosity of η is kept constant at $\eta = 10^{-2}$ poise. The value
$\alpha = 0.63$ is the one obtained from the measurements on Hg (Ref. 3).
The values of α_m and β are chosen in accordance with the calculation
of the surface tension in the structural model for the solid-liquid
interface (Refs. 15 and 16).

211

can be introduced in formula (3), using the Vogel-Fulcher equation:[18]

$$\eta = A \exp[a/(T-T_g)] \qquad (4)$$

The resulting $I(T_r)$ curves for three values of T_{rg} are shown in Fig. 2. It appears that, for $T_{rg} < \frac{1}{2}$, there will be copious homogeneous nucleation in some temperature range at all practically attainable quench rates.

The foregoing results have been derived mainly for pure melts, but they can probably be extended to alloy melts as well. As has been mentioned above, the crystal-melt interfacial tension finds its origin mainly in the large negentropy of localization of the interface, or, equivalently, in the low configurational entropy of the first interfacial layer. This type of localization is, of course, not limited to pure systems, and we therefore expect the interfacial tension in alloy melts also to be entropic in origin and to decrease with T. Furthermore, as Spaepen and Meyer[16] have pointed out, α_m is quite insensitive to considerable variation in the configurational entropy of the interface in the range of interest here. Even though the details of the entropy calculation will change depending on composition and crystal orientation, it is reasonable to expect that the configurational entropy will be of the same order of magnitude in most metallic systems. Therefore, if there are no impurity segregation effects, which seems plausible on geometrical grounds, since the interface has zero density deficit, we expect α_m to be roughly the same in these systems. For this reason, the conclusions from Fig. 2 should hold for alloy systems also. It is significant that the lowest value of T_{rg} observed so far for a metallic alloy is 0.44, which is close to the lower limit estimated from Fig. 2. As predicted, this system also requires high quench rates ($\sim 10^6$°C/sec) for glass formation.

212

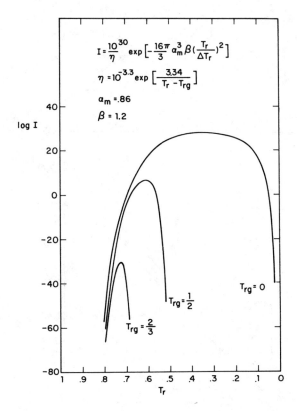

Figure 2.

The variation of the nucleation frequency (I, in cm.$^{-3}$ sec.$^{-1}$) with reduced temperature (T_r), as calculated according to formula 3 for the values of T_{rg}. The temperature dependence of the viscosity is given by the Fulcher equation for a typical metallic liquid. α_m and β are chosen in accordance with the calculation of the temperature dependent surface tension (Refs. 15 and 16). 213

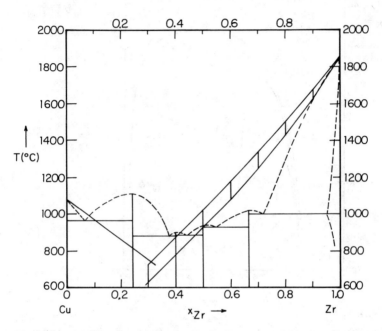

Figure 3.

Comparison between the observed liquidus in the Cu-Zr systems
[see M. Hansen, "Constitution of Binary Alloys", McGraw-Hill (1958)]
of an ideal solution melt in equilibrium with pure crystalline Cu
or Zr. The heats of crystallization selected for Zr ranged from 3.74
(lower branch) to 4.5 kcal/gm. atom (upper branch). (See text for
discussion on choice).

214

Crystal Growth in Amorphous Metals

Even if homogeneous nucleation is completely suppressed, crystallization may still occur by growth from heterogeneous nuclei, "seeds", which are usually present inadvertently, if not by deliberate addition, in a normal size system. Therefore, the formation and persistence, below T_g, of a metallic glass requires conditions such that the crystal growth rate, u, in the amorphous metal does not become large enough to effect crystallization by growth from these seeds.

The isothermal rate of advance of a plane crystal front in a pure melt or glass should be determined by the undercooling, ΔT_r, the crystal ledge-melt tension, σ_e, and the time constant, τ_u, for the attachment of an atom from the melt to a crystal surface site.[1,2] Experience indicates[19] that there is little resistance to crystal-ledge formation, even at small undercooling, in metal and other melts where the entropy of fusion $\tilde{<}$ 2 k per atom. This behavior, which implies that, in metals, σ_e is much less than σ_{LS}, appears quite consistent with Spaepen's[2] model[15] for the interface. More particularly, ledge formation would need little energy and its negentropy contribution also should be small because of the highly pronounced localization of DRP positions and the low configurational entropy attending formation of the interface.

In view of the foregoing, it seems that τ_u may be the most important factor in the resistance of an amorphous metal to crystal growth. We have noted that in non-metallic melts it often scales rougly with η. The time constant, τ_D, for atomic or molecular transport also scales as η, at least for $\eta < 10^7$ poise, which implies proportionality of τ_u to τ_D. The microscopic justification of these relations has been discussed elsewhere.[20]

Impurity redistribution, either ordering or segregation, should proceed with a time constant of order, τ_D. If crystal growth must be attended by such redistribution, then $\tau_u \tilde{>} \tau_D$ and it may be indexed roughly by the viscosity behavior.

These conditions should apply in the crystallization of most alloys so their glass forms ought to persist for long periods at temperatures well below T_g. We note that the scaling relation gives $\tau_D \sim 10^3$ secs at T_g.

If impurity redistribution is not required, the kinetic resistance in crystal growth can be much smaller than in atomic transport. Indeed, it has been suggested[21,22] that the attachment process in crystal growth into pure monatomic melts is simply the collision of an atom of the melt with the crystal surface. If so, the upper limiting crystal growth rate can be expressed as:

$$u \mathrel{\tilde{<}} \frac{\lambda}{\tau_c} \frac{\beta \Delta T_r}{T_r} \tag{5}$$

where λ is the distance advanced by the interface per attachment, $\tau_c = \frac{1}{z_c}$ and z_c is the impingement frequency of atoms from the melt on a crystal surface site. Precise evaluation of z_c is extremely difficult but a rough estimate may be made by applying the mean field model.[22] In this calculation, it is assumed that the only forces acting within the system, including the interface, are hard sphere-repulsive in nature. Then the impingement frequency may be expressed:

$$z_c = z_c^o \frac{P}{P^o} \tag{6}$$

where P is the "internal" pressure of the melt and z_c^o and P^o are, respectively, the impingment frequency and pressure calculated by extrapolating the dilute fluid results to their values at the density, n atoms/cm^3, of the melt. From kinetic theory, we have:

$$z_c^o = \frac{1}{4} n <c_{rel}> \frac{v}{\lambda}$$

216

where $<c_{rel}> = <c> + u$, $<c>$ is the average speed of the impinging atoms in the laboratory frame and v is the volume/atom. Substitution of (6) and (7) into (5) leads to:

$$u \lesssim \frac{n <c_{rel}>}{4} (\frac{P}{P_o}) (v \beta) (\frac{\Delta T_r}{T_r}) = \frac{<c_{rel}>}{4} (\frac{P}{P_o}) (\frac{\beta \Delta T_r}{T_r}) \qquad (8)$$

and:

$$\tau_c = (\frac{P^o}{P}) \frac{4}{<c_{rel}>} \qquad (9)$$

There is no evaluation of the ratio P/P^o. However, for the purpose of making a rough estimate, it seems plausible to assign it the value, obtained by molecular dynamics computations, for the uniform hard sphere fluid at its thermodynamic crystallization point.[23] In this way, Turnbull and Bagley[22] calculated upper limiting crystal growth rates in pure undercooled melts of Ni and Co, which were larger than, but only by factors of 2 to 5, the maximum values measured experimentally.

In the dilute fluid regime, τ_D, the time constant for atomic transport, equals τ_c. However, as the density approaches that of the melt at T_m, τ_D falls substantially below τ_c owing to the onset of backscattering, which reduces the displacements to values well below those predicted by the random walk analysis. This effect is usually specified by the correlation factor: $f = \tau_D/\tau_c$.

With increasing undercooling, the fluid becomes more dense and the backscattering steadily increases, as manifested by falling f. This trend should continue until $f \to 0$ and the atomic motions are almost wholly oscillatory. It is this transition from translatory to primarily oscillatory motion which is associated with the onset of solid-like behavior at $T \sim T_g$ in the Cohen-Turnbull[24] model for the melt$\to$glass transition.

217

In sharp contrast with τ_D, τ_c should not increase sharply[21] with falling T
in the glass transition range and it may remain small to temperatures far below
T_g. Consequently, in systems where τ_u scales with τ_c, rather than with τ_D, the
crystal growth rates may remain high enough to permit crystallization from a small
number of nuclei at temperatures far below T_g. This behavior is possible for pure
metals and, rather than low T_g, may account for their tendency to crystallize
during or after vapor quenching on extremely cold substrates.

Our discussion has been based in part on the assumption that τ_D scales with η.
However, this scaling has been tested only for some molecular systems to viscosities
no higher than 10^7 to 10^8 poise. It seems quite plausible that the required atomic
or molecular rearrangements will be similar in flow and diffusive transport. However, in the high viscosity or glass regime it is possible that the correlation
distances for flow will greatly exceed the atomic or molecular dimension and under
this condition we have to be alert to the possibility that τ_D will fall considerably below the values calculated from viscosity scaling. Indeed, this is indicated
by the recent measurements of Gupta et al.[25] on the diffusivity at $T < T_g$ of Ag
into $Pd_{81}Si_{19}$ glass.

The Melt ↔ Glass Transition in Metals

1. Pure Metals.

Cohen and Turnbull predicted[24] that all pure monatomic melts, including
metals, would form glasses if quenched rapidly enough. This prediction was based
on the result of their transport model that $f \to 0$ at some $T \to T_g > 0$ and the
supposition that crystallization of these systems requires topological reconstruction of the SRO. However we have noted that the quench rates needed to form glasses
from normal size pure metal melts may be practically impossible to achieve and that,

218

even if formed, pure metal glasses might not withstand crystal growth from already existing seeds, excepting at extremely low temperatures, $T \ll T_g$.

Nevertheless, there is a growing body of indirect evidence indicating that metal melts in pure form have, at least, the capability of forming glasses. We note first that the heat capacity of the melt diverges positively[26,27] from that of the crystal as the temperature falls in the undercooled regime. The rather limited data[26] indicate that the coefficients of thermal expansion of the melt and crystal would also diverge similarly in the undercooled regime. These trends are normally terminated by nucleation and growth crystallization. However, if such crystalliza-tion were bypassed by rapid quenching, the divergencies would have to disappear at $T > 0$ either by transition to a glass[24,28] or by a "spinodal" crystallization.[29]

More important, perhaps, are the findings that some metals can be vapor quenched on to cold substrates, with amorphous, but manifestly solid, forms with quite small impurity admixtures.[30,31] It seems plausible that the small amounts of impurity could be removed from the metal without significant alteration of its structure or viscosity. The alloys which have been melt quenched to glasses contained much larger amounts, 15 at. % or more, of impurity.

There are few indications of what the glass temperatures of pure metals may be. We suppose that they must equal or exceed the _kinetic_ crystallization tempera-tures T_c of deposited amorphous films, with or without small impurity admixture. Chen and coworkers,[32,33,34] by extrapolating the measured concentration trends of the T_g's, have estimated some pure metal T_g's of melt-quenched alloys. However, this procedure is quite uncertain since the measurements span only rather narrow concentration ranges.

It appears that the T_g's of pure metals should increase with the rearrange-ment potentials of the constituent metal ions.[35,21] The results of pseudopotential and other theoretical calculations of these potentials are still quite limited and

219

there is little experimental information on them. In the mean field models,
atomic rearrangements in transport would be limited mainly by the interatomic
repulsive interactions. These arise from the forces opposing ion-ion overlap
and the limitations on the electronic redistribution required for screening the
charges of ions approaching each other. From these considerations, it seems that
the rearrangement potential and T_g should generally increase with the ratio of
ion core/atomic radii and with the metallic valence.[35]

2. Compositional Effects.

The effects of alloying on glass forming tendency have been reviewed[36,21]
and interpreted[35,36,37,21] in several earlier papers. It was pointed out that
the alloys which melt-quench most readily to glasses are composed by combining
late transition elements (A) with metalloid and/or early transition elements (B).
Alloying of A and B melts is generally accompanied by relatively large enthalpy
releases and often by substantial volume contractions. From this we expect that
SR compositional ordering may play an important part in atomic transport and glass
formation.

Of the variables determining glass formation, it is the reduced glass temperature, T_{rg} – now defined as T_g/T_ℓ, where T_ℓ is the liquidus temperature – which
seems likely to be most sensitive to alloying. We note (Table 1) that the reported
T_{rg}'s of melt-quenched alloy glasses range from 0.44 to 2/3. The lowest of these
values lies near the estimated lower limit on $T_{rg}(\gtrsim 1/2)$ for glass formation by pure
metals at quench rates 10^6°C/sec.

Provided T_g of A(B) alloys were not strongly dependent on the nature and
atom fraction, x, of B, T_{rg} would be mainly determined by $T_\ell(x)$. While this proviso is not well established, it is found that for a given A host the likelihood
of glass formation in melt quenching tends, indeed, to be greater the smaller is

220

$T_\ell(x)$. Thus in a given system it is the alloys having compositions in the vicinity of eutectics which are most readily quenched to glasses.[38,39] Further, as Turnbull noted,[21] alloys with compositions near "abnormally deep" eutectics should quench to glasses more readily than those at "normal" eutectics. By an abnormally deep eutectic, we mean one where the actual liquidus lies far below that, $T_\ell^o(x)$, calculated by supposing that the liquid alloy behaves ideally. Marcus and Turnbull[40] report that there is, indeed, some correlation between departure from ideality $(T_\ell^o - T_\ell)(x)$ of binary A rich alloys and their glass forming tendency. In particular, they noted that alloys with near eutectic compositions generally form glasses, when quenched to below T_g at rates $\overset{\sim}{>} 10^6$ °C/sec., provided the departure ratio, $(T_\ell^o - T_\ell) (0.15)/T_m^A \geq 0.09$; T_m^A is the thermodynamic crystallization temperature of pure A. For example, the departure ratio is 0.22 for Au(Si) and 0.29 for Pd(Si). However, there are certain binary P alloys with departure ratios substantially larger than 0.09 which fail to form glasses in rapid melt quenching. Marcus and Turnbull[40] attributed this failure of the correlation to turbulence and composition changes attending evaporation of the volatile P from the melt.

There are some alloy systems which exhibit glass formation[36] over composition ranges extending well beyond the eutectic on either side. For example, Giessen, Grant and coworkers[41,42] have shown that all Cu-Zr alloys with compositions in the range 33 to 50 at. % for Zr can be melt-quenched to glasses. However, the liquidus temperatures over the glass forming range are relatively low, as may be seen in Fig. 3 in which the Hansen equilibrium diagram is compared with a simple eutectic one calculated on the assumptions that the melts (1) are in equilibrium with pure crystalline phases of either Cu or Zr; (2) behave as ideal solutions. The range of liquidus temperatures on the Zr branch reflects our estimate that the unknown entropy of fusion at T_m of Zr will lie between the corresponding values of b.c.c.

221

Na and W. Of course, our calculation ignores any stabilization of the crystalline phases by AB interactions, as manifested by the Hansen diagram, which would raise the ideal liquidus temperatures well above our calculated values. Even so, we note that the actual liquidus temperatures of the glass forming alloys lie within or near our calculated range.

Recently Nagel and Tauc[43] pointed out an energetic contribution to the stability of metallic alloy melts. They emphasized the metallic nature of glassy alloys and noted that the amorphous phase should be stabilized, relatively, if its Fermi energy were to correspond with the energy at the minimum in the electronic density of states. They noted that this condition seems to be fulfilled by the compositions of the alloys most prone to glass formation. However, the electronic properties are computed on the basis of the rigid band model only. They do not try to explain the atomic transport behavior leading to glass formation and their discussion of crystallization resistance seems quite ambiguous in view of the heterogeneous nature of the crystallization process. Nevertheless, they have identified an electronic contribution to amorphous phase stability which might lead to a substantial lowering of the liquidus temperature relative to T_ℓ^o.

3. Microscopic Theory of Glass Formation.

Experience indicates that energy minimization by AB association is, normally, much greater in crystalline intermetallic or solution phases than in melts. However, the advantage of the crystalline over the amorphous milieu for such minimization apparently is much less for good glass formers, as indicated by their relatively low T_ℓ's, than for other alloy systems. This viewpoint is further supported by measurements, noted earlier[21,44] which indicate that the crystallization enthalpies of metallic glasses (see Table 1) are only about $\frac{1}{3}$rd to $\frac{1}{4}$th those of the constituent masses in their pure melt forms at T_m. Explaining such small enthalpy differences in systems where the energy driving AB ordering is quite high becomes then a central problem for the microscopic theory of glass formation. In the fol-

222

TABLE 1

Alloy	Ref.	Liquidus $T_\ell(^{\circ}K)$	glass T $T_g(^{\circ}K)$	T_g/T_ℓ	ΔH_c kcal/gm. atom	ΔH_m^o kcal/gm. atom	$\Delta H_c/\Delta H_m^o$
$Au_{81}Si_{19}$	(50)	636	290	0.46	0.76	2.7	0.28
$Au_{77}Ge_{14}Si_9$	(32)	625	294	0.47			
$Pd_{80}Si_{20}$	(33)	1220	655	0.54	1.13	3.6	0.31
$Pd_{82}Si_{18}$	(33)	1170	648	0.55	0.95	3.65	0.26
$Cu_{66}Zr_{34}$	(42)	1263	762	0.60	0.86	3.45	0.25
$Cu_{50}Zr_{50}$	(42)	1213	680	0.56	1.3	3.6	0.36

Table 1: Selected values of reduced glass temperatures, T_g/T_ℓ (see Chen[34] for more complete compilation) and heats of crystallization of glasses at their T_g. ΔH_m^o is the heat of crystallization of the constituents in their pure _metallic_ forms at T_m. In this calculation we have set the crystallization enthalpy of Si equal to 1.7 kcal/gm.atom (value for pure Sn) and that of Zr equal to 4.1 kcal/gm.atom (see text). These enthalpies are from compilation of K.A. Gschneider, Solid State Phys. <u>16</u>, 275, Academic Press, N.Y. (1964).

lowing, we will survey various approaches to this problem.

In the simplest model, the relative stability of the amorphous alloys is attributed to differences in the "sizes" of A and B atoms. The effects of size differences on the structures of amorphous alloys have been explored in a number of papers and surveyed critically in Cargill's recent review.[13] (Also see this review for references). There is evidence that the degree of space filling in the random packing of hard spheres is substantially higher for two sizes of spheres at certain admixtures and size ratios than for uniform spheres. This result implies that such amorphous admixtures will be relatively stabilized when interatomic attractive forces act. Actually, the concept of atom size in metals needs some clarification. For example, the partial atomic volume of P is roughly constant[21] in both crystalline Ni_3P and the nickel rich Ni-P glasses and it is equal to that of pure crystalline Ni, within 2%. Thus the Goldschmidt radius of P is about equal to that of Ni in the glassy alloy. However, the actual "radius" of P in both the glassy and intermetallic phases can be substantially smaller than the Goldschmidt radius, if the coordination number of P is less than that of Ni. This relation is, indeed, indicated by diffraction studies on both crystalline Ni_3P and glassy Ni(P) alloys.

Polk[37,45] developed a model for the relative stabilization of amorphous alloys based on the DRP structure and the assumption that the preferred AB nearest neighbor spacing is substantially less than that of AA. The DRP structure of uniform spheres can be viewed as composed of tetrahedral holes (delineated by center positions of neighboring atoms) in combination with a small number of larger holes which must be incorporated for space filling. Polk pointed out that the structure should become relatively more stable by siting small, in the sense explained above, atoms into the largest of these holes. Also he noted that the atomic configurations surrounding the largest holes are similar to those around the B atoms in

224

certain crystalline phases, such as Ni_3P, of the AB glass formers. Actually, in the Ni(P), as well as in the other systems Polk considered, the AB spacings required to fit B into the holes of the A DRP structure, without its relaxation, were still substantially less than those in the crystalline AB compounds.

Chen[46] suggested that strongly directed A-B covalency plays an essential role in alloy glass formation. This idea emphasizes the important role of developing C-SRO in alloy glass formation but it has not led to clear predictions of the nature of the C-SRO. It is not sharply distinguished from Polk's concept[37] that the C-SRO will be determined by a preferred r_{AB} in conjunction with some specified density of conduction electrons.

While accepting Polk's structural model, Turnbull[21] suggested that the relative stability of AB alloys in the amorphous state is due to the softness of the repulsive potential between AB pairs in comparison with that between either AA or BB pairs. Consequently, at a given electronic density, the near neighbor AB spacings, r_{AB}, could vary considerably, as they are likely to do in an amorphous structure, with little change in the potential energy of the system. Then, there would be only a small contribution to the energy decrease in crystallization from the attending shifts of the r_{AB} to their energetically most preferred values.

These ideas seem to be supported[21] by the occurrence of "fast" diffusion in AB alloy systems where an element of the B type is the crystalline host. It is well established that the diffusion of A type impurities in, at least, some of these systems occurs by an interstitial-type mechanism.[44,47] The interstitial siting and motion of A in these systems would require, unless there is extensive distortion of the B environment, that r_{AB} take on values which are far less than the sum $(r_{AA}^o + r_{BB}^o)/2$, of the Goldschmidt radii of the constituents in their pure form. The accompanying energy increases cannot exceed the relatively small activation energies observed for transport of A in B, which implies that the AB repulsive potentials are, indeed, quite soft or that the energies for distorting B are small. The reciprocal connection between glass forming tendency and fast diffusion is well illustrated by systems composed of early transition (B) elements, such as Zr and the rare earths, and late transition (A) elements, such as Cu and Co.

If the AB repulsive potentials are, indeed, soft, it follows that the atomic transport rates in disordered and relatively concentrated AB alloys should remain high to quite low temperatures. This means that the glass temperatures might be reduced by random AB admixture, thus offsetting the effects of falling T_ℓ on glass forming tendency. We have noted, however, that the thermal behavior clearly indicates that a high degree of compositional SRO (C-SRO) develops in AB melts as the glass transition is approached. If the concentration of B is no greater than 20 to 25 at. %, and if its coordination number is small, the nature of the C-SRO will be such that every B atom will be "caged" by A neighbors with low probability of B-B contact. Polk and Turnbull[48] noted that flow may be much more difficult in an alloy so ordered than in one compositionally disordered at the same concentration. This concept was the basis of their model for flow disordering and softening to account for the ductility of glassy alloys. Spaepen and Turnbull[49] pointed out that both the topological and compositional SRO may decrease markedly with the isothermal application of high tensile stresses. Consequently, such stresses may cause flow and atomic transport at temperatures far below T_g. It follows from these concepts that the temperature dependence of the atomic transport properties should be determined, mainly, by that of the SRO. If the configurations determining the SRO are nearly frozen at T_g, the transport properties will change at relatively slow rates with T decreasing below T_g.

These ideas are supported by Chen and Turnbull's[32] finding that the shear viscosity of a Au-Si-Ge alloy varies sharply with _isothermal_ configuration

226

changes in the glass transition range. Further the recent measurements of Gupta et. al[25] indicate that the diffusivity of Ag into a $Pd_{81}Si_{19}$ alloy glass is, in the temperature range well below T_g, governed by an activation energy (26 kcal/gm. atom) which is smaller by a factor of 3 to 4 than the apparent activation energy for viscous flow in the T_g range. The model predicts, also, that the atomic transport resistance should be least at the highest AB pair concentration. This prediction seems consistent with Kerns et. al's[42] recent findings that T_g of glassy Cu-Zr alloys decreases steadily by more than 70°C as the Zr content increases from 34 to 50 at. %.

Acknowledgement

Our research in this field has been supported in part by grants from the Office of Naval Research under Contract N00014-67-A-0298-0036 and from the National Science Foundation under Contract DMR-72-03020.

REFERENCES

1. Turnbull, D., (a) <u>Contemp. Phys.</u> <u>10</u>, 473 (1969); (b)"Physics of Non-crystalline Solids", 41-56, J.W. Prins, Editor, North Holland, Amsterdam (1965).
2. Turnbull, D., <u>Solid State Physics</u> 3, 225-306, Acad. Press, New York (1956).
3. Turnbull, D., <u>J. Chem. Phys.</u> <u>20</u>, 411 (1952).
4. (a) Davies, H.A., J. Aucote, and J.B. Hull, <u>Scripta Met.</u> <u>8</u>, 1179 (1974).
   (b) Davies, H.A., <u>J. Non-Cryst. Solids</u> <u>17</u>, 266 (1975).
   (c) Davies, H.A. and B.G. Lewis, <u>Scripta Met.</u> <u>9</u>, 1107 (1975).
5. Zachariasen, W.H., <u>J. Am. Chem. Soc.</u> <u>54</u>, 3841 (1932).
6. Frank, F.C., <u>Proc. Roy. Soc.</u> <u>215A</u>, 43 (1952).
7. Hoare, M.R. and P. Pal, <u>Adv. Phys.</u> <u>20</u>, 161 (1971).
8. Burton, J.J., (a) <u>J. Chem. Phys.</u> <u>52</u>, 345 (1970); (b) <u>Nature</u> <u>229</u>, 335 (1971).
9. Bernal, J.P., (a) <u>Nature</u> <u>185</u>, 68 (1960); (b) <u>Proc. Roy. Soc.</u> <u>280A</u>, 229 (1964).
10. Finney, J.L., <u>Proc. Roy. Soc.</u> <u>319A</u>, 479, 495 (1970).
11. Bennett, C.H., <u>J. Appl. Phys.</u> <u>43</u>, 2727 (1972).
12. Adams, D.J. and A.J. Matheson, <u>J. Chem. Phys.</u> <u>56</u>, 1989 (1972).
13. Cargill, G.S., III, (a) <u>Solid State Phys.</u> <u>30</u>, 227-320 (1970);
    (b) this volume.
14. Turnbull, D., <u>Trans. AIME</u> <u>221</u>, 422 (1961).
15. Spaepen, F., <u>Acta Met.</u> <u>23</u>, 729 (1975).
16. Spaepen, F. and R.B. Meyer, <u>Scripta Met.</u> [in press].
17. Miller, W.A. and G.A. Chadwick, <u>Acta Met.</u> <u>15</u>, 607 (1967).
18. (a) Vogel, H., <u>Phys.Z.</u> 22, 645 (1921).
    (b) Fulcher, G.S. <u>J. Am. Ceram. Soc.</u> <u>6</u>, 339 (1925).
19. Jackson, K.A., "Growth and Perfection of Crystals," 319-325, R.H. Doremus, et al., Editors, Wiley, New York (1958).
20. Turnbull, D., <u>J. Phys. Chem.</u> <u>66</u>, 609 (1962).
21. Turnbull, D., <u>J. de Physique</u> <u>35</u>, Colloque-4, C-4.1—4.9 (1974).
22. Turnbull, D. and B.G. Bagley, "Treatise on Solid State Chemistry", <u>5</u>, 513-554, Hannay, N.B., Editor, Plenum Press, New York (1975).
23. Alder, B.J., W.G. Hoover, and D.A. Young, <u>J. Chem. Phys.</u> <u>49</u>, 3688 (1968).
24. (a) Cohen, M.H. and D. Turnbull, <u>J. Chem. Phys.</u> <u>31</u>, 1164 (1959).
    (b) Turnbull, D. and M.H. Cohen, <u>J. Chem. Phys.</u> <u>52</u>, 3038 (1970).
25. Gupta, D., K.N. Tu, and K.W. Asai, <u>Phys. Rev. Let.</u> <u>35</u>, 380 (1975).
26. Borelius, G., <u>Sol. State Phys.</u> <u>15</u>, 1-51, Acad. Press, N.Y. (1963).
27. Chen, H.S. and D. Turnbull, <u>Acta Met.</u> <u>16</u>, 369 (1968).
28. Cohen, M.H. and D. Turnbull, <u>Nature</u> <u>203</u>, 964 (1964).
29. Kauzman, W., <u>Chem. Rev.</u> <u>43</u>, 219 (1948).
30. (a) Hilsch, R., "Non-Crystalline Solids", 348-373, Fréchette, R.D., Editor, Wiley, New York (1960).
    (b) Buckel, W., <u>Z. Phys.</u> <u>138</u>, 136 (1954).
31. (a) Mader, S. and A.S. Nowick, <u>Appl. Phys. Letters</u> <u>7</u>, 57 (1965).
    (b) Mader, S. and A.S. Nowick, <u>J. Vac. Sci. Techn.</u> <u>2</u>, 35 (1965).
    (c) Nowick, A.S. and S. Mader, <u>IBM J. Res. & Dev.</u> <u>9</u>, 358 (1965).

32. Chen, H.S. and D. Turnbull, *J. Chem. Phys.* 48, 2560 (1968).
33. Chen, H.S. and D. Turnbull, *Acta Met.*, 17, 1021 (1969).
34. Chen, H.S., *Acta Met.* 22, 1505 (1974).
35. Bennett, C.H., D.E. Polk, and D. Turnbull, *Acta Met.* 19, 1295 (1971).
36. Giessen, B.C. and C.N.J. Wagner, "Physics and Chemistry of Liquid Metals", 633-695, Beer, S.Z., Editor, Dekker (1972).
37. Polk, D.E., *Acta Met.* 20, 485 (1972).
38. Rawson, H., "Proc. IV Int. Conf. on Glass", 62 Imprimerie Chaix, Paris (1956).
39. Cohen, M.H. and D. Turnbull, *Nature* 189, 131 (1961).
40. Marcus, M. and D. Turnbull, "Proc. 2nd International Conference on Rapidly Quenched Metals", 82, N.J. Grant and B.C. Giessen, Eds., *Mat. Sci. Eng.* 23 (1976) [in print].
41. Ray, R., B.C. Giessen and N.J. Grant, *Scripta Met.* 2, 357 (1968). Recolevski, A. and N.J. Grant, *Met. Trans.* 3, 1545 (1972).
42. Kerns, A., R. Ray, and B.C. Giessen [to be published].
43. Nagel, S.R. and J. Tauc, *Phys. Rev. Let.* 35, 380 (1975).
44. Warburton, W.K and D. Turnbull, *Thin Solid Films* 25, 71 (1975).
45. Polk, D.E., *Scripta Met.* 4, 117 (1970).
46. Chen, H.S. and B.K. Park, *Acta Met.* 21, 395 (1973).
47. Warburton, W.K. and D. Turnbull, *Diffusion in Solids: Recent Developments*, 171-229, Nowick, A.S. and J.J. Burton, Editors, Academic Press, New York (1975).
48. Polk, D.E. and D. Turnbull, *Acta Met.* 20, 493 (1972).
49. Spaepen, F. and D. Turnbull, *Scripta Met.* 8, 563 (1974).
50. Chen, H.S. and D. Turnbull, *J. Appl. Phys.* 38, 3646 (1967).
51. Dr. Spaepen is an I.B.M. Post-Doctoral Fellow, 1975-1976.

229

STRUCTURAL CHANGES AT THE LENNARD-JONES GLASS TRANSITION

Charles H. Bennett

IBM Thomas Watson Laboratories
Yorktown Heights, New York

John L. Finney

Department of Crystallography
Birkbeck College, London

Random close-packings of equal spheres were developed by Bernal and others as first order structural models of simple liquids.[1-3] Although the essential liquid structure is embodied in these models,[3] their high limiting density suggests they are more directly applicable to hard sphere glasses.[4,5] To investigate this suggestion, a set of molecular dynamics calculations were performed on the inner 1000 centres of the large dense hard sphere packing,[6] scaling to a Lennard-Jones 6:12 potential for several densities. All centres closer than r_0 to the centre of mass were allowed to move, the remaining centres forming an irregular, fixed confining shell. The mean square displacement $<r^2 (t)>$ of the moving atoms was monitored to keep track of the diffusion behaviour. Fuller details of the calculations are given in the Appendix to this paper.

Data for three assemblies of particular interest are given in Table 1. For the two lower temperatures (sets 1 and 2) $<r^2 (t)>$ quickly reaches a small value, after which it does not increase further. The two assemblies are still non-crystalline, with diffusion coefficients of effectively zero, indicating their glassy nature. The small, residual $<r^2 (t)>$ is essentially a measure of the Debye-Waller factor of each (high pressure) glass. For the higher temperature (and volume) array (set 3) $<r^2 (t)>$ continues to increase, indicating liquid-like diffusion behaviour, with an estimated diffusion coefficient of about $0.8 \times 10^{-5} cm^2 sec^{-1}$.

Thus, it appears that somewhere between data sets 2 and 3, during which both temperature and volume have increased slightly, a glass transition has occurred in the confined droplet. As the starting point for the calculation was the hard sphere coordinates, this transition cannot be equated immediately with the glass transition expected on reheating an instantaneously vitrified Lennard-Jones liquid. Comparisons between hard sphere packings and artificially hardened Lennard-Jones liquids[3] suggest differences should be small, though not necessarily insignificant.

231

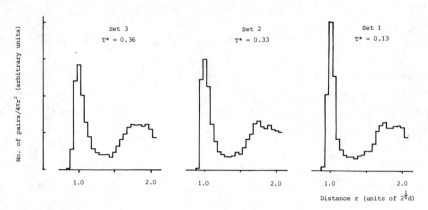

Figure 1. Radial Distribution Functions

Although the calculations were performed to pass through the transition in the direction glass to liquid, the following discussion of structural changes is in terms of the more familiar reverse liquid to glass process.

The main point of interest in the radial distribution functions (Figure 1, supra) is the behaviour of the second peak. This peak is single for liquids, but split for amorphous solids.[7,8] For the present arrays, there is a suggestion of a slight splitting for the $T^* = 0.33$ glass, a splitting which increases as temperature is reduced to 0.13. It is tempting to associate this splitting with the actual glass transition and the associated drop in diffusion coefficient, although the statistics prevent us from asserting this too strongly. The position of the first component of the split peak is about the same as in the hard sphere packing, but the second component moves to a lower r value which is more consistent with amorphous metal data.[8]

Local structure can be investigated through the associated set of Voronoi polyhedra.[3,6,9] The immediate neighbourhood of each atom is uniquely described by the size and shape of that smallest closed polyhedron made up of planes perpendicularly bisecting interatomic vectors.[6,9] Selected average data are shown in Table 2. Also shown for comparison are data from (a) Monte Carlo and Molecular Dynamics calculations on

232

TABLE 1. Selected Data for Three MD Calculations

	Set 1	Set 2	Set 3
Reduced temperature T*	0.13	0.33	0.36
Scaling (σ/hard sphere diameter)	1.110	1.110	1.123
No. of time steps	698 3000	1000 2000 3000	500 1000 1500
Mean square displacement $\langle r^2(t)\rangle$	0.014 0.014 NOT increasing	0.045 0.060 0.052 NOT increasing	0.111 0.161 0.239 increasing
Mean energy E/N	−6.48	−5.95	−5.63

TABLE 2. Selected Polyhedron Data

System	MD[10] (periodic boundary)	MC[6]	This work [c] (confined boundary)			Energy minimised[11] (free boundary)
T*	0.71	0.71	0.13	0.33	0.36	∿0
Average volume [a] per centre $\bar{V}$ (units of σ^3)	1.209	1.183	1.112	1.118	1.159	1.106
Standard deviation of normalised local volume $\sigma(V*)$ [b]	0.086	0.082	0.054	0.062	0.081	0.037
Average no. of faces per poly-hedron $\bar{N}$	14.45	14.46	14.09_7	14.10_3	14.14_6	13.99

(a) For this work, average volume is an independent variable, being fixed by the scaling procedure.
(b) This is equivalent to a measure of volume fluctuation Δ defined by
$\Delta^2 = [\bar{V^2} - (\bar{V})^2] / (\bar{V})^2$
(c) Those centres within or adjacent to the confining shell are excluded.

liquid argon near the triple point,[3,10] and (b) an energy-minimised array of 999 Lennard-Jones atoms with the same starting configuration as the present calculations, but with <u>a free boundary</u>.[11]

[A]   The standard deviation $\sigma(V^*)$ of the local volume distributions shows a large fall through the transition which is relatively greater than that which occurs on further cooling (Table 2). The two periodic boundary liquid argon simulations fit the same picture, $\partial\sigma(V^*) / d\overline{V}$ being much smaller above the transition point than below. The statistics are inadequate for conclusions to be drawn concerning the changes in shape of the local volume distributions.

[B]   The average number of faces per polyhedron $\overline{N}$ also shows a larger fall through the transition than occurs on further cooling, suggesting a small but significant reduction in complexity of the neighbourhood of the average molecule upon losing both volume and temperature as the assembly passes from the liquid to the glass. The much lower value of $\overline{N}$ for the energy-minimised aggregate may reflect the non-existence of thermal fluctuations, just as the much higher values for near-triple point argon may reflect the much greater polyhedron fluctuations arising from thermal effects, in addition to the availability of more free volume. However, the wide difference in boundary conditions may also affect the statistics.[12]

[C]   Similar conclusions can be drawn from the distribution of number of faces per polyhedron (not shown). A large increase in the number of 14-faced polyhedra as we pass down through the transition, together with a smaller increase in 13-faced cells, occurs mainly at the expense of higher-N polyhedra. The low temperature glass data are similar to those of the energy-minimised array, although an additional bias to simpler (smaller-N) polyhedra found in the latter aggregate agrees with the interpretation given above for the relatively low average value of $\overline{N}$ for this array.

[D]   The polyhedron topology statistics (Table 3) reflect a reduction in polyhedron complexity — and hence of average neighbourhood — as we pass from near-triple point liquid, through the glass transition, and towards zero temperature glass. The population of polyhedra with n_3 three-edged faces (Table 3) shows how the marginal reduction of free volume and/or temperature squeezes out the three-edged faces. Again the changes are particularly noticeable across the transition. Classifying polyhedra in terms of $(n_3 n_4)$ — the numbers of 3- and 4-edged faces — shows how the polyhedron types change within the $n_3 = 0$ group (Table 3b). Once more, there are significant changes across the glass transition, lower temperature adjustments within the glass being relatively minor. A point of particular interest is the behaviour of (00) and (01) groups, which refer in the main to (distorted) icosahedra. The molecular dynamics glass assemblies point to reductions in these groups, whereas the energy-minimised array shows significantly higher values. The icosahedron is a particularly efficient way of packing together 13 soft spheres in a low energy configuration, and the higher population in the energy-minimised array suggests there is additional freedom of movement, presumably

234

TABLE 3. Selected Polyhedron Topology Statistics

(a) Percentage of Polyhedra Classified by numbers of 3-edges faces (n_3)

System		$n_3 =$	0	1	2	3	4	≥ 5
MC	T* = .71		41	36	16	5	1	<1
MD	T* = .36		58	31	8	3	<1	-
	T* = .33		69	24	6	1	<1	-
	T* = .13		67	27	5	<1	<1	-
Energy minimised (T* $\sim$0)			75	19	5	<1	-	-

(b) Percentage of polyhedra classified by nos. of 3- and 4-edged faces ($n_3 n_4$) for $n_3 = 0$

System	$n_3 n_4 =$	00	01	02	03	04	05	06
MD	T* = .36	1.8	11.1	12.8	16.4	10.6	4.4	0.9
	T* = .33	1.3	7.3	11.6	24.9	18.5	3.4	1.7
	T* = .13	0.4	5.9	11.8	23.1	19.3	3.8	2.5
Energy minimised (T* $\sim$0)		2.7	8.2	12.8	26.4	21.1	2.8	1.1

from the existence of the free boundary,[12] which allows these local arrangements to form. The confining shell present in the molecular dynamics calculations would seem to have prevented their formation. Thus, the differences already noted above between the energy-minimised and low temperature glass arrays seem to arise from differences in both boundary constraints and thermal fluctuations.

APPENDIX

A Lennard-Jones 6:12 potential was used, with parameters σ and ε, and a cut-off at 2.5σ. The innermost 1000 spheres of the 7934 dense random packing[6] were selected and scaled so that σ was equal to the hard sphere diameter d. r_0^2 was initially fixed at 22.000σ, giving 627 moveable atoms within a fixed confining shell of 473 atoms. The elementary time step used was $\tau = 0.01\sqrt{m\sigma^2 / \varepsilon}$, where m is the mass of an argon atom. Pressure was not monitored.

A first run of 300 time steps with removal of kinetic energy resulted in a glass at T* = 0.00005 (the reduced temperature is given by T* = kT/ε, where k is Boltzmann's constant and T the absolute temperature). For this array, $\langle r^2(t) \rangle$ was 0.007 averaged over the moving atoms only, and did not increase. Further runs at T* = 0.05 and 0.15 were made, the system remaining a high pressure confined glass.

The array was then linearly expanded by a factor of 1.11 (σ = 1.11d) and the system run at T* = 0.13 for a total of 4370τ. This is data set 1: as shown in Table 1, $\langle r^2(t) \rangle$ showed non-increasing behaviour, indicating the array was still glassy. Taking as starting point the T* = 0.13 array after 670τ, a run of 4109τ at T* = 0.33 was performed. Again, Table 1 shows the $\langle r^2(t) \rangle$ behaviour to be still glassy. This is data set 2.

To obtain the liquid aggregate at T* = 0.36 (data set 3) the T* = 0.33 array after 537τ was expanded by a factor of 1.0121, heated to T* = 0.45, and run at constant energy for 2843τ. This gave the T* = 0.36 data (set 3), whose increase $\langle r^2(t) \rangle$ behaviour demonstrated its liquid nature. Because this step involved both a volume and a temperature increase, it is not possible to say which factor was responsible for the transition. We know only that it occurred, and that it could be due to either or both of these factors.

ACKNOWLEDGEMENTS

The work of Charles H. Bennett was done in the Solid State Science Division, Argonne National Laboratory, Ill. 60439, under the auspices of the USERDA.

REFERENCES

1. Bernal, J.D., _Proc. Roy. Soc._ _A280_, 299 (1964.)
2. Scott, G.D., _Nature_ _188_, 908 (1960); _194_, 958 (1962); _201_, 382 (1964).
3. Finney, J.L., _Proc. Roy. Soc._ _A319_, 495 (1970).
4. Cohen, M.H. and D.J. Turnbull, _Nature_ _203_, 964 (1964).
5. Bennett, C.H., _J. Appl. Phys._ _43_, 2727 (1972).
6. Finney, J.L., _Proc. Roy. Soc._ _A319_, 479 (1970).
7. Cargill, A.S., _J. Appl. Phys._ _41_, 12 (1970).
8. Leung, P.K. and J.G. Wright, _Phil.Mag._ _30_, 185 (1974); _30_, 995 (1974).
9. Bernal, J.D. in _Liquids: Structure, Properties, Solid Interactions_ ed. T.J. Hughel, Elsevier, Amsterdam, 25 (1965).
10. Rahman, A., _J. Chem. Phys._ _45_, 2585 (1966).
11. Barker, J.A., M.R. Hoare, and J.L. Finney, _Nature_ _257_, 120 (1975).
12. Finney, J.L., Proc. Second Int'l Conf. on Rapidly-Cooled Metals, M.I.T., 1975, SII, Mat. Sci. Eng. _23_ (1976).

INFLUENCE OF LIQUID STRUCTURE
ON GLASS FORMING TENDENCY

Piotr G. Zielinski and Henryk Matyja

Materials Science Institute
Warsaw Technical University, Warsaw

INTRODUCTION

A method for rapid quenching from the liquid state (splat cooling) makes it possible to obtain new non-crystalline phases in many easily crystallizing systems.[1,2] There is no absolute criterion for glass formation, and it has been suggested by Cohen and Turnbull[3] that any liquid will form a glass if cooled at a rate sufficiently high to prevent crystallization. However, some undercooled liquids have a greater tendency towards glass formation than other ones. To explain this phenomenon, several models for the structural or kinetic aspects of the transition between the melt and the amorphous solid have been developed (for reviews see References 1 and 4).

According to the kinetic models, the cooling rate, concentration of nuclei, rate of crystal growth, etc., are important factors which affect the glass forming tendency of different liquids.[5-7]

At high cooling rates one can expect the liquid structure to exert an effect on the kinetics of crystallization and thus on glass formation.

This is consistent with experimental findings indicating that, in some materials, glass formation is related to the electrical conductivity in the liquid state.[8]

In this paper, glass formation is considered mainly from the standpoint of nucleation and growth processes, as well as their dependence on liquid structure.

LIQUID STRUCTURE

In order to discuss liquid alloys from a fundamental standpoint, three partial interference functions are required: a_{AA}, a_{AB} and a_{BB}.[9] To ob-

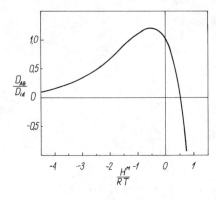

Figure 1. Ratio of the inter-diffusion constants for regular and ideal solutions as a function of the heat of mixing (explanations in the text).

tain these functions, three independent diffraction experiments have to be performed, e.g., using neutrons, on isotopically enriched samples of the same alloy.[10] Therefore, detailed diffraction studies of the liquid structure in binary systems are scarce. However, some information about the structure of liquids can be deduced from measurements of certain structure-sensitive physical properties, e.g., electrical conductivity, Hall effect, viscosity and thermodynamic relationships.

A number of liquid binary alloys ($A_{1-c}B_c$) show abnormal transport properties as a function of c at the actual compound composition A_mB_n.[9,11,12] At this composition, the electrical resistivity and Hall coefficient have maximum values and the thermoelectric power vanishes between the maximum and minimum. Hodgkinson[13] and Cohen and Sak[14] have pointed out the importance of the formation of A_mB_n-type clusters in this phenomenon. According to this model, it is assumed that, in liquid alloys, some of the atoms (or, rather, ions) are distributed randomly, while other atoms associate into small groups. The composition of these groups is assumed to be nearly the same as that of compound A_mB_n. Clusters, i.e., associated groups of atoms, can be considered to be similar to a molecule of composition A_mB_n, with a "flexible" structure in which the interatomic distance between unlike ions is kept nearly constant, whereas the bonding angle is variable over a wide range.[15] Thus, in the first approximation, the system can be described as consisting of liquid metal A or B, containing A_mB_n-type clusters. On the basis of this model, Cohen and Jortner[16] have obtained good agreement between the experimental data and the inhomogeneous transport regime interpretation of electronic transport properties of liquid Hg and Te, as well as of two component systems (e.g., some metallic alloys and metal-ammonia solutions).

GLASS FORMING ABILITY

Glass forming ability (GFA) is generally defined as the ratio of glass-transition temperature T_g, to melting temperature T_m. However, the real glass-transition temperature characterizes the system only under specific, defined experimental conditions; it increases with a rise of the cooling rate.[17] Thus, expressions describing the GFA in terms of T_g (e.g., T_g/T_m; $(T_x-T_g) / (T_m-T_x)$,[18] where T_x = temperature of onset of crystallization of glass) cannot readily be used to compare the tendency of different liquids to form glasses.

For the purpose of this work it was found necessary to express the GFA in terms of the time required to produce a barely detectable crystallized fraction X. This approach, developed by Uhlmann[6] and Davies,[19] is based on the theories of crystal nucleation, growth, and transformation kinetics.

A formula for expressing X as a function of time and temperature, assuming diffusion-controlled growth and constant nucleation and growth rates is given by:[20]

$$X = \frac{4}{3} \pi \int_0^t r^3 \, I \, dt, \tag{1}$$

where I is the rate of nucleation, $r = SD^{1/2}t^{1/2}$ from the theory of diffusion-controlled growth,[21] S = supersaturation term, D = diffusivity, and t = time during which the nuclei grow to size r.

The stationary nucleation rate is generally expressed as:

$$I = Z\beta S*N* \tag{2}$$

where β is the impingement flux of atoms on the critical nucleus surface S*, N* = number of critical nuclei per unit volume, and Z = Zeldovich factor which expresses the departure of the actual concentration and the concentration gradient of critical nuclei from the equilibrium values.

Assuming that β represents the total flux of atoms diffusing through the surface of a sphere of radius $r = SD^{1/2}t^{1/2}$, the following expression can be obtained, according to the theory of diffusion-controlled growth:[21]

$$\beta = 2\pi q S^3 D^{3/2} t^{1/2}, \tag{3}$$

where q is a constant.

Thus it follows that the time t is given by:

$$t = \left(\frac{9X}{8\pi^2 q S^6 ZS*}\right)^{1/3} \left(D^3 N*\right)^{-1/3} \equiv K\left(D^3 N*\right)^{-1/3}. \tag{4}$$

Concerning the GFA in binary systems, the following model of freezing is assumed to simplify the calculations: Independently of its composition, a liquid can be undercooled to any constant temperature $T < T_S$ (T_S = solidus temperature), at which the crystallization of the compound $A_m B_n$ is assumed to take place as a single phase. $S*$ and Z are composition-independent constants, D is replaced by the inter-diffusion constant D_{AB}.

With these assumptions, it is possible to analyze the GFA theoretically, provided that changes in the inter-diffusion constant and density of nuclei as a function of the composition are known, this is discussed further below.

Diffusion

According to the Darken analysis of diffusion processes,[22] the inter-diffusion constant can be expressed as:

$$D_{AB} = \left[c_A D_B(c) + c_B D_A(c)\right] \left[1 + \frac{\partial \ln \gamma_A(c)}{\partial \ln c_A}\right], \tag{5}$$

where $D_i(c)$ is the diffusion coefficient and γ_A the activity coefficient. For a regular solution, equation (5) can be transformed[23] as follows:

$$D_{AB} = D_{id} \left[1 - 2 \frac{H^M}{RT}\right] \exp \frac{H^M}{RT}, \tag{6}$$

where D_{id} denotes the inter-diffusion constant for an ideal solution, H^M = heat of mixing, T = absolute temperature and R = gas constant. Considering this equation to be applicable to liquid binary systems, on the assumption that D_{id} is composition-independent (D_{AB} / D_{id} versus H^M / RT is plotted in Figure 1), it is evident that the inter-diffusion constant proceeds towards minimum values in different ways depending on the sign of the heat of mixing, i.e., for repulsive or attractive interactions between unlike ions in the liquid.

Liquids with repulsive interactions ($H^M > 0$), undergo a spontaneous phase decomposition within a certain temperature range. The critical temperature of spinodal decomposition T_c = $2H^M$ / RT corresponds to the composition at which the sign of the inter-diffusion constant is inverted, with occurrence of up-hill diffusion. For such systems, glass formation is described in the literature[24] and theoretical predictions agree with the experimental data for oxide systems.[25]

For liquids with attractive interactions between unlike ions ($H^M < 0$), the inter-diffusion constant attains a maximum value at H^M / RT = -0.5 and tends to zero at H^M / RT $\rightarrow -\infty$. In the case of a binary system, the minimum heat of mixing appears at the cluster composition, and at this composition the inter-diffusion constant reaches a minimum.

Nucleation
***

The role of fluctuations in liquids and their contribution to nucleation has not been fully assessed so far, mainly because there are no experimental means of testing the fluctuation theory in liquids under normal conditions. However, if the cluster model of liquid structure is adopted, the clusters can be considered to result from compositional fluctuations. Statistically, cluster lifetimes are not appreciable; however, they increase with a drop in temperature.

Therefore, if the liquid is in metastable equilibrium under conditions of supercooling, clusters can be regarded as embryos.

Concerning the nucleation in a binary system, with cluster formation in the liquid state, two pseudobinary systems $A-A_mB_n$ and $B-A_mB_n$ can be considered for the sake of simplicity. In the first of these

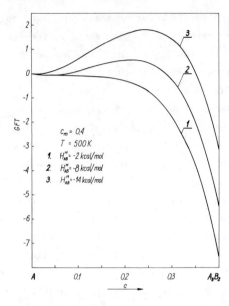

Figure 2. Glass forming tendency (GFT) of the pseudobinary system $A - A_3B_2$ for different values of of the heat of mixing of the liquid compound A_3B_2 $\left[H_{AB}^M\right]$.

systems, nuclei of two kinds can be formed: element A and A_mB_n compound. Dilute alloys exhibit predominant nucleation of the A-phase, and concentrated alloys show nucleation of A_mB_n. Since the A_mB_n-embryos exist in the liquid state, nucleation of this phase is easier. At high cooling rates, only the nucleation of A_mB_n has to be considered, particularly when element A is assumed to increase the glass-forming tendency.

It is very difficult to determine the concentration ranges within which the individual kinds of nucleation are dominant. Therefore, it is assumed that the A_mB_n nucleus concentration changes exponentially as a function of the composition:

$$N^* = a \ \exp(bc^{m+n}), \qquad (7)$$

where a and b are constants. This approximation indicates that the

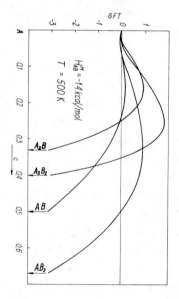

Figure 3. GFT for various cluster types.

nucleation rises within the composition range $p \leq c \leq n/(m+n)$, where $c = p$ is the composition at which the nucleation of $A_m B_n$ compound begins to predominate.

If one takes the boundary conditions in equation (7) as $N* = 1$ for $c = 0$ and $N* = N/(m+n)$ for $c = n/(m+n) = c_m$ (in this case clusters contain only B atoms), the GFA defined in terms of the time t required to produce a barely detectable crystallized fraction has the following form [equations (4), (6), (7)]:

$$t = \frac{K}{D_{id}} \left[1 - 2\frac{H^M}{RT} \right]^{-1} \exp \left[-\frac{H^M}{RT} - \frac{1}{3} \left(\frac{c}{c_m} \right)^{m+n} \ln \left(\frac{N}{m+n} \right) \right], \qquad (8)$$

If it is assumed for the sake of simplicity that D_{id} is constant and composition-independent, and that the heat of mixing changes linearly with composition from zero at $c=0$ to $H^M = H^M_{AB}$ at $c = c_m$, the following equation is obtained:

$$\log t_o \equiv \log \frac{t D_{id}}{K} = -\log \left[1 - 2\frac{c H^M_{AB}}{c_m RT} \right] - 0.434 \frac{c H^M_{AB}}{c_m RT} - 0.145 \left(\frac{c}{c_m} \right)^{m+n} \ln \left(\frac{N}{m+n} \right) \quad (9)$$

In this case, t_o expresses the glass forming tendency (GFT) which represents the GFA (defined in terms of the time required to produce a barely-detectable crystallized fraction) of a liquid alloy referred to the GFA of liquid element A. This results from the assumption that (constant) K/D_{id}, which cannot be determined experimentally, is replaced by the GFA of the pure element. Plots of $\log t_o$ (GFT) versus c, calculated for various values of the parameters H^M_{AB} and $c_m = n/(m+n)$, are shown in Figures 2 and 3. In all calculations, the temperature T to which a liquid can be undercooled is assumed to be constant ($T = 500K$). This value was chosen in view of the further discussion of the GFA as applied to binary tellurium systems.

From equation (9) and Figures 2 and 3 it follows that glass formation is easier when:

    a] the energy of interaction between unlike ions in the liquid state increases, as expressed by a rise of H^M_{AB} (cf. Figure 2), and b] the undercooling of the liquid increases.

The above considerations also explain the slightly lower GFA of dilute alloys as compared to the pure elements A or B. For these alloys, the GFA mainly depends on the rate of crystal growth, and hence on the inter-diffusion constant. Moreover, it can be assumed that the GFA curve exhibits two maxima on both sides of the cluster composition (if two pseudobinary systems $A-A_mB_n$ and $B-A_mB_n$ are considered); the height of these maxima depends on the GFA of the pure elements A and B.

The model presented above is in agreement with Chen[26,27] who has shown that the short-range order which is due to the interaction between unlike ions, influences the glass-transition temperature, as well as with the experimental finding that nearly all liquids quenched to glassy alloys are near low-lying eutectics.[2,28]

EXPERIMENTAL VERIFICATION OF THEORETICAL ANALYSIS

To verify the theoretical analysis of the GFT given above, experimental data concerning the glass formation and electrical conductivity in liquid binary tellurium systems are presented. These systems were chosen because many experimental results are available in the literature.

Data on regions of glass formation in binary tellurium alloys, obtained by splat-cooling (gun technique) or water-quenching, are presented in the Table. Cluster types determined from the minimum electrical conductivity are also recorded in the Table.

The electrical conductivity data for the liquid systems under consideration are in a good agreement with the values calculated using an inhomogeneous transport regime model,[16] except for the Te-Cu system, for which the existence of clusters also has not been confirmed by neutron diffraction experiments.[4,42] It is suggested that in this system clustering is complicated and cannot be described by a simple model of liquid tellurium containing Cu_2Te clusters.

In all systems which exhibit clustering in the liquid state glass formation is observed at high cooling rates. The glass formation regions are the composition range between pure tellurium and the cluster composition. In the case of the Te-In system, t_o can be calculated

Table. Experimental Data Concerning Clustering in Liquid Tellurium
Binary Alloys and Their Structure after Rapid Quenching

System	Quenching method[x]	Glass formation region	Cluster type defined by electrical conductivity measurements		
Te-A	–	at % A	$A_m B_n$	c_m (at % A)	
Te-Cu	SQ	20-40 [29]	Cu_2Te	67	[30]
Te-Ag	SQ	33-40 [31]	Ag_2Te	67	[30]
Te-Au	SQ	π phase only [31]	no clustering		[32]
Te-Ga	SQ	10-30 [33]	Ga_2Te_3	40	[34]
Te-In	SQ	10-30 [33]	In_2Te_3	40	[35]
Te-Tl	SQ	15-40 [36]	Tl_2Te	67	[37]
Te-Ge	WQ	15-20 [38]	$GeTe_2$	33	[34]
	SQ	10-25 [38]	$GeTe_2$	33	[34]
Te-Sn	SQ	10-25 [39]	$SnTe$	50	[30]
Te-Pb	SQ	7-25 [39]	$PbTe$	50	[34]
Te-As	WQ[xx] a. 20-70 ⎫ b. 25-35; 45-70 [40] c. 45-55 ⎭		As_2Te_3?	no data	

as follows. According to electrical conductivity measurements in liquid
Te-In alloys,[35] the clusters occur in the form of In_2Te_3-pseudomole
cules, hence $c_m = 0.4$. If it is assumed that the heat of mixing is
temperature-independent, then the value $H^M_{AB} = -8kcal / mol$[43] can be accepted. Assuming that a liquid can be undercooled about 200K below the
solidus temperature ($T_s = 700K$), t_o can be calculated from equation (9).
This curve is presented in Figure 2, showing that the GFT for the Te-
In system appears to increase in the composition range of 10-28at.%In;
this is in good agreement with the experimentally determined glass formation region of 10-30at.%In.[33] Similar calculations can be performed
for all tellurium systems, if the values of H^M_{AB} are known. These considerations indicate that the real glass formation regions can be pre-

dicted using the model presented.

In the case of the Te-Au system, no glass formation is observed[31] ;
by splat quenching only the metastable crystalline π phase (α - Po -
type structure) can be obtained. In this system no clustering is observed,[32,44] and therefore glass formation requires cooling rates higher
than those necessary to obtain glassy tellurium.[45]

The GFA of the Te-As system can also be explained in terms of the
presented model. If we assume that in liquid Te-As alloys As_2Te_3 -
type clusters occur (data concerning the properties of the liquid
Te-As system are not yet available in the literature), the observed
minimum of the GFA near the As_2Te_3 compound[40] results from the action
of clusters as embryos, especially at low cooling rates.

According to the theoretical analysis of the GFT, glasses obtained
by rapid quenching must exhibit a structure corresponding to the liquid structure, i.e., in binary systems the glass can be regarded as
consisting of non-crystalline element A or B containing disordered
clusters. This conclusion is supported by measurements of the radial
distribution function in the liquid and glassy alloy Te + 17.5at.%Ge.[46]
These measurements indicate that the RDF remains unchanged, if the
density changes between liquid and glass are taken into consideration.
Moreover, certain conclusions about the structure of glass can be drawn
from investigations of the glass crystallization, namely that glass
must crystallize in two stages, during which element A crystallizes
first, and the remaining matrix subsequently. The reciprocal process
is also possible, depending on the thermodynamic conditions of crystallization. This type of crystallization has been observed in splat-
cooled Te-Ge[38] and Cu-Zr[47] alloys.

The presented model can also be used to describe the GFT in some
metallic systems, which will be presented in later papers.

CONCLUSIONS

1. In easily crystallizing binary systems, and at high cooling
   rates, glass formation results from strong attractive interactions between unlike ions which are responsible for clustering in the liquid state.

2. In liquid binary alloys clusters lead to a decrease in the rate of crystal growth and act as embryos; thus the GFA has two maxima within a composition range between the pure element and the cluster composition.

3. Binary glasses obtained by rapid quenching from the liquid state should be "di-phasic".

ACKNOWLEDGEMENTS

The authors are greatly indebted to Dr. M. Lasocka for drawing our attention to this exciting problem. We are also grateful to Professor N.J. Grant for valuable remarks and to J. Jezierska, M.Sc., and F.P. Dabkowski, M.Sc., for helpful discussion. This work was supported by the U.S. National Science Foundation under Grant GF 42176.

REFERENCES

1. Jones, H, Rep. Prog. Phys. 36, 1425 (1973).
2. Giessen, B.C., C.N.J. Wagner in Liquid Metals, Chemistry and Physics, ed. S.Z. Beer, Dekker, p. 633 (1972).
3. Turnbull, D., M.H. Cohen in Modern Aspects of the Vitreous State, Butterworth, Vol. I, p. 1 (1960).
4. Turnbull, D., J. Physique 35, C4-1 (1974).
5. Turnbull, D., Contemp. Phys. 10, 473 (1969).
6. Uhlmann, D.R., J. Non-Cryst. Solids 7, 337 (1972).
7. Lasocka, M. and H. Matyja, Arch. Hutnictwa 18, 247 (1973).
8. Krebs, H., J. Non-Cryst. Solids 1, 455 (1969).
9. Faber, T.E., An Introduction to the Theory of Liquid Metals, Cambridge University Press (1972).
10. Enderby, J.E., D.M. North, P.A. Egelstaff, Phil. Mag. 14, 453 (1966).
11. Allgaier, R.S., Phys. Rev. 185, 227 (1969).
12. Enderby, J.E. in Proc. Internat. Conf. Band-Structure Spectroscopy of Metals and Alloys, ed. D.J. Fabian & L.M. Watson, Academic Press, p. 609 (1973).
13. Hodgkinson, R.J., Phil. Mag. 23, 673 (1971).
14. Cohen, M.H., J. Sak, J. Non-Cryst. Solids 8-10, 696 (1972).
15. Takeuchi, S., O. Uemura, S. Ikeda, Sci. Rep. RITU A25, 41 (1974).
16. Cohen, M.H., J. Jortner, J. Physique 35, C4-345 (1974).
17. Turnbull, D. in Physics of Non-Crystalline Solids, ed. J.A. Prins North-Holland, p. 41 (1965).
18. Hruby, A., Czech. J. Phys. B22, 1187 (1972).
19. Davies, H.A., J. Non-Cryst. Solids 17, 266 (1975).
20. Hammel, J.J. in Nucleation, ed. A.C. Zettlemoyer, Dekker, 489 (1969).
21. Frank, F.C., Proc. Roy. Soc. A201, 586 (1950).
22. Darken, L., Trans. Met. Soc. AIME 175, 184 (1948).
23. Borovskiiy, I.B., K.P. Gurov, I.D. Marchukova, Yu.E. Ugaste, Protsessy vzaimnoiy diffuzii v splavakh, Nauka, 305 (1973).
24. Cahn, J.W. and R.J. Charles, Phys. Chem. Glasses 6, 181 (1965).

25. Levin, E.M in Phase Diagrams, ed. A.M. Alper, Academic Press, III, 144 (1970).
26. Chen, H.S., J. Non-Cryst. Solids 12, 333 (1973).
27. Chen, H.S., Acta Met. 22, 1505 (1974).
28. Duwez, P., Trans. Met. Soc. ASME 60, 605 (1967).
29. Calka, A., M. Wloczewska, H. Matyja, Met. Obr. Cieplna 9, 14 (1974).
30. Dancy, E.A., Trans. Met. Soc. AIME 233, 270 (1965).
31. Luo, H.L., W. Klement, J. Chem. Phys. 36, 1870 (1962).
32. Lee, D.N., B.D. Lichter in Liquid Metals, Chemistry and Physics, ed. S.Z. Beer Dekker, 81 (1972).
33. Anantharaman, T.R. and C. Suryanarayana, J. Mater. Sci 6, 1111 (1971).
34. Valiant, V.C. and T.E. Faber, Phil. Mag. 29, 571 (1974).
35. Ninomiya, Y., Y. Nakamura, M. Shimoji, Phil. Mag. 26, 953 (1972).
36. Anseau, M.R., J. Appl. Phys. 44, 3357 (1973).
37. Mott, N.F., E.A. Davis, Electronic processes in non-crystalline materials, Clarendon Press (1971).
38. Takamori, T., R. Roy, G.J. McCarthy, Mat. Res. Bull. 5, 529 (1970).
39. Kaczorowski, M. [private communication].
40. Cornet, J. and D. Rossier, J. Non-Cryst. Solids 12, 61 (1973).
41. Enderby, J.E., J. Physique 35, C4-309 (1974).
42. Hawker, I., R.A. Howe, J.E. Enderby in Proc. 5th Internat. Conf. on Liquid and Amorphous Semiconductors, 85 (1974).
43. Maekawa, T., T. Yokokawa, K. Niwa, J. Chem. Thermodynam. 4, 153 (1972).
44. Predel, B., J. Piehl, Z. Metallkde 63, 63 (1972).
45. Davies, H.A. and J.B. Hull, J. Mater. Sci. 9, 707 (1974).
46. Nicotera, E., M. Corchia, G. de Giorgi, F. Villa, and M. Antonini, J. Non-Cryst. Solids 11, 417 (1973).
47. Vitek, J.M., J.B. Vander Sande, and N.J. Grant, Acta Met. 23, 165 (1975).

COMPARISONS BETWEEN
VAPOUR- AND LIQUID-QUENCHED
AMORPHOUS ALLOYS

M.G. Scott

Applied Sciences Laboratory
University of Sussex
Brighton, U.K.

R. Maddin

Department of Metallurgy and
Materials Science
University of Pennsylvania
Philadelphia, U.S.A.

INTRODUCTION

Amorphous metallic phases may be prepared by rapid solidification (splat-quenching),[1] thermal evaporation,[2] sputtering,[3] chemical (electroless) deposition,[4] and electrodeposition.[5] Despite the inherent differences between these techniques, few critical comparisons have been made between the structure and properties of the resultant phases and suggestions for a classification of amorphous solids[6] according to mode of preparation have gone unheeded. In this paper we concentrate on the two most used routes, liquid → solid and vapour → solid, by comparing the structure, formation criteria and crystallisation behaviour of representative phases. With one possible exception, no case has been reported of the same amorphous alloy obtained by both liquid and vapour quenching. We therefore describe briefly the results of some attempts to produce similar phases by two routes.

FORMATION CRITERIA

Many pure metals and semi-metals form non-crystalline phases by vapour-deposition onto substrates at cryogenic temperatures[3] although the stability of these phases is frequently the result of small quantities of trapped gaseous impurities. By contrast, rapid quenching of the liquid until recently had been unsuccessful in producing analogous phases; the observation of glassy regions in splat quenched Ni foils[7] may change this, although again, impurity stabilisation cannot be ruled out. Many splat-quenched amorphous alloys consist of one or more noble or transition metals and about 15–25at.% of metalloid(s), e.g., Pd-Si, Fe-P-C.[1]

249

These compositions invariably are close to deep eutectics in the corresponding equilibrium diagram. However, the presence of a metalloid is clearly not essential since similar phases can be obtained in alloys of an early and a late transition metal, e.g., Cu-Zr, Ni-Nb.[1] The wide glass-forming ranges in these alloys (25 to 65at.% Zr in Cu) extend well away from the eutectic and even include the stoichiometry of intermediate phases. Non-crystalline phases stable at room temperature may also be prepared by the co-evaporation of many noble and transition metals, e.g., Co-Au,[2] Au-Ni.[8] These phases form most readily near the equiatomic composition and, as suggested by Mader,[2] the essential criteria for their formation appear to be a solid state immiscibility and a large (>10%) difference in the atomic radii of the constituents. Splat-quenching of these same alloys usually produces only metastable solid solutions. Additionally, several metal-metalloid systems may be produced, in particular Ni-P[9] which yields a non-crystalline phase by evaporation (as well as by electro- and chemical-deposition), but, despite the deep eutectic, fails to do so on splat-quenching.

To form a glassy phase by liquid-quenching the alloy must cool through the temperature range T_m-T_g (where T_m is the melting point and T_g the liquid-glass transition temperature) without detectable crystallisation. Assuming the maximum possible cooling rate the factors which promote glass formation will be those which either minimise the interval T_m-T_g or reduce the crystallisation rate. It is not surprising, therefore, that such phases predominate near eutectics since the associated stabilisation of the liquid depresses T_m and so reduces the interval T_m-T_g. The effects of alloy composition on T_m and T_g are complicated, but in general the existence of strong bonding between the constituent atoms tends to increase T_g and reduce T_m.[10] The role of atomic sizes is less well defined; the increased entropy of mixing associated with hard spheres of different sizes would be expected to raise T_g, but much larger reduction of T_m will lead to an overall decrease in T_m-T_g.[10] Probably the most important role of atomic size differences is in the stabilisation of amorphous phases, in both liquid- and vapour-quenched alloys. From kinetic considerations, Uhlmann[11] suggested that for a low rate of crystal growth (and therefore a low critical cooling rate for glass formation) the liquid alloy should have a high activation energy for viscous flow, E_η, and a rapidly changing viscosity with temperature, $d\eta/dT$. It is significant that those metals which have the highest E_η, namely the noble and transition metals, are those on which many splat-quenched amorphous phases are based. Moreover, the criterion of a rapid $-d\eta/dT$ is exactly the condition at a deep eutectic.

Vapour-quenched amorphous phases are prepared by deposition onto a substrate at some temperature well below T_g. Since at substrate temperatures below $1/3T_m$ it is doubtful whether the liquid phase is ever involved[12] the condensation process is strictly vapour $\rightarrow$ solid. Consequently at no time does the liquid alloy exist in the temperature range T_m-T_g and the arguments discussed above for liquid-quenching are inapplicable. When a stream of vapour atoms arrives at the substrate, temperature T_s, an equilibrium is set up between adsorption and desorption. Since for metal atoms the thermal equilibration time is considerably shorter than the average desorption time virtually all the atoms are adsorbed. After equilibration an adsorbed atom is free to diffuse on the substrate until it is jammed in position, presumably by the arrival of other atoms. The average distance $d(t)$ moved by the atom in time t may be given by[3]:

$$d(t) = \sqrt{2t\, D_s}$$

where D_s, the surface diffusion coefficient, is given by[3]:

$$D_s = a^2\, \nu\, \exp - \left(\frac{Q_d}{kT_s}\right)$$

where a is the atomic jump distance $(\sim 2\text{Å})$, $\nu = \frac{h}{k_B T}$ and Q_d is the activation energy for surface diffusion. It has been suggested[24] that for a deposition rate R Å / sec an adsorbed atom is free to move for a time a/R sec. Typical values of $d(t)$ at liquid nitrogen and room temperatures for an evaporation rate of 5 Å / sec are given in the Table on the following page.

The value of $Q_d \sim 0.2\text{eV}$ is typical for metals on NaCl and might represent the situation during the nucleation stage of a film; that of 0.6eV for metals on metals would be more representative of the later growth.

Table.

Q_d (eV)	T_s (K)	d (Å)
0.2	77	7.0×10^{-1}
0.2	300	2.7×10^{5}
0.6	77	5.7×10^{-14}
0.6	300	4.5×10^{1}

The important observation is that in both cases at 77K the adatom is able to diffuse less than an interatomic distance before it is jammed in position and the formation of a disordered structure is highly likely. It is in this context that the relative atomic sizes of the atoms may be important; the presence on the substrate of different size adatoms might both reduce the mobility and increase the distance that an individual atom would need to diffuse to form an ordered array.

STRUCTURE
___

Amorphous materials are characterised by diffraction patterns which consist of a few diffuse haloes. X-ray diffraction studies of numerous liquid-quenched amorphous alloys have shown that the interference functions calculated from coherent scattering intensity data have a first peak which is somewhat higher and narrower than those obtained from liquid metals, and a second peak which exhibits, to a varying extent, a shoulder on the high angle side.[1] The associated radial distribution functions show a splitting of the second peak which varies from alloy to alloy and in some cases, e.g., Au-Si, is barely detectable. The x-ray diffracted intensities from thin (usually <1000Å) films of vapour-deposited alloys are too low for structural studies. A problem of electron diffraction is that a high proportion of the incident radiation is scattered inelastically and accurate coherent diffraction data is difficult to deduce. However, the use of energy filters and scanning devices has enabled accurate RDF's to be determined for a number of vapour-deposited metals and alloys.[1] The most striking feature is their similarity to those obtained from liquid-quenched alloys. The splitting of the second peak, which is more pronounced in pure metals than alloys is invariably present. The ratios $r_1/r_2 = 1.65$

252

and $r_3/r_1 = 1.90$ (for cobalt), where r_n is the position of the nth peak in the distribution function, are in good agreement with those from splat-quenched alloys. Further comparisons may be possible by neutron diffraction which enables coherent diffraction data to be obtained to higher diffraction angles. The RDF's for sputtered Fe_2Tb[13] and splat-quenched $Fe-P-C$[14] are identical to those described above.

Electron micrographs of amorphous alloys are featureless. However, by forming images from the transmitted beam and part of the first dif-fracted halo (lattice imaging) it has been claimed that no coherently diffracting regions greater than 5Å are present in both splat-quenched $(Cu-Zr)$[15] and vapour-deposited $(Au-Ni$ and $Pd-Si)$[8] films.

The major test of structural models for amorphous metals and alloys is the generation of a split second peak in the atomic distribution function. Of the two extreme approaches, the microcrystallite and the continuous random, the latter appears to have gained most acceptance. Although Polk's[16] model, in which a dense random packing of metal atoms is stabilised by about 20% of smaller interstitial atoms, is a good ap-proximation for the metal-metalloid systems it is clearly inapplicable to the metal-metal systems obtained by both liquid- and vapour-quench-ing. More suitable to the vapour case seem to be those models based on a serial deposition of hard spheres. The model of Dixmier, Sadoc, and Guinier,[17] for example, gives an RDF which not only has a split second peak but correctly predicts the relative intensities of the subpeaks. Moreover, by considering the packing of non-equal sized spheres, the experimentally observed variation of the peak splitting with both al-loy composition and relative atomic size is predicted. The model fits very well the experimental RDF's from vapour-deposited transition metals at low values of $\sin \theta/\lambda$, although at higher angles the Bernal model is better. Clearly, a compromise model is needed. Models based on the 13 atom icosahedron may be particularly useful since not only does the RDF of just one cluster show the split second peak but it can be shown to be preferred growth form.[18]

Roy and co-workers[19] noted that the annealing behaviour of some chalcogenide glasses is particularly dependent on the preparation mode. We might expect to find a similar dependence in amorphous metals. A significant point is the absence of a glass transition phenomenon vapour-quenched films. It is perhaps worth noting that in several binary splat-quenched alloys the presumed T_g is obscured by crystallisation. Since the vapour-deposited alloys are usually binaries it is possible that a similar situation exists in their case. The absence of a T_g need not, therefore, be sufficient grounds for the rejection of these materials as glasses.

Except at very high heating rates the amorphous → crystalline transformation occurs through a series of metastable intermediates (Ostwald's rule). An exception is vapour-deposited alloys having the composition of a stable intermediate phase in which crystallisation occurs directly to that phase.[2] The resistivity vs temperature plots of vapour- and liquid-quenched phases are similar. In both cases the amorphous phase has a negative temperature coefficient of resistivity and crystallisation is marked by a sharp fall in the resistance to about 1/3 of its initial value. The rise in resistance prior to crystallisation often observed in splat-quenched alloys and claimed to arise from clustering[1] has not been reported in the vapour-quenched case. Whilst the value of T_c for vapour-deposited alloys is of the order of $0.3T_m$ the liquid-quenched phases seem to show a much wider scatter from about 0.45 to $0.70T_m$. Moreover, whereas the absolute crystallisation temperatures of the liquid-quenched alloys are quite sensitive to small changes of composition those of the vapour-quenched phases vary little over wide composition ranges. The existence of a well-defined crystallisation temperature is inconsistent with the continuous grain growth of an initially microcrystalline material and in both cases the crystallisation kinetics have been claimed to fit those calculated for crystallisation from a supercooled liquid[2,20]; diffusion appears to be the rate limiting process.

Three distinct crystallisation morphologies are possible: separation into two amorphous phases of different compositions which then crystallise independently; the homogeneous transformation of the amorphous matrix to a fine-grained single crystalline phase; the nucleation and growth of individual crystals in the amorphous matrix. The first, which is analogous to phase separation in oxide glasses, is displayed only by a few liquid-quenched ternary alloys on annealing close to T_g.[21] The second is exhibited by most vapour-quenched alloys irrespective of the heating rate. For example an amorphous Au-Co alloy first forms a homogeneous fcc phase which subsequently separates into the equilibrium crystal phases by a process resembling spinodal decomposition.[2] A similar morphology is obtained when liquid-quenched alloys are annealed for long times at temperatures well below their T_c; Fe-P-C alloys form a homogeneous bcc solid solution.[22] Although such a morphology is symptomatic of the coarsening of a microcrystalline structure it could equally well occur from growth following copious nucleation. High temperature (relative to T_c) isothermal annealing and continuous heating of liquid-quenched alloys usually gives rise to the growth of discrete crystals in an otherwise unchanged amorphous matrix. A typical sequence is that of $Pd_{80}Si_{20}$ in which small crystals of a Pd-rich phase are precipiated.[22] The amorphous matrix is then replaced by a complex silicide and ultimately the equilibrium Pd and Pd_3Si phases appear. Such behaviour, which is inconsistent with anything other than a nucleation and growth mechanism, is also displayed by vapour deposited alloys of composition close to that of an intermediate phase; for example at compositions close to 25% Mg isolated Au_3Mg crystals nucleate and grow in amorphous Au-Mg films.[2] Similarly, depending on composition, the Ni_3P phase grows from Ni-P films either as polyhedral crystals or with a dendritic morphology.[9]

CO-EVAPORATION OF Cu-Zr AND Pb-Au

Pb-Au alloys containing between 25 and 75 at.% Au and Cu-Zr alloys with from 40 to 70 at.% Zr were prepared by co-evaporation onto liquid nitrogen-cooled sapphire substrates. Pb, Au, and Cu were evaporated from

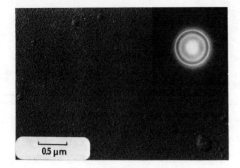

Figure 1: Cu$_{60}$Zr$_{40}$ deposited at 100K.

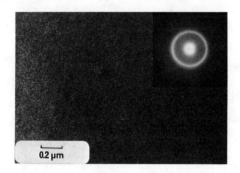

Figure 2: Pb$_{50}$Au$_{50}$ deposited at 100K.

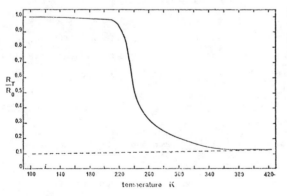

Figure 3: Resistance vs temperature curve for Pb$_{50}$Au$_{50}$ deposited at 100K.

256

resistance heated boats, whilst an electron-gun was used for Zr. Total evaporation rates were about 5Å/sec. Splat-quenching of these alloys yields non-crystalline phases at $Pb_{75}Au_{25}$[23] and $Cu_{75}Zr_{25}$ to $Cu_{35}Zr_{65}$.[1] They were chosen for evaporation in anticipation that their large atomic size differences (21.5 and 25% respectively) might promote amorphous phase formation.

In no cases were amorphous phases found at room temperature. The Cu-Zr films frequently showed a featureless microstructure (Figure 1a) but in some cases precipitates were present. In all cases the diffraction patterns consisted of two broad haloes and a set of outer rings which could be indexed on the basis of a fcc arrangement (Figure 1b). The microstructure of the Pb-Au films was dependent on composition; near 25 and 75at.% Pb the equilibrium phases were present; near the equiatomic composition, however, a fine-grained fcc phase was obtained (Figure 2). The resistance of the Pb-Au films was measured during their annealing from the deposition temperature to room temperature (Figure 3). Whereas the resistance of the $Pb_{75}Au_{25}$ film fell steadily with increasing temperature that of the equiatomic alloy remained almost constant and then fell suddenly at about 230K. This behaviour was taken as evidence of an amorphous phase in as-deposited $Pb_{50}Au_{50}$ alloys. The crystallisation temperature is approximately one third of the average melting point of the alloy, a result consistent with other vapour-deposited amorphous phases. It is interesting to note that the composition at which the amorphous phase is formed is different from that for splat-quenching, supporting the earlier observation that the formation criteria are different in the two cases.

CONCLUSIONS

Although the products appear structurally similar, the criteria for formation of amorphous phases by vapour- and liquid-quenching are different. The morphology of the crystallisation products depends on the preparation method. Equiatomic lead-gold alloys give an amorphous phase at liquid nitrogen temperatures.

ACKNOWLEDGMENTS

This work was supported by the N.S.F. through the Laboratory for Research on the Structure of Matter, University of Pennsylvania. We are grateful to Drs. C. Briant (Pennsylvania) and B. Cantor (Sussex) for useful discussions.

REFERENCES

1. Giessen, B.C. and C.N.J. Wagner: Liquid Metals, 633, S.Z. Beer, editor, Marcell Dekker, Inc., N.Y. (1972).
2. Mader, S.: J. Vac. Sci. Technol. 2, 35 (1965).
3. Chopra, K.L.: Thin Film Phenomena, McGraw-Hill, New York (1969).
4. Bagley, B.G. and D. Turnbull, J. Appl. Phys. 39, 5681 (1968).
5. Schlesinger, M. and J.P. Martin, J. Phys. Chem. Sol. 29, 188 (1968).
6. Roy, R., J. Non-Cryst. Sol. 3, 33 (1970).
7. Davies, H.A., J. Aucote, and J.B. Hull: Nature [Phys. Sci.] 246, 13, (1973).
8. Herd, S.R. and P. Chaudhari: Phys. Stat. Solidi 26(a), 267 (1974).
9. Bagley, B.G. and D. Turnbull, Acta Met 18, 857 (1970).
10. Chen, H.S.: Acta Met. 22, 1505 (1974).
11. Uhlmann, D.R.: J. Non-Cryst. Sol. 7, 337 (1972).
12. Behmdt, K.H.: J. Appl. Phys. 37, 3841 (1966).
13. Rhyne, J.J., S.J. Pickart, and H.A. Alperin: Amorphous Magnetism, 373, edited by H.O. Hooper and A.M. de Graaf, Plenum Press, New York (1973).
14. Mizoguchi, T., K. Yamauchi, and N. Numura, Japan J. Appl. Phys 14, 711 (1975)
15. Revcolevschi, A. and N.J. Grant: Met. Trans 3, 977 (1972).
16. Polk, D.: Acta Met 20, 485 (1972).
17. Sadoc, J.F., J. Dixmier, and A. Guinier, J. Non-Cryst. Sol. 12, 46 (1973).
18. Briant, C.L. and J.J. Burton [private communication].
19. Messier, R., T. Takamori, and R. Roy: J. Non-Cryst. Sol. 8-10, 816 (1972).
20. Srivastava, P.K., B.C. Giessen, and N.J. Grant: Met. Trans. 3, 977, (1972).
21. Chou, C.P. and D. Turnbull: J. Non-Cryst. Sol. 17, 169 (1975).
22. Masumoto, T. and R. Maddin, Mat. Sci. Eng. 19, 1 (1975).
23. Predecki, P., B.C. Giessen, and N.J. Grant: Trans. Met. Soc. AIME 233, 1438 (1965).
24. B. Cantor and R.W. Cahn, this volume.

THE KINETICS OF FORMATION OF METALLIC GLASSES AND ITS PRACTICAL IMPLICATIONS

H.A. Davies and B.G. Lewis

Department of Metallurgy
University of Sheffield, U.K.

INTRODUCTION

The recent development of glassy alloys based on iron and nickel, having potentially useful combinations of properties, and the parallel development of techniques for high-speed casting of the alloys as continuous glassy filament focusses particular attention on the problem of the critical cooling rate R_C for the formation of a glassy alloy. The lower R_C, the thicker the maximum possible section of glass for given conditions of heat removal; in some cases at least, the range and flexibility of applications of glassy alloys might be enhanced by increasing the maximum thickness.

The kinetic treatment of metallic glass formation, based on theories of homogeneous nucleation, crystal growth and transformation kinetics has been shown[1,2] to yield satisfactory theoretical estimates of R_C for several glassy metallic phases. In this approach the reduced glass transition temperature T_g/T_m, where T_m is the melting or liquidus temperature, is of crucial importance in determining the magnitude of R_C, as has also been emphasized previously.[3,4] In this paper the functional dependence on T_g/T_m of the theoretical R_C and corresponding maximum thickness of glassy phase are quantitatively derived and discussed.

DEPENDENCE OF CRITICAL COOLING RATE ON T_g/T_m

The time t in seconds for a fraction X crystallised as a function of reduced temperature T_r ($= T/T_m$) and reduced undercooling ΔT_r [$= (T_m-T)/T_m$] can be expressed thus:[2]

$$t \sim \frac{9.3\eta}{kT} \left\{ \frac{X\, a_o^9}{N_v^o\, f^3} \frac{\exp\left\{\dfrac{1.024}{Tr^3 \Delta Tr^2}\right\}}{\left[1 - \exp\left\{\dfrac{-H_f^m\, \Delta Tr}{RT}\right\}\right]^3} \right\}^{1/4} \tag{1}$$

where a_o is the mean atomic diameter, ΔH_f^m the molar heat of fusion, N_v^o the average number density, η the liquid viscosity at temperature T, and f is the fraction of sites at the crystal/liquid boundary where atoms are preferentially added or removed.

Using assumed viscosity-temperature curves for the undercooled liquid regime between T_m and T_g, most or all of which, depending on the metal or alloy, is not accessible to experimental measurement, Time-Temperature-Transformation curves can be constructed from Equation (1) expressing the time for a fraction crystallised $X = 10^{-6}$ as a function of T. $\left(X = 10^{-6}\right.$ is an arbitrarily small value below which crystallinity is considered to be undetectable.) R_c is then given approximately by the tangent to the nose of the T-T-T curve represented by temperature T_n and time t_n. Theoretical estimates of R_c were found to be in reasonable accord with corresponding experimentally derived estimates for Ni, Pd-18%Si, Pd-6%Cu-16.5%Si and for Au-13.8%Ge-8.4%Si[2] spanning a wide range of T_g/T_m (or T_c/T_m where no T_g is observed).

To investigate the effect of T_g/T_m on R_c a model alloy is chosen for convenience, having the following properties, considered to be typical average values for metallic glass forming systems: $a_o = 0.26\,\text{mm}$, $\Delta H_f^m = 12500$ J mole^{-1}, density of 9.0gcm^{-3} and a gram formula weight of 70. Two different viscosity models are considered:

[1] the liquid viscosity at T_m and the activation energy for fiscous flow assumed equal to the corresponding value for liquid Ni at its melting point. T_g/T_m is varied by keeping T_m fixed and changing T_g. This has an equivalent effect to varying T_m while keeping T_g and η at T_m fixed. See Figures 1(a) and 1(b).

[2] the liquid viscosity at T_m assumed to be a linear extrapolation of the Arrhenius plot for liquid Ni, T_g/T_m being varied by keeping T_g fixed and varying T_m, as has been assumed in the absence of experimental data for Pd-Si alloys.[1] See Figure 1(c).

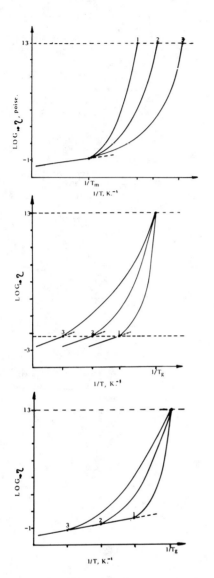

[A] Model [1]
The viscosity at T_m is held constant and T_g/T_m is varied by changing T_g.

[B] Model [1]
Variations in T_g/T_m are effected by changing T_m and keeping the viscosity at T_m constant. This has an equivalent effect to [A] (supra).

[C] Model [2]
Variations in T_g/T_m are effected by changing T_m but here the viscosity at T_m is a linear extrapolation of the Arrhenius plot for Ni above its T_m.

Figure 1. Viscosity-Temperature Curves for the Model Alloy (The sequence $1 \rightarrow 2 \rightarrow 3$ shows the effect of reducing T_g/T_m).

261

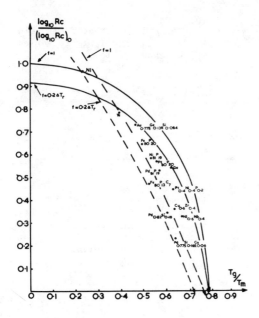

Figure 2. Variation of critical cooling rate with T_g/T_m for the model alloy. The solid lines are for model [1] and the broken lines for model [2]. The $\log_{10} R_c$ is normalised to $\log_{10} R_c$ (= 10.04) for model [1] as $T_g/T_m \rightarrow 0$. Also included are normalised values for selected real glass-forming alloys.

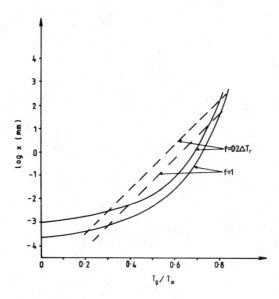

Figure 3. Variation of the theoretical maximum thickness of glassy phase for the model alloy. Solid and dashed lines refer to the same conditions as in Figure 2.

Using these two viscosity models, the critical cooling rates are
then calculated as a function of T_g/T_m, using Equation (1) for both
the rough (f = 1) and smooth (f = 0.2ΔT_r) interface conditions. These
curves are shown in Figure 2; they are all normalised to the value of
$\log_{10} R_c$ (= 10.04) for Model [1] for the conditions $T_g/T_m \rightarrow 0$ and f = 1
and hence the ordinate is converted approximately to absolute values
$\log_{10} R_c$ by using a scale factor 10. The viscosity model [2] gives a
lower cooling rate for a given $T_g T_m$ since for a given T_g/T_m and hence
T_n the viscosity is greater in case [2] than in case [1]. The normalised theoretical R_c values for some real metals and alloys are also
given in Figure 2; for most of the alloys, particularly those with
three components, no experimental liquid viscosity data above the respective T_m were available and values have been estimated from available data for pure solvent metals and trends shown in binary systems.
For alloys not having congruent melting temperatures, the liquidus temperature has been used as T_m and in cases where no T_g is observed the
temperature of onset of crystallisation, for the highest heating rate
for which data are available, has been taken as the lowest limit of T_g.
In each case, a value of f has been chosen, depending on the type of
crystal/liquid interface considered to be involved in the nucleation
and growth processes. The maximum thickness x of glassy phase, on the
assumed criterion of glassiness, can be expressed by:

$$x \sim \sqrt{D_\lambda t_n}$$

where D_λ is the thermal diffusivity of the liquid and t_n is the time
corresponding to the nose of the T-T-T curve. Assuming an average D_λ of
10 $mm^2 s^{-1}$, the calculated theoretical variation of x with T_g/T_m is given
for the model alloy in Figure 3 for the two viscosity models and interface conditions.

DISCUSSION

It is appropriate first to discuss errors involved both in the theory
and in the calculations. Some of the uncertainties arising from the
numerous assumptions in the theoretical arguments leading to Equation
(1) and from the general lack of experimental data required to predict

263

R_c for specific alloys have been discussed recently by the present authors.[2,5] Uncertainties arising in the first category affect the calculated values of R_c for both the model alloy and real alloys to the same extent but are rather difficult to quantify. Uncertainties in the experimental data which are mainly determined by uncertainties in the magnitude of liquid viscosity in the temperature range between T_m and T_n and in the type of interface involved in crystal growth relate only to the R_c for the real alloys. The <u>maximum</u> uncertainty in R_c arising from compounded errors in the data only is estimated to be about ±2 orders of magnitude but with a more realistic value being about ±1 order of magnitude.

In spite of the assumptions and the empirical way in which the viscosity curves are interpolated there is remarkably close agreement between the values of R_c predicted theoretically and those obtained from independent experimental estimates for pure Ni[7] $\sim$ 5 x 10^9 K/s and for Pd-6%Cu-16.5%Si[8] $\sim$ 100 K/s at the two extremes of T_g/T_m and cooling rates. The uncertainties in theoretical R_c values is better reflected however, in the intermediate cooling region where the Au-Ge-Si alloy has a theoretical R_c of $\sim 10^7$ K/s but can be splat-quenched completely to a glass by gunquenching whereas gunquenching foils of Ni-20%P, Pt-20%P and Pd-20%P are only partly glassy, in spite of having theoretical R_c values in the range 10^5-10^6 K/s.

The limitations discussed above notwithstanding the trends shown in Figure 2 can be considered to give a useful measure of the ease of glass formation for metallic alloys as a function of their T_g/T_m for various conditions of liquid viscosity and crystal/liquid interface. An approximate prediction of R_c is thus possible for any particular alloy of known T_g/T_m and providing its liquid viscosity is known or can be estimated. These theoretical curves introduce some rationality into the glass-forming behaviour of alloys in general and serve to emphasise that, when assessing quantitatively the critical cooling rate for glass formation of a particular alloy, T_m is equally relevant a parameter to T_g.

The positions of the theoretical points for the real alloys relative to the curves for the model alloy in Figure 2 depend largely on their
264

viscosities and on the interface conditions, i.e., smooth or rough, assumed in each case. In almost all cases considered, the points lie between the two sets of model alloy curves which is an indication that, at least for the alloys for which viscosities could be established with some certainty, the two viscosity models represent approximate limits of behaviour of alloy systems.

It should be emphasised that the theoretical curves relate only to the formation of <u>metallic</u> glasses where the liquid viscosities above their T_m fall within a fairly narrow regime around $1- 10^{-2}$ poise. For network glass formers, the liquid viscosities would generally be much higher and, although the general trend of R_c as a function of T_g/T_m would be similar, the curves would be displaced to much lower cooling rates for corresponding values of T_g/T_m. Critical cooling rates for some of these and other inorganic and organic glass-formers have been investigated by Uhlmann.[6]

It is clear from the foregoing that, when considering whether a metallic alloy is or is not a glass former, it is necessary to specify the thickness. It is interesting to note that for viscosity model [2] $log_{10}x$ is almost a linear function of T_g/T_m though this has no particular theoretical significance. Also, it should be borne in mind that all the curves apply to the model system chosen so that they only act as a guide to the values of x for particular alloys. Thus, by reference to Figures 2 and 3, it can be seen that for $T_g/T_m = 0.25$ where $f = 1$ and viscosity model [2] (approximating to pure Ni) x is $\sim$0.3μm whereas at the other extreme of $T_g/T_m = 0.64$ (approximating to Pd–6% Cu–16.5%Si) x is $\sim$4mm, roughly consistent with experimentally observed estimates of x.[1,7,8] In the intermediate range, Fe–13%P–7%C has a T_g/T_m, or more correctly T_c/T_m, of $\sim$0.52 and this gives x $\sim$300μm which again is very roughly consistent with experimental observation.

The highest value of T_g/T_m or T_c/T_m observed for any metallic glass thus far is $\sim$0.67 for Ni–40%Nb. As T_g/T_m increases beyond 0.7 it is predicted that critical cooling rates of 10 K/s and less and thus glassy thickness of 10mm and upwards would be possible. In network glasses, values of T_g/T_m up to 0.78 can occur but it is the very high liquid viscosties above their melting temperatures that have the greatest effect in promoting very low values of R_c, e.g., 2×10^{-4} K s^{-1} for SiO$_2$.[6] Since such high viscosities are unattainable in metallic systems, lower cooling rates for glassy alloy systems of practical importance would require increasing T_g/T_m by reducing T_m, raising T_g, or at least raising T_g significantly more than T_m. The latter two courses would, in practice, be more useful since not only does a given increase in T_g have a greater effect on T_g/T_m than a reduction in T_m of equal magnitude but an increased T_g also reflects an increased thermal stability of the glassy phase which would be additionally useful.

ACKNOWLEDGEMENTS

The authors are grateful to the Science Research Council for financial support and to Professor B.B. Argent for providing laboratory facilities.

REFERENCES

1. Davies, H.A., J. Aucote, and J.B. Hull, Scripta Met. 8,1179 (1974).
2. Davies, H.A., J. Non-Cryst. Sol. 17,266 (1975).
3. Turnbull, D., Contempt. Phys. 10,473 (1969).
4. Chen, H.S., Acta Met. 22,1505 (1974).
5. Davies, H.A. and B.G. Lewis, Scripta Met., [in press].
6. Uhlmann, D.R., J. Non-Cryst. Sol. 7,337 (1972).
7. Davies, H.A. and J.B. Hull, J. Mater. Sci. [in press].
8. Chen, H.S. and D. Turnbull, Acta Met. 17,1021 (1969).

CRYSTALLIZATION OF AMORPHOUS ALLOYS PREPARED BY ELECTROLESS DEPOSITION

W.G. Clements

Maharishi International University
Switzerland

B. Cantor

University of Sussex
Brighton, U.K.

INTRODUCTION

Crystallization from the amorphous state frequently involves the transient formation of a metastable single crystalline phase prior to decomposition to two or more equilibrium phases.[1,2] This paper describes a study of the complex crystallization behaviour of amorphous electroless-deposited Ni-based alloys, which was undertaken with the initial objective of obtaining each alloy in both amorphous and single-phase crystalline forms. Both forms of the alloys were required in order to compare their magnetic properties, as part of an extensive investigation into the magnetic properties of amorphous alloys.[3-6]

EXPERIMENTAL TECHNIQUE

Amorphous alloys were prepared by an electroless plating process in which metal ions are reduced, and deposited from solution onto a catalytic substrate surface.[3-6] The plating bath was fed continuously with alkaline plating solution at constant pH and temperature for periods of a few days, to produce deposits 0.1-1.0mm thick. The chemical composition of the solution varied considerably depending upon the alloy to be deposited, but usually included:

[1] the metal or metals to be deposited in the form of sulphates or chlorides;

[2] a complexing agent to hold the metal ions in solution prior to catalytic deposition;

[3] hypophosphite ion as a reducing agent, and to promote an amorphous deposit by incorporation of phosphorus;

[4] NH_4OH to provide an alkaline solution, and $(NH_4)_2SO_4$ as a buffer to stabilise the pH.

All chemical components of the solution were ANALAR grade, and the substrate was an etched rod of 99.99% pure Al rotated at 1 rpm during

267

TABLE : Electroless-Deposited Amorphous Alloys

SYSTEM	COMPOSITION (at%)
Ni-P	5P; 7P; 8P; 12P; 13P; 14P; 16P; 17P; 18P.
Ni-Co-P	13Co,20P; 22Co,17P; 26Co,22P; 27Co,21P; 47Co,20P;
	72Co,12P; 94Co,6P(zero Ni).
Ni-Co-Fe-P	35Co,1Fe,24P; 50Co,1Fe,17P; 36Co,4Fe,17P; 40Co,4Fe,18P

plating. After deposition, alloys were removed from the substrate by etching in warm caustic soda. Alloy compositions were determined by energy-dispersive x-ray microanalysis, and alloy structures were determined by x-ray diffraction in a Debye-Scherrer camera. The Table (supra) shows the alloys prepared by this technique and subsequently used to study crystallization behaviour. As-deposited Ni-5P and Ni-7P alloys showed faint traces of crystallinity but all the others were completely amorphous.

To study crystallization behaviour, a number of specimens of each alloy were heated between 300 and 800K at heating rates of 5-160K min^{-1} in a differential scanning calorimeter (DSC). Sample weights were between 1-15 mg. Once the crystallization characteristics had been evaluated, specimens suitable for subsequent x-ray measurements were heated in the DSC to significant temperatures within the range, to determine the structure obtained after a particular stage of the transformation. Some of the alloys were up-quenched at 320K min^{-1} and allowed to crystallize isothermally. The isothermal heat treatments were repeated over a range of temperatures to provide data for the kinetics of crystallization.

CRYSTALLIZATION BEHAVIOUR

Typical continuous heating DSC curves are shown schematically in Figure 1. For all alloys, transformation temperatures were dependent upon heating rate; for a change in heating rate of 40 to 10 K min^{-1}, the temperature of each transformation would fall by ~30K (Figure 2). Binary Ni-P alloys with <15 at.%P crystallized with 3 distinct and irreversible exothermic peaks. The first broad peak extended from ~480-600K and cor-

268

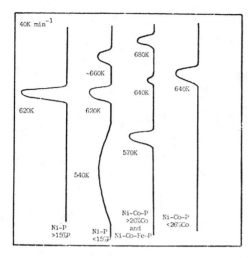

Figure 1. Typical DSC Curves.

responded to formation of a single phase metastable fcc solid solution of P in Ni. The other two peaks were sharper, extending over ~20K; with increasing P content, these peaks occurred at progressively lower temperatures and became closer together. Specimens annealed beyond either of these peaks exhibited complex x-ray patterns, which corresponded to a mixture of fcc Ni and one other phase. For alloys with >15 at.%P, the broad peak disappeared and the other two peaks merged to produce a single crystallization peak at ~620K which corresponded to direct formation of a two-phase structure.

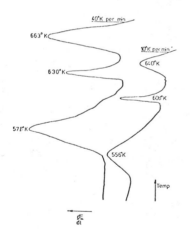

X-ray patterns from specimens annealed just below 620K confirmed that there was no fcc metastable solid solution. The above data are in broad agreement with previous work,[6,7] although the critical composition for the change in crystallization behaviour has been determined more precisely.

Despite their high P content, Ni-Co-P alloys with >26 at.%Co

Figure 2. The Effect of Heating Rate.

269

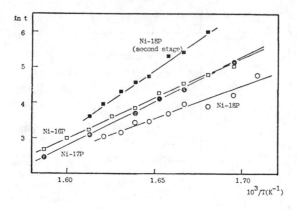

Figure 3. Time to initiate crystallization versus inverse temperature.

and all the Ni-Co-Fe-P alloys exhibited 3 irreversible exothermic peaks, with the first peak sharper than that observed in low P alloys of Ni-P. X-ray analysis showed that the first peak corresponded to an amorphous to single phase crystalline transformation. For these alloys the single crystalline phase was hcp with lattice parameter close to that of hcp Co. Therefore the presence of Co or Fe in Ni-P favours the formation of an intermediate metastable crystalline phase. Presumably, as the system becomes more complex, more diffusion and atomic rearrangement is required to form the equilibrium crystalline structure. The corresponding processes to form the single phase are not affected so much, and the intermediate formation of a metastable single crystalline phase becomes more favorable.

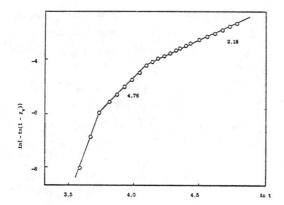

The remaining alloys investigated in the DSC were Ni-rich Ni-Co-P with high P content. In support of previous work,[6] these showed only 1 exothermic peak and transformed directly

Figure 4.
Fraction transformed as a function of time.

to a two phase mixture. As with the Ni-P alloys, one phase was fcc Ni, but the other phase could not be unambiguously identified and did not fit the expected diffraction lines of Ni_3P, Ni_7P_3, Ni_2P, or NiP_2.

CRYSTALLIZATION KINETICS

Most of the Ni-P and Ni-Co-P alloys were annealed isothermally in the DSC to investigate the kinetics of crystallization. For alloys which crystallized initially to a single metastable phase, the transformation was slow and it was not possible to obtain accurate kinetic data. For Ni-P alloys with >15 at.%P, crystallization was more rapid, and the incubation time was obtained as a function of temperature (Figure 3). From these data, the activation energies for initiation of crystallization were found to be 46.8, 49.2, and 38.6K cals mole^{-1} respectively for Ni-16,17 and 18P. For Ni-18P, isothermal crystallisation was in two almost superimposed stages, and the second stage had an activation energy of 70.0 K cals mole^{-1}. For this alloy, the DSC curve at 610K was integrated to obtain the fraction transformed (x) as a function of time. When plotted as $\ln[-\ln(1-x)]$ versus $\ln$ time (Figure 4), the two stages wer found to have exponents of 4.8 and 2.2, respectively. This suggests that the two stages correspond to the initial formation of single phase nuclei followed by diffusion-controlled growth for which the expected exponents would be 4 and 1.5.[8] Deviation from these expected values is probably due to systematic errors in determining the zero and baseline for the DSC curves.

The slow rates of transformation observed in low P alloys could well correspond to crystallization by microcrystallite grain growth as suggested by Bagley and Turnbull.[9] However, the present data indicate that single phase nuclei form as a precursor to dendritic growth in high P alloys.[9]

REFERENCES

1. Mader, S. and A.S. Nowick, <u>Acta Met.</u> <u>15</u>,215 (1967).
2. Masumoto, T. and R. Maddin, <u>Mat. Sci. Eng.</u> <u>19</u>,1 (1975).
3. Simpson, A.W. and D.R. Brambley, <u>Phys. Stat. Sol.</u> <u>43</u>,291 (1971).
4. ibid., <u>49</u> 685 (1972).

5. Clements, W.G., D. Phil. thesis, Sussex (1974).
6. Brambley, D.R., D. Phil. thesis, Sussex (1971).
7. Randin, J.P., P.A. Maire, E. Saurer, and H.E. Hintermann, <u>J. Elec-</u>
   <u>trochem. Sci</u>. <u>114</u>,442 (1967).
8. Burke, J., "Kinetics of Phase Transformations in Metals", Pergam-
   mon (1965).
9. Bagley, B.G. and D. Turnbull, <u>Acta Met</u>. <u>18</u>,857 (1970).

EFFECTS OF ALLOYING ELEMENTS ON STRENGTH
AND THERMAL STABILITY OF AMORPHOUS IRON-BASE ALLOYS

M. Naka, S. Tomizawa, and T. Masumoto

The Research Institute for Iron,
  Steel, and Other Metals
Tohoku University, Sendai, Japan

T. Watanabe

Department of Applied Physics
Tohoku University
Sendai, Japan

INTRODUCTION

Amorphous iron alloys quenched from the liquid state have high static
strength,[1,2] consistently high toughness,[3] and good resistance to chemical and pitting corrosion.[4] It was suggested by Chen[5] that the chemical bonding plays a major role in determining some of the properties
of amorphous alloys. From this point of view, it is thus important to
clarify the effects of alloying elements on the strength and thermal
stability of these alloys. In the present work, the effects of alloying
elements on the hardness and crystallization temperature were studied
as a measure of the atomic bonding force and structural thermal stability of amorphous iron alloys.

EXPERIMENTAL METHODS

Amorphous alloys with the compositions $Fe_{80-x}M_xP_{13}C_7$, where $M = Ti$, V,
Cr, Mn, Co, Ni, and Cu, were prepared by a centrifugal quenching method. The quenched foil specimens were about 30μm thick and 0.5mm wide.
The microhardness was measured using a pyramidal diamond indenter with
a 100 gram load, and the tensile strength using an Instron type machine
at a strain rate of $1.34 \cdot 10^{-4}$ / sec. Calorimetric measurements were carried out with a differential thermal analyzer using a heating rate of
5°C / min in order to examine the crystallization temperature, i.e., the
temperature at which the exothermic peak begins to appear.

RESULTS AND DISCUSSION

Figure 1 shows the relationship between hardness and fracture stress
for the amorphous iron alloys prepared in the present study. A linear

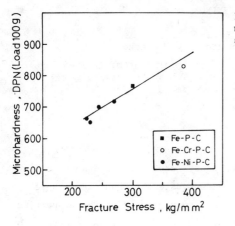

Figure 1. Relationship between microhardness and fracture stress of amorphous iron alloys.

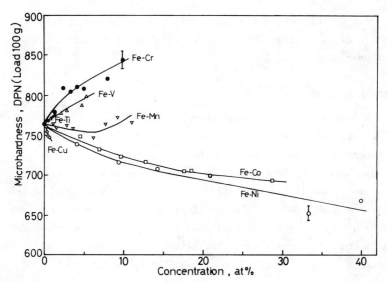

Figure 2. Effect of alloying elements on the microhardness of a-morphous iron alloys.

274

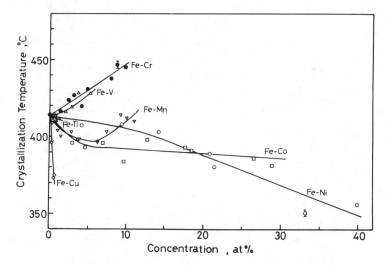

Figure 3. Effect of alloying elements on the crystallization temperature of amorphous iron alloys.

relationship between these quantities is found to exist. In Figures
2 and 3 the hardness and crystallization temperature are plotted a-
gainst the concentrations of the alloying elements. The figures show
clearly that the concentration dependence of both properties exhibits
a very similar behavior, thus suggesting that their main characteristics have a similar origin. In general, the values of both properties
decrease when the atomic number of the alloying elements is larger
than that of iron, and they increase in the opposite case. It seems
that alloying elements from Ti to Cu show a "group number effect",
i.e., an effect of the periodic table position on the hardness as
well as the crystallization temperature.

According to Hume-Rothery,[6] the prime factors determining the properties of metals are the atomic size, electronegativity and valence.
In the following, the possibility is discussed that these factors
play a role in the amorphous iron-base alloys studied in the present
work.

275

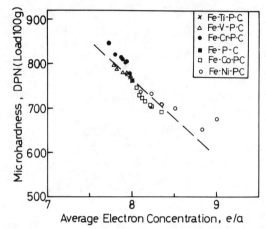

Figure 4. Microhardness plotted against the average outer electron concentration of the metallic atoms (e/a) for amorphous iron alloys.

First, the role of atomic size is considered to be of small importance in determining the mechanical strength and thermal stability of the alloys, since the present measurements show that Cr and Ni affect the hardness and crystallization temperature of the alloys in distinct and opposite ways although the atomic diameters of these elements are nearly equal (2.37, 2.34, and 2.31Å for Cr, Fe, and Ni, respectively). The atomic sizes of the alloying elements differ from that of iron by not more than 15% and the Hume-Rothery 15% size factor rule seems to be applicable to the amorphous iron alloys. Thus, as far as the atomic size is concerned, the effect of alloying elements can be considered to be weakened by the random structure of amorphous iron alloys.

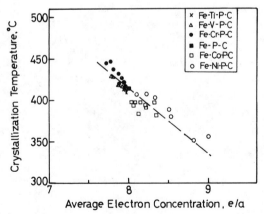

Secondly, we also argue against a possible role of the electronegativity in determining the hardness and crystallization temperature of amorphous iron alloys, using a similar

Figure 5. Crystallization temperature plotted against the average outer electron concentration of the metallic atoms (e/a) for amorphous iron alloys.

reasoning. The electronegativities of Fe, Co, and Ni are given by Pauling[7] as 1.8 for each element. Nonetheless, it is observed that alloying gives rise to significant changes in the hardness and crystallization temperature; therefore, the electronegativity also cannot be a major factor in determining these properties.

The only possible factor left to be considered is the relative valence of the alloying elements. In Figure 4 and 5 the measured values of hardness and crystallization temperature are plotted against the averaged outer electron concentration of the metallic atoms in amorphous iron alloys, which is customarily used as a measure of the valence of transition elements in alloys. By taking the numbers of the outer electrons to be 4, 5, 6, 7, 8, 9, and 10 for Ti, V, Cr, Mn, Fe, Co, and Ni, respectively, the average concentrations are given by the weighted means according to the atomic percentage of the transition elements. Manganese, which often exhibits anomalous features in crystals, is omitted from the analysis. Also, copper is not included in the figures, because it is usually eliminated from arguments where the group number relation among alloying elements is considered. It is clearly seen in the figures that the average outer electron concentration has a significant effect on the hardness and crystallization temperature.

Now the possible role of the outer electrons of the transition elements in amorphous iron alloys are discussed from several points of view.

It is known that the microhardness of the $Fe_3C - Fe_3P$ compound in the crystalline form is considerably higher ($1000 Kg/mm^2$) than that for amorphous Fe-P-C alloys ($760\ Kg/mm^2$) and also that an addition of outer electrons gives rise to an increase of the hardness in dilute solid solutions of Cu.

Furthermore, as shown in Figure 6, an increase of the (P+C) concentration in amorphous $Fe_{80-x}(P+C)_x$ alloys leads to an increase of the hardness of the alloys. These facts suggest that the major role of the outer electrons of the transition elements in amorphous alloys is not to increase the cohesive energy of solids (as in the case of crystalline solids), but to weaken the bond strength associated with P,C atoms which seems to play the dominant role in determining the properties of amorphous alloys. It was found recently that the effect of alloying on the electrical resistivity is not appreciably different in its overall behavior for amorphous and crystalline alloys. This suggests that the alloying effect is mainly confined to the bonding character of the outer electrons of the component elements.

In these alloys, bonds are formed by the overlap of the s-p hybrid orbitals of P and C atoms and the s-p-d orbitals of transition elements including iron. The present measurements suggest that the strength of these bonds is weakened by the increase of the d-electron concentration, and as is indicated in Figure 3 this effect becomes extremely remarkable in the case of alloying by Cu. The same result can be also derived from the change in the saturation magnetization of Fe induced by changes of the P and C concentration in amorphous Fe-P-C alloys.[8]

The strength of bonding between transition elements and C is also reflected in the interaction parameters of the alloying elements with C in liquid iron alloys. These values are -7.88, -5.08, 2.86, 2.85 and 4.06 for V, Cr, Co, Ni and Cu, respectively, at 1600°C, and they change from attractive to repulsive with increasing atomic numbers.

On the other hand, an indication of the strength of bonding between the transition elements and P can be obtained from the eutectic temperature of M-P binary alloys in the metal-rich region, as suggested by Chen.[5] The eutectic temperatures of Cr-, Mn-, Fe-, Co-, Ni-, and Cu-P are 1370, 960, 1050, 1023, 880, and 714°C, respectively. This suggests that the interaction of P with transition elements decreases with increasing atomic numbers.

A correlation of the effects of alloying elements on hardness and crystallization temperature of amorphous alloys exists as shown in Figure 7. We believe that these quantities are predominantly governed by the bonding character of the outer electrons of the component elements.

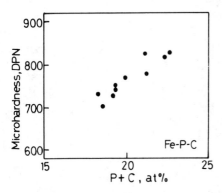

Figure 6. Effects of the concentration of the metalloid elements P+C on the microhardness of amorphous iron alloys.

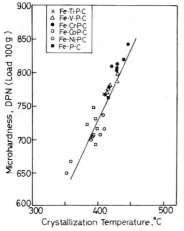

Figure 7. Relationship between microhardness and crystallization temperature of amorphous iron alloys.

REFERENCES

1. Masumoto, T. and R. Maddin; <u>Mater. Sci. Eng</u>. <u>19</u>, 1 (1975).
2. Gilman, J.J., <u>J. Appl. Phys</u>. <u>46</u>, 1625 (1975).
3. Kimura, H. and T. Masumoto, <u>Scripta Met</u>. <u>9</u>, 211 (1975).
4. Naka, M., K. Hashimoto, and T. Masumoto, <u>J. Japan Inst. Metals</u> 38, 835 (1974).
5. Chen, H.S., <u>Acta Met</u>. <u>22</u>, 897 (1974).
6. Hume-Rothery, W., "The Structure of Metals and Alloys of Iron", Pergamon Press (1966).
7. Pauling, L., "The Structure of the Chemical Bond", Cornell U. Press (1960).
8. Watanabe, K., M. Naka, and T. Masumoto [to be published].

279

PHASE TRANSFORMATIONS IN AMORPHOUS Fe-P-C ALLOYS

Helen L. Yeh and Robert Maddin

Department of Metallurgy and Materials Science and
Laboratory for Research on the Structure of Matter,
University of Pennsylvania, Philadelphia, Pennsylvania

INTRODUCTION

The iron base system known to exhibit amorphous structure produces interesting properties and it is necessary to understand the
crystallization and annealing behavior of these alloys. Studies such
as these may provide clues to the initial structure, since direct observation of the structure is not easy. There have been previous
studies of the phase transformations of these alloys, e.g. Rastogi
et al. [1,2] and Masumoto et al. [3,4] The former studied $Fe_{75}P_{15}C_{10}$
and the latter considered the $Fe_{80}P_{15}C_7$ and $Fe_{80}P_{13}C_7$ and $Fe_{80}P_{15}C_5$
compositions. In this paper, studies of the compositional dependence
of the phase transformations in Fe-P-C alloys are presented.

EXPERIMENTAL TECHNIQUES

Different compositions of Fe-P-C alloys were prepared by properly
mixing purified powder to produce intermediate alloys. Amorphous alloys were made by the modified Pond method. [5]   The quenched samples
consisted of strips about 30-50 μm thick and 0.5 mm wide. The samples
were checked by XRD with a multi-channel pulse analyzer equipped with
a silicon detector in a liquid nitrogen cooling system. This technique indicated that the samples were non-crystalline. The surfaces
of the sample were checked by Auger microscopy for oxygen contamination.

For a given composition all experiments were performed on quenched samples from the same production run to avoid any effects due to
quenching rate differences. The DSC [6] was used to check the crystallization behavior. The samples were heated under an argon atmosphere.
A Debye-Scherrer camera or a diffractometer was used to determine

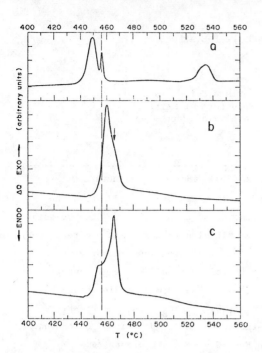

Figure 1:
DSC results at a heating rate of 20°C/min for alloys (a) $Fe_{83}P_{10}C_7$, (b) $Fe_{80}P_{15}C_5$, (c) $Fe_{80}P_{13}C_7$.

the crystalline phases present after annealing. The structures of the strips and their annealing behavior were investigated by a hot-stage TEM operated at 100kV. For thinning purposes, an ion micro-milling instrument was used; typical ion bombardment conditions were: beam current 100μA at 6 kV for at least 8 hours. Usually more than two areas, each about 5 μm long were studied in the same sample in an attempt to obtain representative micrographs. The actual temperature of the specimens in the hot stage was estimated to be about 20-30°C lower than that read from the thermocouple on the basis of comparisons with DSC results.

RESULTS

The DSC results are shown in Figure 1 for different compositions.
The XRD showed that the metastable crystalline phases associated with

282

the first peak of the $Fe_{83}P_{10}C_7$ glass (Fig. 1(a)) were mainly αFe and
Fe_3P; there were also present unidentified lines, probably corresponding to ε carbide and Fe_3C as reported by Tekin et al [7] . After the
second peak additional lines from Fe_3P appeared in the X-ray pattern.
The third peak was associated with the crystalline Fe_3C phase, αFe and
Fe_3P without any unidentified lines. The products for $Fe_{80}P_{15}C_5$
(Fig. 1(b)), and $Fe_{80}P_{13}C_7$ (Fig. 1(c)) after the first peak were the
same- - αFe, Fe_3P and one unidentified line. No further changes in the
crystalline phases were observed during the subsequent heat evolution
except that Fe_3C was indicated by very weak lines. None of these
lines was very sharp, probably due to the small crystallite size.
Note that only $Fe_{83}P_{10}C_7$ has a third peak around 530°C.

The kinetics of transformation of the $Fe_{83}P_{10}C_7$ amorphous alloy
were also studied by isothermal DSC measurements. Using these DSC results and the recrystallization theory due to Avrami [8] , it was found
that the nucleation rates decreased with time for both crystallization
processes corresponding to the first and second peaks.[9]

Since DSC provides only qualitative information about the transformation, TEM was applied. For consistency of both the TEM and DSC
studies, the same samples examined under the TEM were thinned from
their thickness after quenching to 30-50μm required for the DSC studies. This assured that both the DSC and TEM samples had the same
quenching rate.

The maximum size of the as-quenched samples was limited by the
necessity to obtain the non-crystalline alloy by our techniques, especially for the case of $Fe_{83}P_{10}C_7$ which was difficult to produce in
the amorphous form. Since the narrow samples were thinned by ion
bombardment, the effects were checked by dark field examination and
compared with the published transformation results of the non-treated
thin sample.[3]   These effects were negligible if proper selections
of the area were made for the in situ micrographs.[9]   The results
are as follows:

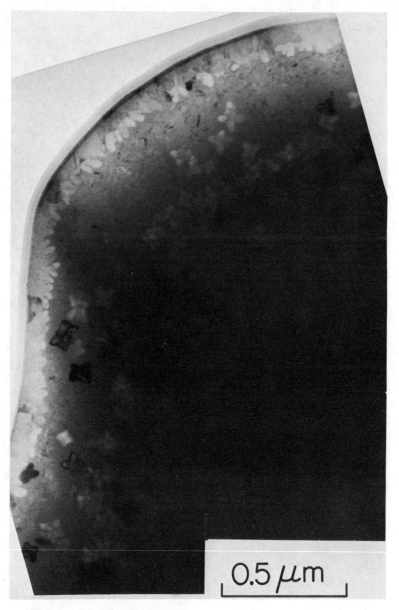

Figure 2. Bcc αFe dendrites and needle-shaped eutectic phases in a $Fe_{83}P_{10}C_7$ alloy held at 425°C for 360 min.

(a) $Fe_{83}P_{10}C_7$ alloy:

No obvious crystallization was observed after 315°C but after
the sample was held at 405°C for less than 5 minutes, the bcc phases
precipitated in a square morphology around 100 Å in size. This form
grew about 1.0 Å/min and the nucleation rate decreased with time. After holding at 405°C for 40 minutes, the sample was heated to 425°C.
The crystallites, now in an obvious dendritic configuration, grew to
about 700-1400 Å length and 600-1000 Å width after 250 minutes and
then stopped growing.

Figure 2 shows a different and needle shaped phase about 80 Å
wide, 100-400 Å long precipitated from the amorphous matrix after 360
minutes at 425°C. This 'phase' was found to be a mixture of αFe, Fe_3P
and iron carbide.[1] The coarsening of these small grains was observed
after the amorphous alloy transformed totally to the crystalline state.
As the sample was heated to 530°C, the coarsening rate increased and
simultaneously the dendritic bcc phases began to dissolve and form a
series of complicated phases along with the coarsening of the eutectic
phases. This may correspond to the third peak of the DSC curve shown
in Fig. 1(a). The corsened grains were about 500-3000 Å in diameter
after being held at 610°C for 185 minutes. The final phases were mainly αFe, Fe_3P and Fe_3C.

(b) $Fe_{80}P_{15}C_5$ alloy:

The amorphous matrix containing some cracks resulting from the
thinning and/or the handling process was examined by TEM. At the crack
tips small αFe nuclei of about 50 Å were observed at 380°C during heating. The sample heated to 410°C and held for 210 minutes showed the
crack tips to be filled with bcc dendrites while only a few of the
dendrites were observed in the amorphous matrix. The density of these
dendrites was much lower than that observed in the $Fe_{83}P_{10}C_7$ alloy.
After 640 minutes, the sample showed no obvious changes and was then
heated to 455°C. Figure 3 taken after 52 minutes at 455°C shows the
growth of lamellar eutectics (area A, B) along with the nucleation
of both the needle shaped and lamellar eutectics (area C). It was not-

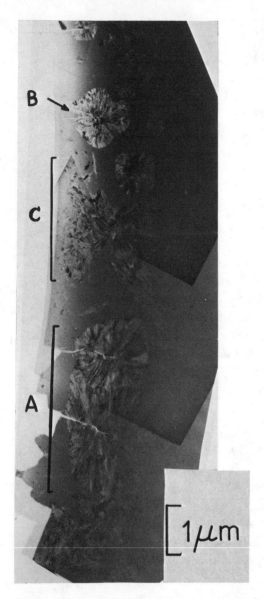

Figure 3. Transformed needle-shaped and lamellar eutectic in a $Fe_{80}P_{15}C_5$ alloy held at 455°C for 52 min.

ed that the nucleation of the eutectic phases decreased with time; the growth rate was about 22 $\overset{\circ}{A}$/min at this temperature. After 105 minutes at 455°C, the entire area was observed to be completely transformed. The final phases and the sizes of the coarsened grains taken at 625°C were the same as for the $Fe_{83}P_{10}C_7$ alloy.

(c) $Fe_{80}P_{13}C_7$ alloy:

After 300 minutes at 400°C, this alloy did not transform to the αFe phase in the matrix but showed some small crystallites at an isolated crack tip. On the other hand, a small amount of equiaxial grains identified as Fe_2P and XFe_2C precipitated at 400°C proceeding the eutectic transformation. In another area only needle shaped eutectic phases occurred.

The nucleation rate of the Fe_2P and XFe_2C phases increased at 400°C followed by a decrease after which no new nuclei appeared. Most of the area remained non-crystalline (shown by bright field and selected area diffraction pattern) after 300 minutes. The sample was then heated to 450°C, at which temperature transformation to the eutectic phases occurred. This transformation showed the earliest nuclei to be about 100$\overset{\circ}{A}$ in size after 30 min at 450°C. The nucleation rate of this eutectic phase again decreased with time. After about 50 minutes, the entire area transformed to both the lamellar and needle shaped eutectics. Fe_2P and XFe_2C remained in the matrix as shown by weak diffraction patterns.

The coarsening process of this alloy was the same as the other alloys discussed.

(d) $Fe_{75}P_{15}C_{10}$ alloy:

From the studies by Rastogi et al.,[1,2] Fe_2P is known to precipitate in the following cases: (i) After a small heat release at a heating rate of 100°C/min, 100–2000 $\overset{\circ}{A}$ granular shaped grains are embedded in amorphous matrix. (ii) Heating at 1°C/min to 375°C produces grains about 50 $\overset{\circ}{A}$. They also report the Fe_2P precipitates first followed by the αFe, Fe_3P and iron carbide phases. Analyzing the X-ray

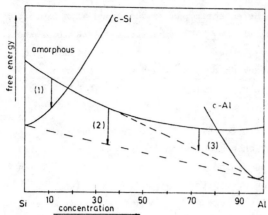

Fig.4 Hypothetical plot of free energy for various phases versus aluminum concentration in Si-Al. The observed crystallization reactions are indicated by arrows: (1) poly-
morphous transformation into a supersaturated Si solid
solution, (2) eutectic crystallization of Si and Al, (3)
pre-crystallization of Al spheres (From Ref. 11).

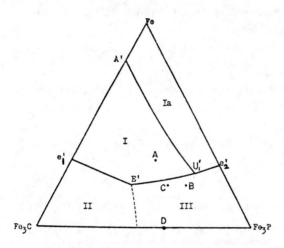

Fig.5 Fields of primary solidification in iron-carbon phosphorus
alloys. Ia:α/δ,I:γ, II:Fe_3C, III: Fe_3P; U_1': peritecto-
eutectic point, E': ternary eutectic point, e_1': ledeburite
eutectic point, e_2': binary ($\delta+Fe_3P$) eutectic point; A:
$Fe_{83}P_{10}C_7$, B: $Fe_{80}P_{15}C_5$, C: $Fe_{80}P_{13}C_7$, D: $Fe_{75}P_{15}C_{10}$.

288

and electron diffraction pattern reported in their report and our TEM results showing the Fe_2P and χFe_2C phases for $Fe_{80}P_{13}C_7$, it is suggested that Fe_2P along with iron carbide precipitated in the $Fe_{75}P_{15}C_{10}$ alloy rather than γFe and Fe_3P.

DISCUSSION

A schematic free energy versus composition diagram first used by Köster, et al, [10,11] to explain the compositional dependence of the transformation processes of vapor-deposited noncrystalline Al-Si and Al-Ge alloys can be extended to explain the observed reactions in the Fe-P-C system. Fig. 4 taken from Ref. 11 illustrates the free energy model so applied. For the systems reported in this paper, and on the basis of the diagram showing the fields of primary solidification for Fe, Fe_3P and Fe_3C in Fig. 5 [12] the following reactions are suggested to occur:

(a) By pre-crystallization of αFe the homogeneous amorphous alloy transforms into a metastable equilibrium mixture consisting of αFe embedded in the amorphous matrix. This reaction was observed in the alloy with 83 atomic percent Fe (alloy A in Fig 5). At higher temperatures, the amorphous phase transforms to the eutectic phases. The stable states are obtained by dissolution of the metastable Fe dendrites and by the coarsening of the eutectic phases.

(b) Fe_2P and iron carbide crystallize first, followed by the eutectic reaction. This was observed in $Fe_{75}P_{15}C_{10}$ (alloy D in Fig 5), the phosphorus rich alloy, and some of the area of $Fe_{80}P_{13}C_7$ (alloy C in Fig. 5).

(c) Eutectic crystallization occurs over the entire observed concentration range. Some areas of the alloys with compositions near the eutectic line produce the eutectic crystallization without the processes discussed in (a) and (b) above, such as $Fe_{80}P_{15}C_5$ and $Fe_{80}P_{13}C_7$ (alloys B and C in Fig. 5).

289

For the compositions near the eutectic line, it is difficult to predict at this time which of the above possible reactions occurs first. However, when the reactions discussed in (a) and (b) above occur, the transformation occurring is predominantly the eutectic reaction. Calculation from the TEM results shows that the nucleation rates for all these three processes (a), (b) and (c) decrease with time. The kinetics obtained from the DSC measurements for the $Fe_{83}P_{10}C_7$ alloy showed that the nucleation rates decrease with time for both the first and the second heat releases. Our results agree with the Avrami analysis at least qualitatively.

CONCLUSIONS

On the basis of these results, we suggest that the crystallization of Fe-P-C alloys proceeds by the following processes: For the iron rich alloy, bcc αFe precipitates first. For the phosphorus rich alloy, Fe_2P plus iron carbide precipitate first. For alloys near the eutectic line, the eutectic reactions dominate although some αFe or Fe_2P plus iron carbide crystals are observed.

ACKNOWLEDGEMENTS

We thank Dr. C. Laird and Dr. H.S. Chen for many helpful discussions, Dr. P. Morgan for help in using the X-ray diffractometer at the Franklin Institute, and Dr. C. Briant for proof reading the manuscript. This work was supported by the Office of Naval Research through Contract No. N0014-67-A-0216 — 19 and the National Science Foundation.

REFERENCES

1. Rastogi, P.K. and P. Duwez, <u>J. Non-Cryst. Sol</u>. <u>5</u>, 1 (1970).
2. Rastogi, P.K., <u>J. Mat. Sci</u>. <u>8</u>, 140 (1973).
3. Masumoto, T. and H. Kimura, <u>J. Jap. Inst. Metals</u> <u>39</u>, 273 (1975).
4. Masumoto, T. and R. Maddin, <u>Mater. Sci. and Eng</u>. <u>19</u>, 1 (1975).
5. Pond, R., Jr., and R. Maddin, Trans. AIME <u>245</u>, 2475 (1969).
6. A Perkin Elmer DSC 2 was used for the kinetics study. A Du Pont 990 Thermal Analyzer with a DSC cell was used for regular checks.
7. Tekin, E. and P.M. Kelly, <u>Precipitation from Iron-Base Alloys</u>, ed. by G.R. Speich and J.B. Clark, p. 173.
8. Avrami, M., <u>J. Chem. Phys</u>. <u>8</u>, 212 (1940).
9. Yeh, H. and R. Maddin [to be published].
10. Küster, U., <u>Acta Met</u>. <u>20</u>, 1361 (1972).
11. Küster, U. and P. Weiss, <u>J. Non-Cryst. Sol</u>. <u>17</u>, 359 (1975).
12. Hume-Rothery, W., <u>The Structures of Alloys of Iron</u>, Pergamon Press, New York, 312 (1966).

291

CURRENT VIEWS ON THE STRUCTURE OF AMORPHOUS METALS

G.S. Cargill III

IBM T.J. Watson Research Center
Yorktown Heights, New York 10598

INTRODUCTION

Questions concerning the structure of amorphous metals, from both experimental and theoretical points of view, can be discussed on many levels, beginning perhaps with discussion of whether the metals and metallic alloys, now being described as amorphous, are (or are not) microcrystalline. If they are not microcrystalline, what are the best methods for characterizing and describing their atomic scale structures? Although most attention has been focussed on short-range atomic structure of amorphous solids, the importance of macro- and microscopic structure must not be overlooked, for structure on these larger scales may be as important as atomic scale structure for many properties of these solids. Possibilities of structural anisotropy in amorphous metals, on either atomic or microscopic scales, have also been receiving increasing attention. In this review on the structure of amorphous metals, I will first discuss "history," which includes developments which took place prior to 1975 and to completion of Ref. 1. I will then point out some more recent developments, both experimental and theoretical.

HISTORY

The earliest report of an amorphous metallic alloy which I have discovered is in the 1950 paper by A. Brenner and co-workers at NBS,[2] "Electrodeposition of Alloys of Phosphorus with Nickel or Cobalt." In x-ray examinations of these alloys they observed that "a specimen of the high nickel phosphorus alloy yielded only one diffuse band, which indicated that the material was amorphous. . .on heat treating the specimen at 800°C, numerous lines appeared." Their "high nickel phosphorus alloy" contained between 24 and 30 at. % P. In the mid-1950's Hilsch and co-workers reported x-ray and electron scattering patterns for amorphous metallic films which crystallized well below room temperature.[3] The next report of an amorphous metallic alloy which could be retained near room temperature was by Klement, Willens, and Duwez in 1960, "Noncrystalline Structure in Solidified Au-Si Alloys."[4]

The first radial distribution function for an amorphous metallic alloy at room temperature was published in 1963 by Dixmier, et al.,[5,6] who demonstrated that the x-ray scattering pattern for an amorphous chemically deposited Ni-P alloy was inconsistent with scattering patterns calculated for small fcc cr hcp crystals.[6] X-ray scattering data and radial distribution functions subsequently became available for several amorphous metal-metalloid alloys (splat cooled Pd-Si, Au-Si, Fe-P-C; electrodeposited Ni-P); Wagner[7] and Cargill[8] pointed out the striking similarities among x-ray scattering patterns and among radial distribution functions for these amorphous alloys. Cargill extended comparisons with microcrystalline models to include strains in fcc microcrystals but was unsuccessful in accounting for the observed scattering patterns in terms of such microcrystalline models.[8] These comparisons and the high densities of the amorphous Ni-P alloys led him to conclude that "the alloys have a continuous structure rather than one in which internal boundaries separate small well ordered regions."[8]

These conclusions were presented at the VIIIth Congress of the International Union of Crystallography, Stony Brook, N. Y., in 1969, and in the same session J. L. Finney reported results on dense random packing models for simple liquids. His model distribution functions showed striking similarities to those seen for the amorphous metal-metalloid alloys, and closer comparisons revealed that his dense random packing of hard spheres (DRPHS) model was a very attractive alternative to microcrystalline models for the amorphous alloys.[9] In 1964 Cohen and Turnbull had suggested dense random packing of hard spheres as the prototype structure of a monatomic glass[10] and this suggestion was supported by detailed comparisons, in 1970, of the Finney DRPHS distribution function with radial distribution functions for amorphous electrodeposited Ni-P alloys.[9,11]

These developments were quickly followed by Polk's suggestion that the simple DRPHS model for these amorphous alloys be modified by allowing the metal atoms to have an arrangement similar to DRPHS, but with most of the metalloid atoms occupying "the larger holes inherent in the random packing."[12] In 1972 Polk generalized his view of the DRPHS void-filling model to allow the metal atoms to occupy random packing structures somewhat less dense than those of Bernal[13] and Finney,[11] which should provide larger numbers of larger voids to accommodate the

Characteristics of the scattering anisotropy for various model structures. The magnitude of the scattering vector corresponds to the first diffraction ring. (From Alben et al.[45])

Model	$\bar{I}$	I_{max}	$I_{max}/\bar{I}$	$(I_{max}/\bar{I})_{random}$
519-random	0.92	7.3	7.9	6.6 $\pm$ 1.5
996-Finney	2.93	38.9	13.3	7.6 $\pm$ 1.5
890-Bennett	2.93	33.5	11.4	7.5 $\pm$ 1.5
864-fcc	5.09	864.0	169.8	7.4 $\pm$ 1.5

metalloid atoms.[14] This point of view is similar to that of Sadoc, et al., who in 1973 proposed for the structure of these alloys a binary dense random packing model in which no small spheres, which represent metalloid atoms, are allowed to be near neighbors.[15]

In parallel with the latter stages of these developments, progress was also being made on other fronts. Structural data on other types of amorphous metallic solids were being obtained. Amorphous films of Co and of Fe, with no intentional impurity additions, were prepared by vacuum evaporation at 4°K and 10^{-8} Torr. These films crystallized at approximately 50°K; crystallization temperatures increased with increases in residual gas pressure during evaporation. In-situ electron scattering measurements by Ichikawa (1973)[16] and by Leung and Wright (1974)[17] were used to calculate interference functions and radial distribution functions. These were very similar to those found for the amorphous metal-metalloid alloys. Amorphous rare earth-transition metal (RE-TM) alloys which are metastable well above room temperature were made by evaporation (1972)[18] and by dc and rf sputtering, (1972)[19] and (1973).[20] X-ray and neutron scattering data from some of these films soon became available.[21,22] Three contributions to the near neighbor radial distribution functions could be attributed to TM-TM, TM-RE, and RE-RE near neighbor pairs.

Dense random packing models for atomic scale structure of amorphous metals have also been examined further. Computer generated dense random packings of single size hard spheres, similar in many ways to those obtained by Finney with ball bearings,[11] have been studied by Bennett[23] and by Adams and Matheson.[24] Both groups used similar criteria in generating their hard sphere structures; densities and distribution functions from these computer generated models are indistinguishable from one another. Binary DRPHS structures, i.e. with two sizes of spheres, have been generated by Sadoc, et al.,[15] to simulate atomic arrangements in amorphous metal-metalloid alloys and by Kirkpatrick and Cargill[1,25] for amorphous RE-TM alloys.

Most comparisons between models and experimental data for amorphous metals or alloys have employed either interference functions or distribution functions. Comparisons of model and experimental densities are also of interest, although densities of the amorphous transition metal films[16,17] are not known. In all other cases, e.g. metal-metalloid alloys and RE-TM alloys, the densities of the DRPHS models (hard spheres per scaled unit volume) are smaller than those of the amorphous alloys.[1] However, density values are not available for the Sadoc, et al.,[15] binary DRPHS metal-metalloid model, and comparisons of metal-metalloid alloy densities with those of single-size-sphere DRPHS models are complicated by uncertainties concerning the locations and effective sizes of the metalloid atoms.[1]

RECENT DEVELOPMENTS: EXPERIMENTS

There has been a great deal of experimental research on amorphous metallic alloys since the preparation of Ref. 1. In this section I will review some of this recent work concerning the structure of amorphous alloys, particularly (1) the role of metalloid atoms in the atomic scale structure of metal-metalloid alloys, (2) the effects of mechanical deformation and of thermal annealing on this atomic scale structure, and (3) the compositional homogeneity and anisotropy of some of these alloys. I will also briefly discuss (4) atomic scale anisotropy in an amorphous RE-TM alloy and (5) recent studies of atomic scale structure in other types of "metal-metal" amorphous alloys.

<u>Where is the P in Amorphous Ni-P and Co-P Alloys?</u> This has been a difficult question to answer, although it appears to be an important one. The metalloid components of the transition metal-metalloid alloys play an important role in stabilizing these amorphous alloys, i.e. in raising their crystallization temperatures to well above room temperature, in comparison to the nominally pure transition metal films, which crystallize at about 50°K. Polk pointed out the possible importance of compositional short range order and atomic size differences in the structure of amorphous Pd-Si and Ni-P alloys, and he has discussed the role of metalloid atoms in "jamming" dense random packings of the metal atoms, by reducing the free volume.[12,14] Both Polk[14] and Chen and Park[26] have speculated that chemical bonding between the metal and metalloid atoms is particularly strong and that the surroundings of metalloid atoms in these amorphous alloys are probably similar to those in the corresponding metal-rich metal-metalloid crystalline compounds. It appears that experimental data to test these hypotheses are now becoming available.

X-ray distribution functions for metal-metalloid alloys are dominated by the contributions of metal-metal pairs, because the metal atoms are both more numerous, typically comprising between 70 and 80 at. %, and have much larger x-ray scattering factors. Weighting factors

$$W_{ij} = c_i c_j f_i f_j / <f>^2 \tag{1}$$

of partial distribution functions $\rho_{ij}(r)/c_j$ in the composite distribution function

$$\rho(r) = \sum_{i,j} W_{ij} \rho_{ij}(r)/c_j \tag{2}$$

obtained by x-ray scattering with Mo K_α radiation for $Co_{78}P_{22}$ are $W_{CoCo}=0.75$, $2 W_{CoP}=0.23$, and $W_{PP}=0.02$. However, Dixmier, Bletry, and Sadoc[27] obtained a distribution function from unpolarized neutron scattering data, for which the corresponding weighting factors W_{ij} are 0.40, 0.46, and 0.13. Comparison of the x-ray and neutron composite distribution functions, shown in Fig. 1, clearly indicates that Co-P near neighbors occur at smaller separations than do the Co-Co near neighbors.

More recently, Bletry and Sadoc[28] have used polarized neutron scattering data from an amorphous ferromagnetic Co-P(16.6 at. % P) alloy to calculate partial interference functions $I_{ij}(K)$, which are just Fourier transforms of the partial distribution functions discussed above,

296

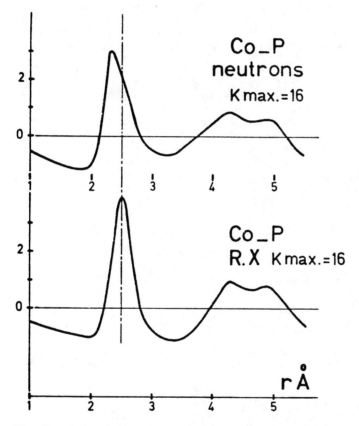

Fig. 1. Reduced distribution functions $G(r)=4\pi r(\rho(r) - \rho_o)$ obtained
with neutrons and with x-rays; displacement of the first maximum indicates that Co-P near neighbor pairs occur at smaller
separations than do Co-Co near neighbors, (from Dixmier, et al.[2]).

$$I_{ij}(K) = 1 + (1/K) \int 4\pi r [(\rho_{ij}(r)/c_j) - \rho_o] \sin Kr \, dr. \qquad (3)$$

The K-dependence of the magnetic scattering factor for Co limited the range of $K(<10\text{Å}^{-1})$ over which the partial interference functions could be obtained. Their results are reproduced in Fig. 2(a).

A different approach has been used by Waseda and Tamaki[29] in obtaining partial interference functions for amorphous Ni-P alloys. They obtained x-ray scattering data with Mo K_α, Cu K_α, and Co K_α radiations. The Cu K_α and Co K_α characteristic wavelengths are near the K absorption edge of Ni, and the Ni x-ray scattering factor is modified by dispersion corrections $\Delta f'$ and $\Delta f''$ which differ for the three radiations used in the scattering experiments. These small differences in $f_{Ni}(K)$ produce differences in the W_{ij} weighting factors, and Waseda and Tamaki[29] were able to extract the three partial interference functions from their careful scattering measurements. The long wavelength of the Co K_α radiation limited the K range over which they could obtain these partial interference functions $(K<7\text{Å}^{-1})$, and the smallness of the dispersion corrections to f_{Ni} limited the precision of the partials. Their results are also reproduced in Fig. 2(b), and they are in encouraging agreement with the partial interference functions obtained by Bletry and Sadoc[28] for the amorphous Co-P alloy. However, the split second peak in the $I_{NiNi}(K)$ is centered on $\sim 6\text{Å}^{-1}$ but the corresponding peak in $I_{CoCo}(K)$ is at $\sim 5\text{Å}^{-1}$. The curious shape of the split first peak of $I_{NiP}(K)$ or $I_{CoP}(K)$ is clearly indicated by both sets of data. This characteristic split first maximum is <u>not</u> reproduced in partial interference functions calculated from the binary DRPHS model of Sadoc, et al.[15] Availability of these experimentally determined partial interference functions should provide critical tests for structural models which incorporate both metal and metalloid atoms.

<u>Effects of Annealing and of Mechanical Deformation.</u> Isothermal annealing experiments by Waseda and Masumoto[30] indicate that detectable changes occur in the atomic scale structure of amorphous $Fe_{80}P_{13}C_7$ before indications of crystallization are detectable by transmission electron microscopy or by appearance of crystalline features in x-ray interference functions. The first two maxima in I(K) become slightly higher after anneals at 300°C for 500, 1000, and 1300 min. The magnitude of oscillations at larger K values also appears to be increased.

It may be useful to distinguish between two regimes of short range order in such amorphous alloys. The distribution functions W(r) or G(r) are dominated by the strong, rather narrow near neighbor maximum at $r=r_1$. Likewise, interference functions I(K) or F(K) are dominated by a first strong, sharp maximum, $K=K_1$. It is tempting to attribute particular features of G(r) to individual peaks in I(K), although evaluation of G(r) for any r value actually involves integration over the complete reduced interference function F(K). However, for amorphous metal-metalloid alloys, the structure in G(r) for $r>2r_1$ is determined almost solely by the position, width, and amplitude of the first maximum in F(K), because this maximum is much sharper and much more intense than other

298

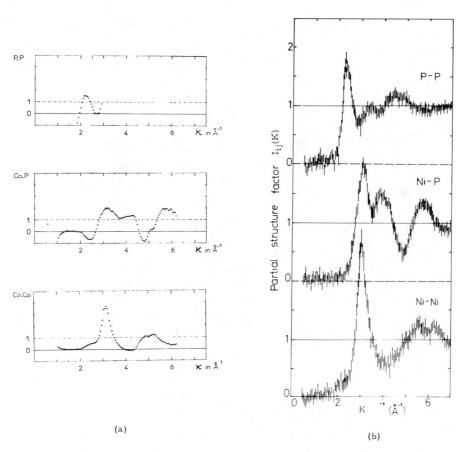

(a)

(b)

Fig. 2. Partial interference functions $I_{ij}(K)$ for metal-metalloid alloys:

(a) polarized neutron scattering for Co-P (16.6 at.%P) from
Bletry and Sadoc.[28]

(b) x-ray scattering, with dispersion corrections, for Ni-P
(16.9 – 26.9 at.%P) from Waseda and Tamaki.[29]

components of $F(K)$. This is illustrated in Fig. 3. A Gaussian has been fitted to the first maximum in $F(K)$ for $Ni_{76}P_{24}$. The Fourier transform of the Gaussian is a damped sine wave of period $2\pi/K_1$. This is shown in Fig. 3(b), together with the complete $Ni_{76}P_{24}$ distribution function. The unusual sharpness of the first maximum in $F(K)$ results from nearly periodic oscillations in $G(r)$ for $r>2r_1$. For the Ni-P alloys, the period of these oscillations $2\pi/(3.11\text{Å}^{-1})=2.02\text{Å}^{-1}$, is only 0.5% less than the (111) plane spacing in fcc nickel. The significance of these oscillations in $G(r)$ remains uncertain. It is interesting that Pang et al.,[43] and Chaudhari and Graczyk[44] have noted the occurrence of "warped sheet," planelike configurations in dense random hard sphere packings. More extensive results on "planes" in DRPHS structures will be given in the last section of this review.

Just as the first maximum in $F(K)$ is associated with periodic damped oscillations in $G(r)$ for $r>2r_1$, the first maximum in $G(r)$ can be associated with (or results in) periodic damped oscillations at $K>2K_1$ in $F(K)$. With these simplifications, effects of annealing on the Fe-P-C atomic scale structure[30] can be described in r-space as follows:

(1)   Increase in the amplitude of the near neighbor maximum in $G(r)$, probably from a narrowing of this peak.

(2)   Increase in the strength of periodic, damped oscillations in $G(r)$, which may be associated with the "planes" to be discussed in the last section of this review.

Masumoto and Maddin[31] have recently reported that mechanical deformation of amorphous $Pd_{80}Si_{20}$ at room temperature "produces a more disordered atomic structure than is present in the as-quenched state by introducing additional irregularities through 'defects'." Effects of cold rolling on the first maximum of the x-ray scattering pattern of amorphous $Pd_{80}Si_{20}$ are shown in Fig. 4.[31] The peak position shifts to lower angles and the width of the peak increases with this mechanical deformation. These changes suggest a reduction of strength and an increase in the period of the damped oscillations in $G(r)$.

Compositional Homogeneity in Amorphous Metal-Metalloid Alloys. Chou and Turnbull[32] have recently reported x-ray small angle scattering (SAS) measurements on roller quenched Pd-Au-Si alloys. They followed the phase separation which preceded crystallization of an amorphous $Pd_{74}Au_8Si_{18}$ alloy. Some of their x-ray SAS data are reproduced in Fig. 5. The flat SAS for the as-quenched alloy suggests that it is quite homogeneous on the scale of $\lesssim 10\text{Å}^{-1}$, but larger scale composition inhomogeneities ($\gtrsim 100\text{Å}^{-1}$) may be present. Also, slower quench rates might have produced phase separation even in the as-quenched materials. Normalized (electron units/atom) SAS data are useful in quantifying the magnitude and extent of composition inhomogeneities in amorphous alloys. Such normalized SAS measurements for amorphous electrodeposited Co-P alloys are discussed below.

300

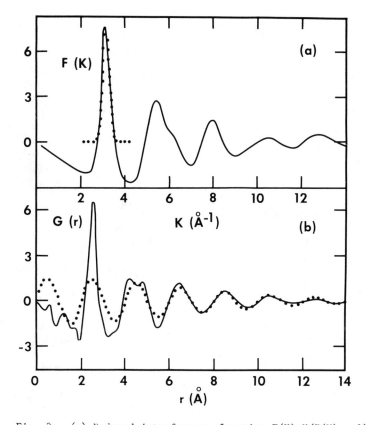

Fig. 3. (a) Reduced interference function $F(K)=K(I(K) - 1)$ for $Ni_{76}P_{24}$ (solid line) and gaussian fitted to its first maximum (points).

(b) Reduced distribution function $G(r)$ for $Ni_{76}P_{24}$ (solid line) and Fourier transform of the gaussian peak shown in (a) (points), (from Cargill[1]).

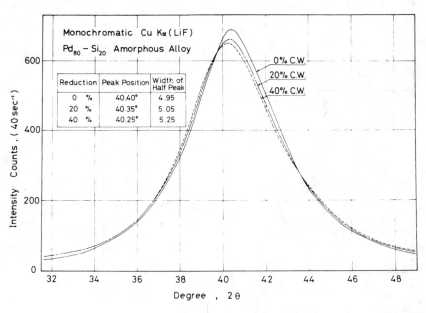

Fig. 4. Changes in the scattering pattern of $Pd_{80}Si_{20}$ after cold rolling thickness reductions of 20% and 40% (from Masumoto and Maddin[31]).

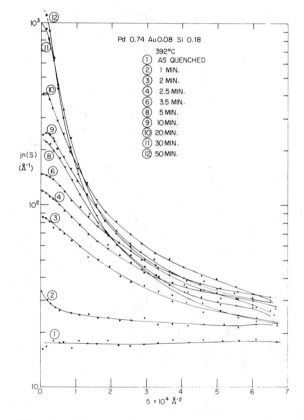

Fig. 5. Guinier plots of x-ray SAS from $Pd_{74}Au_8Si_{18}$ glassy alloys (T_g =394°C) annealed at 392°C for different lengths of time, (from Chou and Turnbull[32]).

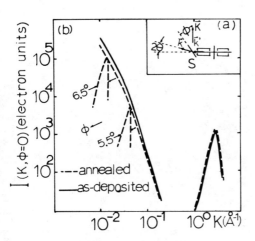

Fig. 6.   (a) SAS geometry with scattering sample S, scattering angle 2θ, and rotation angle ϕ.
(b) Observed dependence of SAS intensity on both $K = 4\pi\sin\theta/\lambda$ and ϕ for electrodeposited Co-P (22.5 at.%P), both as-deposited and annealed, without crystallization, (from Chi and Cargill[34]).

304

<u>Anisotropic Microstructure in an Amorphous Co-P Alloy</u>. In a contributed paper at this conference, Chi and Cargill[33] report x-ray SAS data for electrodeposited Co-P alloy films. Normalized, slit-length corrected data are shown in Fig. 6, together with a schematic of the experimental configuration used to obtain these data. SAS from these films was anisotropic, i.e. it depended strongly on the orientation of the film with respect to the incident x-ray beam. The intensity of the SAS decreased when the samples were annealed at 200°C. These observations indicate that the as-deposited films contained anisotropic composition inhomogeneities, of characteristic dimensions $\sim$ 100-300Å in the film plane and $\gtrsim$ 2000Å normal to the film plane. The magnitude of these inhomogeneities was decreased by annealing, presumably by atomic scale diffusion; this is in contrast to the phase separation behavior seen by Chou and Turnbull[32] in roller-quenched Pd-Au-Si alloys.

<u>Atomic Scale Anisotropy in Amorphous Sputtered RE-TM Alloy Films</u>. Although atomic arrangements in amorphous solids have commonly been assumed to be macroscopically isotropic, some amorphous ferrimagnetic and ferromagnetic alloys are magnetically anisotropic. For example, the electrodeposited Co-P alloys discussed above have weak perpendicular easy-axis magnetic anisotropy. The anisotropic microstructure revealed by the x-ray SAS measurements is thought to be largely responsible for their magnetic anisotropy.[34]

Some amorphous alloys have much larger magnetic anisotropies, particularly some sputtered amorphous ferrimagnetic Gd-Co alloys have anisotropy fields $H_K > 4\pi M_s$.[20] The origin of this magnetic anisotropy has not yet been established, but it is thought to involve anisotropies in short range structural or compositional ordering. It has been proposed that an excess of Co-Co near neighbor pairs in the film plane may be responsible for the perpendicular magnetic anisotropy in amorphous Gd-Co alloy films, with perhaps 1% more pairs "in-plane" than "out-of-plane".[35] Data of the type shown in Fig. 7 could provide direct evidence for such structural anisotropies, if they were several times larger than this estimate.[36] Near neighbor portions of two radial distribution functions for an amorphous sputtered $Gd_{18}Co_{82}$ alloy are shown. One was obtained from x-ray scattering data taken with the scattering vector perpendicular to the film surface; the other, with the scattering vector parallel to the film surface. Results from these two types of scattering measurements, i.e. reflection and transmission, should be identical if local atomic arrangements are macroscopically isotropic. Arrows in Fig. 7 indicate average separations for Co-Co, Co-Gd, and Gd-Gd nearest neighbor pairs; these data indicate that the number of Co-Co pairs aligned perpendicular to the film is very close to the number aligned parallel to any in-plane direction. The difference is certainly less than 5%. However, the data do suggest that there is a deficiency of "in-plane" Gd-Gd near neighbor pairs.

This film was produced with very low rf bias, and may therefore be a poor candidate for seeking evidence for anisotropic Co-Co nearest neighbor distributions.[35] Similar measurements on high anisotropy films

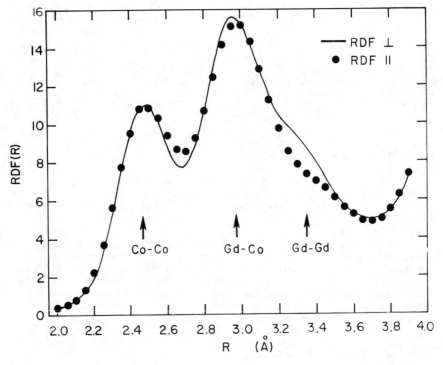

Fig. 7. Comparison of radial distribution functions RDF(r)=$4\pi r^2 \rho(r)$ from scattering measurements with scattering vector $\vec{K}$ perpendicular and parallel to the surface of an amorphous, sputtered $Gd_{18}Co_{82}$ films, (from Cargill[36]).

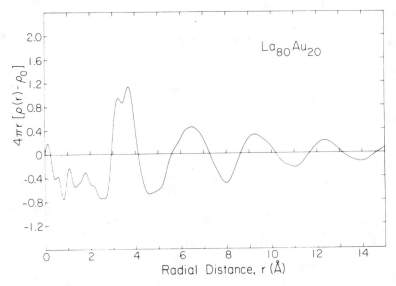

Fig. 8. Reduced distribution function G(r) for an amorphous La$_{80}$Au$_{20}$ alloy, prepared by rapid quenching, (from Logan[38]).

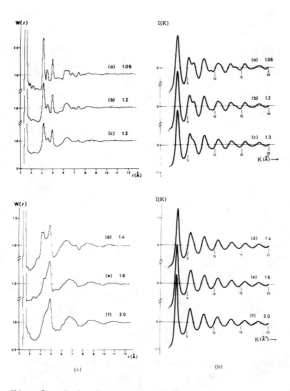

Fig. 9. Pair distribution functions $W(r)=\rho(r)/\rho_o$ and interference functions $I(K)$ for computer generated DRPHS structures. The sequences
(a) to (f) represent decreasing degrees of tetrahedron perfection,
(from Ichikawa[41]).

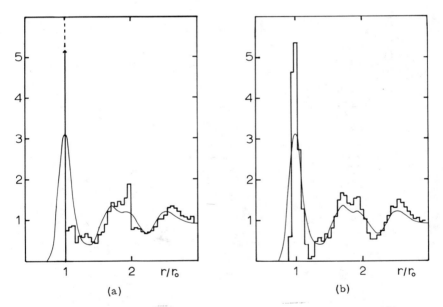

Fig. 10. (a) Pair distribution functions $W(r)=\rho(r)/\rho_o$ for an 888 atom inner section of the 3999 atom Bennett model[32] (histogram) and for amorphous Co[17] (thin line).
(b) $W(r)$ for the relaxed 888 atom Bennett model (histogram) compared with that for amorphous Co (thin line), (from v. Heimendahl[42]).

309

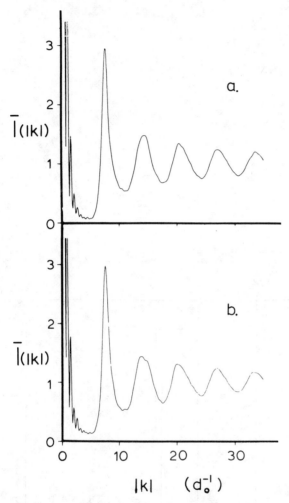

Fig. 11. Directionally averaged interference function $\bar{I}(|K|)$ for two
DRP models. The scattering vector $|K|$ is given in units of
one divided by the hard sphere diameter d_o:

(a) 890 central units from a 3999-unit computer generated
DRPHS of Bennett[23]
(b) 996 central units from a 7934-unit DRPHS model hand built
by Finney[11], (from Alben, et al.[45]).

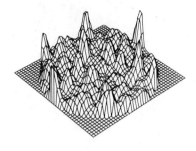

(a)

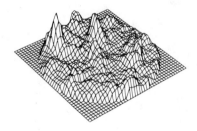

(b)

Fig. 12. Dependence of the interference function $I(\vec{K})$ on the direction
of $\vec{K}$ for $|K|=7.7$ d_o^{-1} for:

(a) the 890 central units of the DRP model of Bennett and
(b) the 996 central units of the DRP model of Finney.

The interference function has been averaged over directions
within 5° of the indicated directions for purposes of illustration, (from Alben, et al.[45]).

311

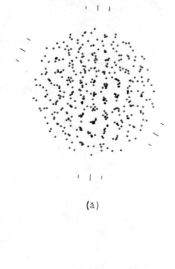

(a)

(b)

Fig. 13.   Projection of 440 central units of DRP models on the planes
containing the two strongest scattering directions for $|K| =$
7.7 d_o^{-1}.   The first few planes of high density and their
spacing are indicated by the tick marks: (a) Bennett model
and (b) Finney model. For case (a) there is a third set of
planes perpendicular to a direction about midway between the
directions perpendicular to the indicated sets of planes. The
planes are most easily seen by tipping the page, (from Alben,
et al.[45]).

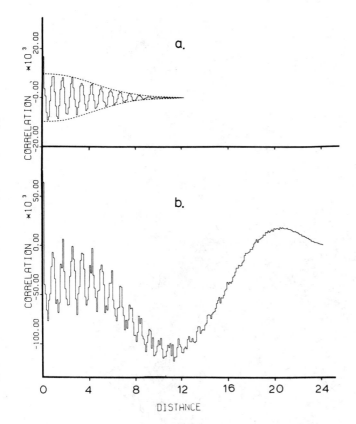

Fig. 14. Autocorrelation of the projected density for the strongest scattering direction of the 996 central units of the Finney DRP
model for $|K|$=7.7 d_o^{-1}. A smooth background corresponding to a
finite sphere has been removed.

(a) 996 central units of Finney model. The broken line envelope
shows the decrease in the autocorrelation due solely to the
finite size of the model.
(b) Full 7934 units of the Finney model. The broad oscillation
is due to asphericity of the model. It is possible to discern
20 oscillations corresponding to a plane spacing of slightly
more than 0.82 d_o, (from Alben et al.[45]).   313

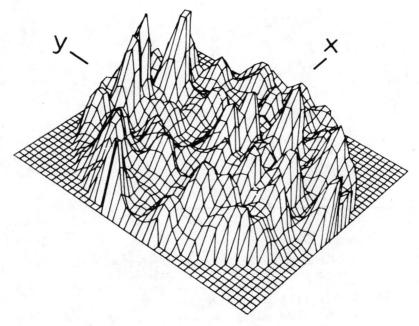

Fig. 15.   Dependence of the interference function $I(\vec{K})$ on the direction
of $\vec{K}$ for a spherical cluster of 519 randomly placed scatterers,
(from Alben <u>et al</u>.[45]).

<u>may</u> provide direct evidence for the atomistic origin of magnetic ani-
sotropy, but expected experimental uncertainties are of the order of 3%.
Wagner, <u>et al</u>.,[37] have recently reported that interference functions
I(K), obtained from high anisotropy, bias sputtered Gd-Co films, using
reflection and transmission configurations, were identical within experimental uncertainties. Nevertheless, such comparisons of reflection
and transmission data may be useful in characterizing anisotropies in
other amorphous alloys.

<u>Atomic Scale Structure in Other Types of Amorphous Metal-Metal Alloys</u>.
Large angle x-ray scattering data have recently been published for two
amorphous metallic alloys which belong to neither metal-metalloid nor
RE-TM alloy families. Logan[38] obtained data for splat cooled $La_{80}Au_{20}$;
his reduced radial distribution function G(r) is shown in Fig. 8. As
for the RE-TM alloys, the near neighbor maximum apparently consists of
more than one discernable contribution. In this case, the contributions
can be associated with La-La pairs at 3.33Å (W_{LaLa}=0.54) and La-Au pairs
at 3.75Å ($2W_{LaAu}$=0.39); these correspond closely to distances expected
on the basis on sums of Goldschmidt radii (3.31Å and 3.74Å). Logan concluded that the compositional short range order of crystalline La_2Au is
maintained in the amorphous alloy.

Waseda and Masumoto[39] recently reported x-ray scattering data for
rapidly quenched $Cu_{57}Zr_{43}$, for which W_{CuCu}=0.24, $2W_{CuZr}$=0.50, and
W_{ZrZr}=0.26. One might expect to see three near neighbor contributions
at 2.56Å, 2.88Å, and 3.20Å, but only a single peak at 2.81Å was found.
The Goldschmidt radii difference for Cu and Zr is 22% while that for La
and Au is 26%. It is difficult to determine whether the failure to resolve individual components in the first peak for $Cu_{57}Zr_{43}$ indicates (a)
that some of these components are not present, i.e. chemical ordering,
or (b) that all components are present, but with overlap which prevents
their being resolved from one another. Data for amorphous binary alloys
in which contributions from both components are important should provide
additional critical tests for binary DRPHS structural models.

RECENT DEVELOPMENTS: MODELS

Interesting developments concerning three aspects of structural
models for amorphous metals have recently occurred. The first of these
involves effects of modifying Bennett's algorithm[23] in computer generation of DRPHS structures. The second involves relaxation of DRPHS
structures for energy minimization with pairwise potentials. A third
area of development is analysis of anisotropies and of long range ordering, both in physically constructed DRPHS and in computer generated
DRPHS structures.

<u>Computer Generated DRPHS Models with Modified Algorithms</u>. As has been
noted by several authors, the Finney[11] and Bennett[23] DRPHS models are in
good, but not perfect, agreement with observed scattering data for amorphous metal-metalloid alloys and for amorphous transition metal films.[1]
Specifically, the second peak in I(K), calculated from either of these
models, does not have the strong shoulder seen for the metal-metalloid

alloys or the splitting seen for the transition metal films
(see Fig. 11). The second maximum in distribution
functions $G(r)$ or $W(r)$, calculated from either of these
models, although split, as in the experimental distribution functions,
has the weight in the two components reversed from that seen experimentally; the positions of the two components also differ slightly from those
seen experimentally (see Fig. 10(a)).

Modifications of the Bennett "global" criterion[23] for adding spheres
to a DRP cluster have been made by Sadoc, et al.,[15] by Mrafko and Duhaj[40]
and by Ichikawa.[41] All of these modifications increase the number of
nearly perfect tetrahedra in the DRPHS structures. Ichikawa[41] has found
that the packing fraction is reduced as the degree of "tetrahedron perfection" is increased. Also observed were both progressive reversal of
weight in the two components of the second peak of $W(r)$ and growth of the
shoulder or splitting in the second peak of $I(K)$, as shown in Fig. 9(a)
and 9(b) taken from Ichikawa.[41] The height of the first peak in $I(K)$ is
apparently reduced as the degree of tetrahedron perfection is increased.
This is particularly true for the Sadoc, et al.,[15] DRPHS, for which
$I(K_1) \sim 2.2$ and $\Delta K_1 \sim 0.09 Å^{-1}$.

Relaxation of DRPHS Structures. L. von Heimendahl[42] recently investigated effects of relaxing a dense random packing of 888 hard spheres with
respect to a realistic interatomic potential. He began with the 888
central atom region of Bennett's 3999 atom model[23] and used a Leonard-
Jones potential for the initial relaxation. He then further relaxed the
structure using a Morse potential, adjusting the interaction range to
extend to the minimum following the near neighbor maximum in $W(r)$. Distribution functions for his initial and final structures are compared in
Fig. 10 with those for amorphous Co at 4.2°K.[17] In both cases, $r_o = 1$ is
the center position of the first peak. This relaxation has clearly increased the similarity between the model and experimental distribution
functions. Similar trends were also obtained in relaxing a 888 atom section of the Finney model,[11] but the changes were smaller than those for
the Bennett model;[23] the atoms in the Finney model are perhaps from the
beginning more "locked in" and less free to move.

Anisotropies and Long-Range Ordering in DRPHS Models. Dense random
packing of hard sphere models have been generally considered to be
"ideally amorphous," with no significant regions of crystalline ordering,
i.e. no periodically repeating structure. However, Pang, et al.,[43] and
Chaudhari and Graczyk[44] have noted the occurance of "warped sheet" planelike configurations in dense random packings. More extensive studies of
planes and of anisotropies in dense random packings have been reported by
Alben, et al.,[45] and some of their rather surprising results will be discussed in this section.

The scattering intensity per atom, for a model structure with just
one type of scatterer, is given by

$$I(\vec{K}) = \frac{1}{N} \sum_{ij} e^{i\vec{K} \cdot (\vec{r}_i - \vec{r}_j)} \tag{4}$$

The isotropic average scattering $\bar{I}(|K|)$, the average of $I(\vec{K})$ over all
directions of the scattering vector $\vec{K}$, is given by
316

$$\bar{I}(|K|) = \frac{1}{N} \sum_{ij} \frac{\sin|K||r_i - r_j|}{|K||r_i - r_j|} \tag{5}$$

In Fig. 11 are shown $\bar{I}(|K|)$ for spherical regions from two DRP models: (a) the central 890 units of the 3999 unit model of C. H. Bennett[23] and (b) the 996 central units of the 7934 unit model of J. L. Finney.[11] The first peak associated with correlations within the models occurs at $7.7\ d_o^{-1}$, where d_o is the hard sphere diameter. The peak indicates the presence of structural periodicities with a spacing of $(2\pi/7.7)d_o = 0.82 d_o$, and the sharpness of the peak indicates that these periodicities persist over considerable range.

In Fig. 12 is shown the behavior of the direction dependent intensity function $I(\vec{K})$ with the direction of $\vec{K}$ for $|K| = 7.7\ d_o^{-1}$ for the two models. Points in the x-y plane correspond to the spherical angles θ and ϕ and the intensity is given by the height z of the "scattering net." For each model there are several directions for which the scattering tends to be quite large. The "planes" responsible for these maxima can be seen in Fig. 13, which shows the projection of 440 central units of the DRP models onto the planes containing the two strongest scattering directions for $|K| = 7.7\ d_o^{-1}$. For the Bennett model[23] there is also a third set of planes perpendicular to a direction about midway between directions perpendicular to the indicated sets of planes, and these three coplanar strong scattering directions are correlated to the symmetry axes of a triangular three unit seed which was used in generating the model.

In Fig. 14 is shown the autocorrelation function for density oscillations for the strongest scattering direction for the 996 central units of the Finney model.[11] This is compared with the uniform sphere autocorrelation function, which dies off simply because fewer pairs of units can be found at larger separations in a finite cluster. Remarkably, the oscillatory correlation for this case falls off no faster than expected from finite size effects alone. Also shown is a similar result for the full 7934 unit model. The autocorrelation function for this larger cluster does not oscillate about zero because the cluster does not have a spherical shape; about 20 correlation peaks can be seen. Even for this large model, the overall decrease in the correlation is not significantly more rapid than that due strictly to finite size.

These observations of apparent anisotropies and planes in DRPHS models can be compared with results for a cluster of characteristic dimension $\bar{R}$ of randomly placed scatters. For this case it can be shown that the expected range for the maximum value I_{max} of $I(\vec{K})$ for fixed $|K|$ is given by

$$2\ \ell n(|K|\bar{R}) - 2.5 < I_{max}(|K|) < 2\ \ell n(|K|\bar{R}) + 0.5 \tag{6}$$

where the probabilities that I_{max} will be above or below this range are each about 0.1. Since the directional average of $I(\vec{K})$, denoted $\bar{I}(|K|)$, is unity in this case, Eq. (6) may also be regarded as the expected range for the ratio $I_{max}/\bar{I}$. The anisotropic scattering for such random structures can be attributed to randomly occurring reinforcement and interference.

317

The overall behavior of anisotropy in $I(\vec{K})$ for a particular 519 atom cluster of scatterers randomly distributed within a sphere is shown in Fig. 15. The ratio $I_{max}/\bar{I}=7.9$ is within the range 6.6 ± 1.5 given by Eq. (6). Thus, even in a model where intrinsic correlations are relatively short ranged and, on the average, isotropic in character, it must be expected that by chance the correlations will persist over much larger range for some directions.

Characteristics of the scattering anisotropy of the Finney[11] and Bennett[23] DRP model structures are given in Table I. The degree of anisotropy $I_{max}/\bar{I}$ exceeds that expected for randomly occurring reinforcement and interference $(I_{max}/\bar{I})_{random}$ for both the Finney[11] and Bennett[23] models. Much smaller differences were found between $I_{max}/\bar{I}$ and $(I_{max}/\bar{I})_{random}$ for tetrahedral continuous random network structural models.[45] This extra anisotropy for DRPHS models, with the associated strongly scattering "planelike" structural periodicities, may be responsible for the strong, sharp first maxima of $I(K)$ for amorphous metal-metalloid alloys, but direct evidence for the presence of such planelike, anisotropic structures in these alloys is not available. It would be interesting to investigate such anisotropies in DRPHS structures generated by other algorithms, e.g. those discussed above in the section "Computer Generated DRPHS Models with Modified Algorithms."

CONCLUSIONS

Significant progress has been made in unraveling the structure of amorphous metals and alloys since the first radial distribution function for an amorphous Ni-P alloy was obtained in 1962.[6a] The growth of research in this area in the last year has been quite amazing. Some of the "recent developments" which I have selected for this review may provide the basis for even further progress.

REFERENCES

1. G. S. Cargill III, in Solid State Physics, Vol. 30 (F. Seitz, D. Turnbull and H. Ehrenreich, eds.), p. 227. Academic Press, New York 1975.

2. A. Brenner, D. E. Couch and E. K. Williams, J. Res. Natl. Bur. Std. 44, 109 (1950).

3. R. Hilsch, Non-Crystalline Solids, ed. V. D. Frechette (Wiley, New York, 1960) p. 348.

4. W. Klement Jr., R. W. Willens and P. Duwez, Nature 187, 869 (1960).

5. J. Dixmier and K. Doi, Compt. rend. 257, 2451 (1963).

6. J. Dixmier, K. Doi, and A. Guinier, Physics of Non-Crystalline Solids, ed. J. A. Prins (North Holland Publishing Co., Amsterdam, 1965) p. 67.

6a. V. P. Moiseev, [Izv. Akad. Nauk. SSSR Ser. Fiz., 378 (1962)] had published a radial distribution function for an amorphous Ni-P alloy in 1962, but its resolution was much poorer than that of Dixmier, et al.[5,6]

7. C. N. J. Wagner, J. Vac. Sci. Tech. 6, 650 (1969).

8. G. S. Cargill III, J. Appl. Phys. 41, 12 (1970).

9. G. S. Cargill III, J. Appl. Phys. 41, 2249 (1970).

10. M. H. Cohen and D. Turnbull, Nature 203, 964 (1964).

11. J. L. Finney, Proc. Roy. Soc. (London) A319, 479 (1970).

12. D. E. Polk, Scripta Met. 4, 117 (1970).

13. J. D. Bernal, Proc. Roy. Soc. (London) A280, 299 (1964).

14. D. E. Polk, Acta Met. 20, 485 (1972).

15. J. F. Sadoc, J. Dixmier and A. Guinier, J. Non-Cryst. Solids 12, 46 (1973).

16. T. Ichikawa, Phys. Stat. Sol. (a) 19, 707 (1973).

17. P. K. Leung and J. G. Wright, Phil. Mag. 30, 185 and 995 (1974).

18. J. Orehotsky and K. Schroder, J. Appl. Phys. 43, 2413 (1972).

19. J. J. Rhyne, S. J. Pickart and H. A. Alperin, Phys. Rev. Lett. 29, 1562 (1972).

20. P. Chaudhari, J. J. Cuomo and R. J. Gambino, IBM J. Res. Dev. 17, 66 (1973).

21. J. J. Rhyne, S. J. Pickart and H. A. Alperin, AIP Conf. Proc. 18, 563 (1974).

22. G. S. Cargill III, AIP Conf. Proc. 18, 631 (1974).

23. C. H. Bennett, J. Appl. Phys. 43, 2727 (1972).

24. D. J. Adams and A. J. Matheson, J. Chem. Phys. 56, 1989 (1972).

25. S. Kirkpatrick and G. S. Cargill III, unpublished.

26. H. S. Chen and B. K. Park, Acta Met. 21, 395 (1973).

27. J. Dixmier, J. Bletry and J. F. Sadoc, J. Phys. (Paris) 36, C2-65 (1975).

28. J. Bletry and J. F. Sadoc, J. Phys. F 5, L110 (1975).

29. Y. Waseda and S. Tamaki, Z. Physik B 23 (1976) in press.

30. Y. Waseda and T. Masumoto, Z. Physik B 22, 121 (1975).

31. T. Masumoto and R. Maddin, Mat. Sci. and Engr. 19, 1 (1975).

32. C. P. P. Chou and D. Turnbull, J. Non-Cryst. Solids 17, 169 (1975).

33. G. C. Chi and G. S. Cargill III, presented at the Second Int. Conf. on Rapidly Quenched Metals, Boston, Mass., November 1975, to be published in conference proceedings.

34. G. C. Chi and G. S. Cargill III, to be presented at 21st Annual Conf. on Magnetism and Magnetic Materials, Philadelphia, Pa., December 1975, to be published in conference proceedings.

35. R. J. Gambino, P. Chaudhari and J. J. Cuomo, AIP Conf. Proc. 18, 578 (1974).

36. G. S. Cargill III, AIP Conf. Proc. 24, 138 (1975).

37. C. N. J. Wagner, N. Heiman, T. C. Huang, A. Onton and W. Parrish, to be presented at 21st Annual Conf. on Magnetism and Magnetic Materials, Philadelphia, Pa., December 1975, to be published in conference proceedings.

38. J. Logan, Scripta Met. 9, 379 (1975).

39. Y. Waseda and T. Masumoto, Z. Physik B 21, 235 (1975).

40. P. Mrafko and P. Duhaj, J. Non-Cryst. Solids 17, 143 (1975).

41. T. Ichikawa, Phys. Stat. Sol. (a) 29, 293 (1975).

42. L. v. Heimendahl, J. Phys. F 5, L141 (1975).

43. T. W. S. Pang, U. M. Franklin and W. A. Miller, *Mater. Sci. and Engr.* 12, 167 (1973).

44. P. Chaudhari and J. Graczyk, *Bull. Am. Phys. Soc.* 19, 317 (1974), and to be published in *J. Non-Cryst. Solids*.

45. R. Alben, G. S. Cargill III and J. Wenzel, to be published in Phys. Rev. B.

CLOSE-PACKED HARD SPHERE MODELS OF METALLIC GLASS ALLOYS

D.S. Boudreaux and J.M. Gregor

Materials Research Center
Allied Chemical Corporation
Morristown, New Jersey

In this paper, computer simulation of metallic glass structures is discussed. Large clusters of spheres are geometrically assembled according to the global criterion of Bennett[1]; however, the work focuses on modeling systems containing more than one type of atom. In particular, two structures of the TM-M (transition metal-metalloid) type will be analyzed. Previous theoretical work on binary systems has been restricted either to consideration of rare-earth-transition metal glasses[2] or, to limited statistical summaries (in the one case where TM-M systems were studied).[3]

The primary difficulty with the computational task is the growth of time usage with model size. The following simple analysis serves to illustrate the point. The simulation proceeds by adding atoms (represented by spheres) one at a time to a seed cluster. Consider the point at which the model contains a large number, say n, atoms; the atom being added represents an increment, dn. To avoid sphere overlap, the distance must be calculated between the addition site and each of the other n sites. The total number of distance calculations for an N atom structure is then

$$N_d = \int_o^N n\,dn \propto N^2.$$

If each distance calculation requires a time K, then the total time is at least KN^2. Thus computer time, and hence cost, vary quadratically with the number of atoms desired in the simulation.

One can avoid the run-away cost if the number of operations at each addition step can be kept fixed or at least upper bounded. The present work accomplishes this by using a trade-off between computer time and information storage capacity. Basically, one considers space to be partitioned into both cubically symmetric and radially symmetric bins. Every atom or addition site belongs to one and only one bin of each symmetry and a cross-reference between bins is kept.

Overlap calculations need only be done between atoms and/or addition sites in neighboring bins, and since there is a maximum occupancy of each bin, the desired limitation of operation is effected. A more detailed explanation[4] would require excessive space in this publication.

323

To simulate multi-component alloy systems one uses different size spheres to represent the various elemental alloy components. Each different size sphere used in the simulation requires a continually updated list of sites which satisfy Bennett's global criterion, namely, that the site touch three atoms in the existing cluster without overlapping any other atom. Atoms are serially designated to occupy addition sites. The type of atom (size of sphere) to be added is first determined by comparing a uniformly distributed random number, r: $0 < r \leq 100$, to the desired composition. The list of addition sites for that type atom is then searched for that site closest to the geometrical origin of the cluster. Since the smaller atoms tend to be placed nearer to the center of the model this procedure permits only approximate control of the desired composition.[4]

The clusters so generated are spherical in nature. As the construction proceeds, the radially symmetric bins of data become devoid of addition sites one-by-one, and are called <u>filled</u>. Analysis of the contents of the filled bins yields packing fraction, density and composition as a function of cluster size. Analysis of the data from a completed cluster by a separate computer program yields coordination numbers, pair correlation functions, radial distribution functions, and interference functions (or structure factors). The latter program was carefully written to expose any possible anisotropies in the cluster. Basically, the computation is concerned with the distances of all atoms from a given atom considered as center; in addition to classifying centers and atoms by type, the centers are classified by the radial shells to which they belong and the atoms are classified by the orientation of the atom to center vector relative to a cluster radius.

The first two structures constructed and analyzed as described above will be discussed in this paper. The Fe-P system was chosen since it was the subject of recent investigations by Logan.[5] The Fe radius was taken to be 1.36 Å and the P radius to be 0.98 Å. The composition used in conjunction with the random number selection procedure was 80% Fe and 20% P (atomic percent). The initial cluster consisted of two Fe and one P atom in contact with one-another. Two structures containing 2500 atoms each were generated; they differ only in that one allows P-P near neighbors and the other does not.

Analysis of composition with model size exposes the fact that the algorithm being used at present permits some initial variation in the ratio of the number of Fe to the number of P atoms. By the time the model contains 1000 atoms, however, the ratio levels off to approximately 77% Fe and 23% P.

Packing fraction, p, variations with model size are a puzzling feature of Bennett's model[1] and a similar trend is observed in this work. p is larger for our binary system as expected from packing considerations, but a decrease in p with model size is seen. So as not to over-emphasize this point, the decrease from 1000 atoms to 2500 atoms is from 0.640 to 0.636 and is nearly the same for the two models discussed herein.

Coordination of both Fe and P sites is higher in the center of the structure as observed previously.[1] Averaged over the entire models the calculated coordination numbers are:

	P-P forbidden	P-P allowed
Fe by Fe:	6.71	6.69
Fe by P:	1.59	1.51
P by Fe:	5.29	4.87
P by P:	--	0.69

These numbers are lower than those measured by Logan,[5] and it is unlikely that any application of the present construction algorithm could produce very much higher coordinations. Further, Logan measured all the area out to the first minimum in

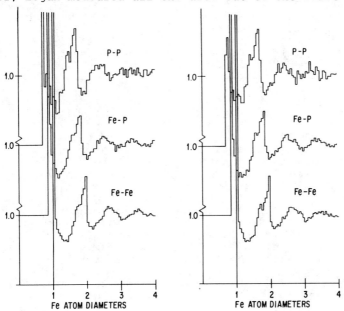

Pair correlation functions broken down as labeled. On left, P-P near neighbors were allowed; on right, they were not. Note the splitting of the second Fe-Fe peak on the right.

the RDF which does not strictly correspond to the numbers cited above which are calculated from the first bar only of a histogram; also Logan is measuring some "composite" coordination number since he does not resolve the various pair contributions.

Pair correlation functions (PCF), broken down into the 3 pair types, are shown in Figure 1. Unfortunately, one cannot compare these data with the work of Sadoc et al.[3] because the latter authors combined them into a radial distribution function weighing each part equally (which is incorrect). Bennett[1] observed the split second peak as have others[2]; it illustrates an ordering in the structure of the hard sphere packing not found in simple liquids.

In the model allowing P-P near neighbors, the inner half of the split second peak is suppressed. This effect is even more evident in the PCF's separating the center atoms into three radial classes. In the outermost class the second peak sharpens substantially to the outer parts.

In the model where P-P near neighbors are not allowed, the suppression of the first peak seems to move the intensity into the inner half of the second peak of the Fe-Fe and Fe-P PCF's. This corresponds to an "increase of order" in this glass as compared to the other. In the overall radial distribution of atoms, "removing" P atoms from the P-P nearest neighbor shell forces them outward to establish a shell which is not prominent when the P atoms are allowed farther in.

Logan[5] measured radial distribution functions for some Fe-P alloys and observed the split second peak. Both the splitting and intensity ratio are in rough agreement with the calculated values, although he observes them slightly closer in (8%).

Another observation worth noting is that there is an orientational dependence to the width of the first peak, particularly in the case of the Fe-Fe PCF. The PCF's calculated using atoms located in a 60° cone "looking" outward or "looking" inward from a given center are statistically indistinguishable. However, the atoms in the remaining space (oriented tangentially with respect to the center atom) illustrate a much less intense first peak and a broad shoulder on the high side. This is apparently connected with the fact that atoms are being added under geometries which emphasize that they touch those above and below rather than to the side. The effect diminishes somewhat with radial class, but the statistics are not sufficiently accurate to permit relating this phenomenon to the decreasing packing fraction with radial class. Further, we have noted that the number of counts under the first bar of the Fe-P PCF histogram increases in the tangential directions. It is felt that this is due to the smaller P atoms being able to fit in surface pockets between the larger Fe species.

In this paper, the initial applications of a new method of simulating metallic glass structures have been discussed. Plans are to study compositional variation of multicomponent alloys and to isolate and remove any items in the present

326

theoretical algorithms which are causing unphysical anisotropies in computer simulations.

REFERENCES

1. Bennet, C.H., J. Appl. Phys. 43, 2727 (1972).
2. Cargill, C.S. III, Solid State Phys. 30, 227 (1975).
3. Sadoc, J.F., J. Dixmier, and A. Guinier, J. Non-Cryst. Solids 12, 46 (1973).
4. Gregor, J.M. and D.S. Boudreaux [to be published].
5. Logan, California Institute of Technology Report #CALT-822-65 (1974).

ELECTRONIC STRUCTURE OF METALLIC GLASSES

M. Fischer, H.-J. Güntherodt, E. Hauser, H.U. Künzi, M. Liard, and R. Müller

Experimental Physics of Condensed Matter
Institute of Physics
The University of Basel, Switzerland

INTRODUCTION

In recent years, metallic alloys in the amorphous[1] and in the liquid[2] state have received much attention. Transition metals and their alloys, which are of current interest in both fields, provide a unique opportunity for a link between the amorphous and the liquid state.

There are certain metallic alloys of transition metals which are good glass formers and remain relatively stable in the amorphous state. Examples of such glass systems are $Pd_{80}Si_{20}$[3] and the Metglas 2826A $(Fe_{32}Ni_{36}Cr_{14}P_{12}B_6)$.[4] Their mechanical, magnetic, electrical, and chemical properties have unexpected features, which qualify them for many applications.

On the other hand, progress has been made in understanding the electronic structure of liquid transition metals and their alloys. In particular, the theory of the electrical resistivity[5] has been developed to give a qualitative and sometimes even quantitative explanation of the resistivity data in the liquid state. It seems therefore worthwhile to look for an application of these theories to the amorphous state.

Such an extension is based on the strong experimental similarities in the electronic properties of amorphous and liquid alloys of transition metals.

EXPERIMENTAL

The different samples used in our experiments were taken from commercially produced Metglas 2826A (Allied Chemical). The electrical resistivity was measured by the four-point-probe method with an ac current of 85cps. An ac current—ac magnetic field technique has been used to determine the Hall coefficient in a magnetic field of 0.1 Tesla. The

329

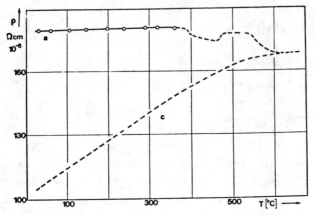

Figure 1. Electrical resistivity of Metglas 2826A as a function of temperature.

magnetic susceptibility was measured by a Faraday balance.

The electrical resistivity as a function of temperature is shown in Figure 1 (supra). In the amorphous (a) state, the electrical resistivity is 180μΩcm.

This value remains temperature independent within the relative experimental error of 0.1%. Then follows the irreversible transition into the crystalline (c) state. The resistivity of the crystalline sample varies linearly with temperature up to 400°C. For higher temperatures the resistivity tends to bend away from linear T behavior toward the temperature axis. The absolute value and the temperature dependence of the resistivity show drastic differences between the amorphous and the crystalline state. Due to the high vapor pressure of P it is not easy to heat the Metglas sample up to the melting point and compare directly the amorphous and the liquid state of the same alloy.

The Hall coefficient [Figure 2] shows in the amorphous state at room temperature a drastic decrease due to the effects associated with the Curie temperature of 250K. The plot of $1/R_H$ versus temperature shows a deviation from a straight line in the amorphous state. Figure 3 indicates that the Hall coefficient in fact obeys a $1/R_H^2$ versus temperature law. Above 300°C the Hall coefficient approaches the value of $+27 \cdot 10^{-11} m^3$/As. Then follows the transition into the crystalline

330

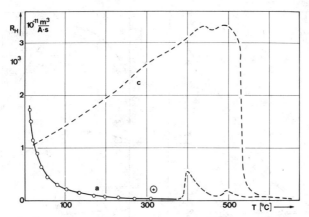

Fig.2 Hall coefficient of Metglas 2826A
as a function of temperature.

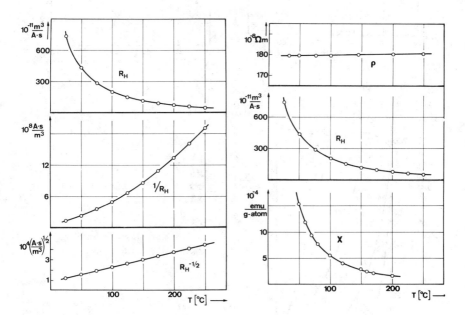

Fig.3 Different plots of
the Hall coefficient.

Fig.4 Summary of the properties of Metglas
2826A.

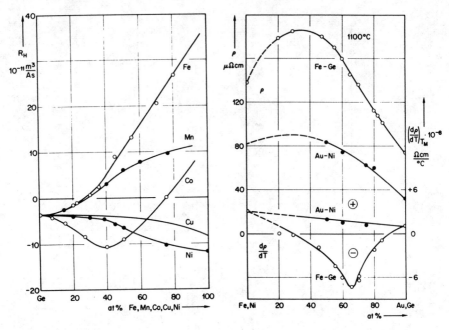

Fig. 5

Fig. 6

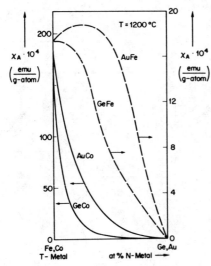

Fig. 5
Hall coefficient of liquid transition metal alloys.

Fig. 6
Electrical resistivity of liquid transition metal alloys

Fig. 7
Magnetic susceptibility of liquid transition metal alloys.

state. The Hall coefficient at 300°C in the crystalline state is two
orders of magnitude larger than in the amorphous state.

Figure 4 shows the magnetic susceptibility in the amorphous state.
The susceptibility follows the same temperature dependence as observed for the Hall coefficient. Figure 4 also contains as a summary
for the discussion the relevant values of the resistivity and the Hall
coefficient in the amorphous state.

DISCUSSION

First of all, we will discuss the strong experimental similarities
between the amorphous and the liquid state. The observed values of the
electrical resistivity, the Hall coefficient, and the magnetic susceptibility in the amorphous state are in good agreement with typical values observed for liquid transition metal alloys. Such an agreement is
surprising in view of the different components forming Metglas 2826A.
For a detailed comparison, we present such data of the electrical resistivity [Figure 5], the Hall coefficient [Figure 6], and the magnetic susceptibility [Figure 7] for different liquid alloys. We can conclude that the typical values of the amorphous alloy: the electrical
resistivity of $180\mu\Omega cm$, the zero temperature coefficient of the resistivity, the Hall coefficient of $+27 \cdot 10^{-11} m^3/As$ and the magnetic susceptibility of $+15 \cdot 10^{-4} emu/g$-atom are characteristic for a liquid
$Fe_{80}Ge_{20}$ alloy. Therefore we believe, that the Metglas 2826A
($Fe_{32}Ni_{36}Cr_{14}P_{12}B_6$) can be treated in a simple model as a pseudobinary
alloy $T_{82}N_{18}$ containing roughly 82at.% of an effective transition metal
and 18at.% of a polyvalent metal. This means that the Fe, Ni, and Cr
components act as a 3d configuration close to Fe. Support of this
idea is also given from the recently established relation[6] between the
composition range of metallic glasses and the wave numbers $2k_F$ (Fermi-
wave number) and K_p (position of the first peak of the liquid structure factor). The observed similarities of the amorphous and liquid
state suggest that the explanations and the theories for the resistivity, the Hall coefficient, and the susceptibility in the liquid state,[2]
should be applicable to the amorphous state.

333

As an example we deal with the ideas of the theory of the electrical resistivity. Such a theory explains why the resistivity values of 180μΩcm in amorphous and liquid alloys are considerably larger than the resistivities of pure liquid Fe and Ni.[7]

In our simple model we assume the Metglas to consist of statistically distributed T and N ions and conduction electrons. The T and N atoms are represented by their nonoverlapping muffin-tin potentials. The electrical resistivity is due to the scattering of conduction electrons by the muffin-tin potentials. For pure transition metals, the main contribution arises as resonance scattering from the 3d states lying in the conduction band. On alloying with polyvalent metals the contribution of the structure factor to the resistivity increases. The electrical resistivity in a single-site approximation is given by

$$\rho = \frac{3\pi\Omega}{4e^2 v_F^2 \hbar k_F^4} \int_0^{2k_F} |U(K)|^2 K^3 dK$$

with

$$
\begin{aligned}
|U(K)|^2 &= c_T |t_T|^2 (1 - c_T + c_T a_{TT}) \\
&+ c_N |t_N|^2 (1 - c_N + c_N a_{NN}) \\
&+ c_T c_N (t_T^* t_N + t_T t_N^*)(a_{TN} - 1)
\end{aligned}
$$

where c_T, c_N are the concentrations, t_T, t_N the t matrices of the components T and N, and a_{TT}, a_{TN}, and a_{NN} the partial structure factors of the alloy. The relevance of the spin-disorder contribution to the resistivity in disordered systems is still under investigation.

CONCLUSIONS

This paper shows how fruitful the fields of amorphous and liquid metals can interact. We hope that the first values of the electrical resistivity of liquid Pd[7] make it possible in the near future to compare a simpler alloy such as $Pd_{80}Si_{20}$ in the amorphous and in the liquid state.

334

ACKNOWLEDGEMENTS

One of the authors, H.-J. Güntherodt, is very much indebted to IBM for the opportunity of an introduction into the field of amorphous metals at the IBM Watson Research Center, Yorktown Heights. Many stimulating discussions by Dr. P. Chaudhari and Dr. C.C. Tsuei are gratefully acknowledged. We would like to thank the Schweizerische Nationalfonds zur Förderung der wissenschaftlichen Forschung and the Research Center of Alusuisse for financial support.

REFERENCES

1. G.S. Cargill III, Solid State Physics 30, 227, H. Ehrenreich, et al., Editors, Academic, New York (1975).
2. Busch, G. and H.-J. Güntherodt, Solid State Physics 23, 235, H. Ehrenreich, et al., Editors, Academic, New York (1974).
3. Duwez, P., et al., J. Appl. Phys 36, 2267 (1965).
4. Gilman, J.J., Phys. Today, p. 46 (May 1975).
5. Dreirach, et al., J. Phys. F2, 709 (1972).
6. Nagel, S.R. and J. Tauc, Phys. Rev. Let. 35, 380 (1975).
7. Güntherodt, H.-J., et al., Phys. Let. (1975).

STABILITY OF METALLIC GLASSES: PHOTOEMISSION STUDIES AND A NEW ELECTRONIC MODEL FOR GLASS STABILITY[16]

S.R. Nagel and J. Tauc

Division of Engineering and Department of Physics
Brown University, Providence, Rhode Island

This paper will summarize the results of two recent investigations on the stability of metal glasses. The first of these consisted of a series of photoemission measurements on $Pd_{.775}Cu_{.06}Si_{.165}$ in its glassy and crystalline forms[1]; the second was the development of a model which arrives at a criterion for stability based on the properties of the nearly free electron gas[2]. Additional evidence will be presented which lends support to this electronic theory. Finally there will be a brief discussion of how this electronic theory may be extended to treat other classes of alloys than those previously considered.

Our photoemission studies were undertaken to look for evidence of strong local chemical bonding in the particularly stable[3] metallic glass $Pd_{.775}Cu_{.06}Si_{.165}$. Chen and Park[4] had previously suggested that such bonds could play an important role in the glass forming tendency. Evidence for their presence would have been large chemical shifts in the core levels of the various elements. We looked for such shifts between the x-ray photoemission spectra of pure Pd and of the glass and crystal alloy. No large shifts were found. Fig. 1 shows a comparison of the palladium 3d levels for the three samples. The spectra for the crystal and glass are virtually identical. That for the pure palladium is only slightly shifted to lower binding energy as compared to the alloy. This can be understood after examining Fig. 2, the ultraviolet photoemission spectra for the valence bands of the three samples. In

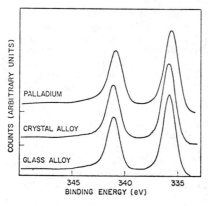

Fig. 1) The XPS spectra ($h\nu = 1253.6$ eV) for the palladium 3d levels for the three samples: 1) palladium, 2) crystal and 3) glass Pd-Cu-Si alloy. Binding energies are measured relative to the Fermi level.

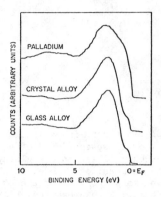

Fig. 2) The UPS spectra
(hν = 21.2 eV) of the valence
band for the three samples:
1) palladium, 2) crystal and
3) glass Pd-Cu- Si alloy.

this figure we see again that the glass and crystalline alloy are very
similar whereas the pure palladium is significantly different: the
density of states at the Fermi level is much larger than in the alloys
and the width of the valence band is larger as well. This result
implies that a rigid band model is applicable near the Fermi surface.
The four valence electrons of silicon fill the d levels of palladium
and the Fermi level is now above the top of the d-bands. This explains
the slight shift of the palladium core levels on alloying since the
Fermi level will be raised with respect to these core levels as well.
This shift of the Fermi level is consistent with other studies of this
alloy, particularly the positron annihilation studies[5] and the magnetic
susceptibility measurements[6]. (Our samples were prepared in an identical
manner to the samples used in the latter experiment.) Although these
measurements do not rule out the possibility of strong chemical bonds
playing a significant role in the glass formation they do make such an
explanation less convincing at the same time, giving additional information
on the electronic structure of the alloy. Based on this information
as well as on the properties of liquid transition metal alloys we have
developed an alternative theory for metal glass stability. This theory
deals with alloys of the form $M_{1-x}X_x$ where M is a noble or transition
metal and X is an element from group IVA or VA. In particular our model
shows why increased stability should occur with the metal content around
80%. This electronic theory need not rule out the importance of chemical
effects or the geometrical ones[7,8] based on Bernal random packing of
hard spheres. It seems that the various theories are not mutually
exclusive.

It has long been known that the properties of the electron gas,
in particular the geometry of the Fermi surface, can influence the atomic
structure of a metal. Three examples are immediately apparent: 1) the
prediction of a Peierls instability in one-dimensional system[9] 2) the
observation of charge density waves in layered compounds[10] and 3) the
success of the Hume-Rothery rules for predicting the crystal structure

of some alloys[11]. These three examples all involve the interaction of the electrons at the Fermi surface with the reciprocal lattice. The reciprocal lattice, which is derived from a Fourier transform of the pair correlation function, is responsible for opening up gaps in the band structure at the zone boundary (i.e., at a value of $\vec{q} = \vec{G}/2$ where $\vec{G}$ is any reciprocal lattice vector). When the wavevector at the Fermi surface lies at $k_F = \vec{G}/2$, the occupied states will be lowered while only the unoccupied states will be raised. The system will be relatively stable against distortion or change of crystal structure. For a larger value of k_F some states will also be raised by the opening of a gap and the system may transform to another configuration with larger reciprocal lattice vectors and lower energy. In a system where k_F is much less than any G the crystal may distort to produce a reciprocal lattice vector at a value of $2k_F$, the "spanning vector" of the Fermi surface. This is what occurs in a charge density wave or Peierls instability.

For an amorphous metal a set of reciprocal lattice vectors no longer exists, but an analogous quantity does exist and is the structure factor, $S(q)$. $S(q)$ is also related, via a Fourier transform, to the pair correlation function. In a glass or liquid both the Fermi surface and the structure factor are spherically symmetric whereas in a crystal $S(\vec{q})$ depends on the direction as well as on the magnitude of $\vec{q}$. The essential point of our electronic model is that there will be a similar behavior in the amorphous metal as in the crystalline cases mentioned in the previous paragraph. That is, when the value of $2k_F$ equals q_p, (where q_p is the value of the wavevector at the first peak in $S(q)$) the system will be in a relatively stable state. We will now show why this is true.

As mentioned above, the photoemission measurements show that for the alloy, the Fermi level lies above the palladium d-bands. Thus one might expect that the Fermi level electrons can be treated, as in the noble metals, as if they are nearly free, interacting only weakly with the ionic cores. The energy levels of the system and the density of states $D(E)$ can be calculated in perturbation theory:

$$E = E_k^o + v(o) + \frac{\Omega}{8\pi^3} \int \frac{|v(q)|^2 \, S(q)}{E_k^o - E_{k+q}^o} \, d^3q$$

where $v(q)$ is the pseudo-potential of the ions, $E_k^o = \hbar^2 k^2/2m$ and Ω is the volume. We have neglected the influence of the d-bands except in so far as they alter $v(q)$. This expression fails when $|\vec{k}+\vec{q}_p| = |\vec{k}|$ just as it does in the crystal where a gap occurs in the energy band at the zone boundary. Although for the liquid we do not expect a gap, we do expect a similar decrease in $D(E)$ at the energy $E = (\hbar^2/2m)(q_p/2)^2$. We expect a substantial change since in the liquid or glass $S(q)$ is spherically symmetric so that <u>all</u> the states at $|k| = q_p/2$ (which also all correspond to the same unperturbed energy) are affected. In addition the larger is the value $v(q)$ or $S(q)$ at q_p, the larger will be the perturbation and the deeper we expect the minimum in the density of states to be.

A system for which $2k_F = q_p$ should be more stable against crystallization than a system for which $2k_F \neq q_p$. This condition, $2k_F = q_p$, implies that the Fermi level E_F lies at the minimum of the density of states. Even though the electronic energy of the glass may be higher than that of the crystal, the glass is still at a metastable minimum of energy. If the system starts to crystallize, due to a fluctuation in which the system develops long range order, $S(q)$ is perturbed and is no longer spherically symmetric. The argument that all the states with a given magnitude of $|q| = q_p/2$ are affected is no longer valid. The value of q_p will now depend on direction and will consequently act to lower the density of states at a different value of the energy for each direction. The deep minimum at the Fermi level is no longer expected. Thus, as $S(q)$ is perturbed the total energy of the system increases as more electron states are moved up to E_F where they have higher energy. The glass is in a metastable state since any perturbation will destroy the spherical symmetry of $S(q)$.

The glass forming tendency will be enhanced for two reasons. The first is that the total energy of the liquid will be decreased at the composition where $2k_F = q_p$ and the eutectic which occurs at this concentration will be deep. The second is that there will be a barrier set up against crystallization due to the perturbation of $S(q)$ just mentioned.

For atoms with valence $z = 1$, $2k_F$ is less than q_p while for divalent atoms $2k_F$ lies slightly higher than q_p. One effect of alloying a monovalent metal with an element of higher valence is to shift the effective valence, $z_{eff} = z_1(1-x) + z_2 x$. This in turn shifts the value of $2k_F$ through q_p for some value of x. That this shift is a valid picture of what happens upon alloying is well substantiated from resistivity measurements. To get an idea of what concentrations are needed for the situation $2k_F = q_p$ we take as a typical value $z_{eff} = 1.7$. If we alloy a monovalent atom with one of valence four we get a value for x of .23 ($M_{.77}X_{.23}$). It is just at such concentrations, when $2k_F = q_p$, that the Fermi energy will lie at a minimum of the density of states curve and the glass will be stable against crystallization. The choice of $z_{eff} = 1.7$ is somewhat arbitrary but it is a reasonable estimate. Using this value we find that we can indeed predict increased stability in the same concentration range as that observed experimentally. Of course depending on the details of $S(q)$ in each alloy system, q_p will coincide with $2k_F$ at different values of z_{eff}. One example of this will be presented below. As distinct from a model based on local chemical bonds no prediction is made in this theory of what is the exact arrangement of nearest neighbors; e.g., whether two metalloid atoms can be next to each other. The one prediction is that the structure factor be sharply peaked. Atomic arrangements which have a large $S(q_p)$ will be favored.

340

As we have argued the pseudo-potential is also important for determining the glass forming tendency of an alloy system. This can be clearly seen in the case of the alkali metals which do not form glasses. In these metals $v(q_p)$ is small whereas for transition metals like copper, due to the influence of the d-bands, it is much larger[12]. The d-bands thus do play a central role in the glass formation. For example, calculations of D(E) for Hg show a pronounced minimum whereas for Al they show none at all[13].

In reference 2, evidence was presented from resistivity and magnetic susceptibility measurements which helped support this model. These measurements showed that group VIII transition metals act like the noble metals upon alloying. Further support can be found from examining the alloy system Au : Si which forms a glass at 30 at.% Si. This is a higher metalloid content than one finds in many other glasses. However, if one compares the structure factors for pure Au and for the $Au_{.70}Si_{.30}$ glass one finds that the value of q_p is significantly greater in the alloy[14]. In order to have the value of $2k_F$ coincide with the value of q_p more than the normal amount of metalloid must be added. This is one instance where the strict rigid band approximation would break down: the motion of q_p on alloying, though often small, must be taken into account in calculating the correct concentration for glass formation. The anomalously large Si concentration in this glass can therefore be understood in terms of this electronic model.

Another piece of evidence again comes from the magnetic susceptibility measurements on Pd, liquid Si[15] and the $Pd_{.775}Cu_{.06}Si_{.165}$ glass[6]. Whereas both Pd and liquid Si are paramagnetic the glass is slightly diamagnetic. One possible reason for this is that the Pauli paramagnetism, which is proportional to $D(E_F)$, is a minimum for the glass. This is in agreement with the predictions of our model.

The model that we have presented above can be generalized to include other glass forming systems such as those that contain two transition or noble metals and no metalloids. It can be extended to any situation where there is a minimum in D(E) at the Fermi level and where this minimum is most pronounced for a truly amorphous system. That is, the model should remain applicable as long as any perturbation which sets up long range order makes the minimum more shallow. One possible way in which this situation might occur is if on alloying two transition metals the electrons filled the band structure up to the hybridization gap[12]. For this to happen one of the elements must have a low valence and the other a high valence (e.g., Cu:Zr). It is conceivable that the energy for the hybridization gap would be better defined in an amorphous system than for a crystalline one. In a crystal each direction will have a distinct value for the energy at which this gap occurs. If the d-states are influenced by more than just the nearest neighbors, the strong variation in direction produced by the short range order may be smeared out. This particular model also depends strongly on assuming that the electrons from one element can fill, in a rigid band fashion, the d-bands of the other. It is not clear that this is always the case.

In conclusion we have presented the results of photoemission measurements on a stable glass. These results do not show evidence for strong chemical bonds. They are consistent however with a nearly free electron theory of the glass which is also supported by much other experimental evidence from alloy systems. One distinguishing feature of this model is that it is formulated in reciprocal lattice space instead of real space. This enables one to see from the outset the effect that long range order will have on the system.

REFERENCES

1. Nagel, S.R., G.B. Fisher, J.Tauc, and B.G. Bagley [to be published].
2. Nagel, S.R. and J. Tauc, Phys. Rev. Lett 35, 380 (1975).
3. Chen, H.S. and D. Turnbull, Acta Met. 17, 1021 (1969).
4. Chen, H.S. and B.K. Park, Acta Met. 21, 395 (1973).
5. Chuang, S.Y., S.J. Tao, and H.S. Chen [to be published].
6. Bagley, B.G. and F.J. DiSalvo in: Amorphous Magnetism, ed. by H.O. Hooper and A.M. deGraaf (Plenum Press, New York), 143 (1973).
7. Polk, D.E., Acta Met. 20, 485 (1972).
8. Turnbull, D., J. de Physique 35, Colloque 4, 1 (1974).
9. Peierls, R., Quantum Theory of Solids, Oxford (1955).
10. Wilson, J.A., F.J. DiSalvo, and S. Mahajan, Phys. Rev. Lett 32, 882 (1974).
11. Hume-Rothery, W., Electrons, Atoms, Metals and Alloys, Dover (1963).
12. Harrison, W.A., Solid State Theory (McGraw Hill, New York), Chap. 2.9.3 (1970).
13. Ballentine, L.E., Can. J. Phys. 44, 2533 (1966); L.E. Ballentine and T. Chan, in Proceedings of the Second International Conference on the Properties of Liquid Metals (Taylor & Francis, London), 197, (1973).
14. Giessen, B.C. and C.N.J. Wagner in: Physics and Chemistry of Liquid Metals, ed. S.Z. Beer (Dekker), 633 (1972).
15. Busch, G. and H.-J. Güntherodt in: Solid State Physics, ed. H. Ehrenreich, F. Seitz, and D. Turnbull, Vol. 29, 235, Academic Press, New York (1974).
16. Work supported by the National Science Foundation and the NSF-MRL program at Brown University. We would like to thank G.B. Fisher and B.G. Bagley for collaboration on the photoemission experiments.

MEASUREMENTS OF THE SHORT RANGE ORDER OF GLASSY METALLIC ALLOYS
QUENCHED FROM THE MELT BY THE X-RAY AND THE MÖSSBAUER TECHNIQUE
AND COMPARISON WITH THE ATOMIC DISTRIBUTION OF THE MELT

G.E.A. Bartsch, P. Glozbach, and T. Just
Technische Universität Berlin, West Germany

Since the x-ray investigations of amorphous metallic alloys only yield
information about the total number of the next neighbour atoms, and
since it is impossible to separate the radial distribution function
(RDF) (See Figure 1), in order to obtain the atomic arrangement of either the metal atoms or the nonmetal atoms, the Mössbauer technique is
suggested as a means of obtaining detailed information about the environment of one metal atom in these alloys. This technique has the advantage
that it is in principle possible to distinguish between the nonmetal
and the metal neighbours of a metal center atom.

MÖSSBAUER EXPERIMENTS

The following amorphous alloys were investigated[1]: $Pd_{79}(Fe_1)Si_{20}$,
$Fe_{80}P_{13}C_7$, $Au_{76}(Fe_1)Ge_{13}Si_{10}$, and $Pt_{60}Ni_{15}P_{25}$. The Mössbauer experiments
with the alloys Pd(Fe)-Si, Fe-P-C, and Au(Fe)-Ge-Si were performed us-
ing the 14.4keV radiation of a 10mCi Co_{57}/Pt source. The amorphous alloy Pt-Ni-P was investigated using the 99keV radiation of Pt 195 of a
5mCi Au_{195}/Pt Mössbauer source. The non-crystalline samples were obtained using the piston and anvil quenching technique. While the experiments discussed in this work using the Fe 57 Mössbauer isotope were
performed at room-temperature, the experiments on the amorphous Pt-Ni-
P alloy using the Pt 195 Mössbauer isotope had to be done by cooling
source and absorber down to 4.2K.

Figure 2 shows the characteristics Mössbauer spectrum of the non-
crystalline alloys investigated. The spectra obtained with the Fe 57
Mössbauer isotope show evidence of a quadrupole interaction. In the
case of the amorphous $Fe_{80}P_{13}C_7$ alloy, an additional magnetic hyperfine interaction was observed. However, the spectrum of non-crystalline Pt-Ni-P shows a single line having a line width FWHM (free width

343

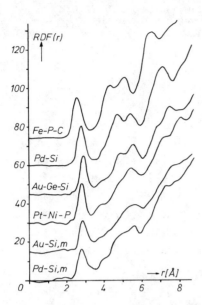

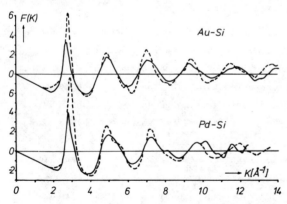

Figure 1. The RDF's of metallic
glasses compared with those of
liquid $Pd_{84}Si_{16}$ (T = 900°C) and
$Au_{81}Si_{19}$ (T = 410°). The zeros
are displaced.

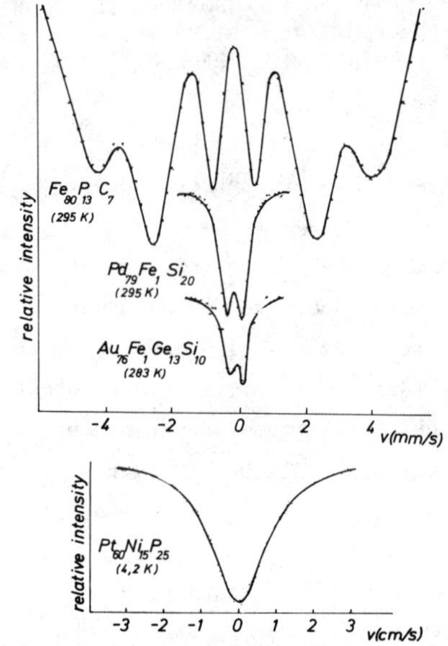

Figure 2. Mossbauer spectra of non-
crystalline Fe-P-C,Pd(Fe)-Si, Au(Fe)
-Ge-Si, and Pt-Ni-P.

Figure 3. The reduced interference functions F(K)=K[I(K)-1] of glass (---) and
molten (—) $Pd_{84}Si_{16}$(T=900°) and of glassy (---) $Au_{77}Ge_{13}Si_{10}$ and molten (—)
$Au_{81}Si_{19}$ (T=410°C).

at half maximum) that shows no broadening when compared with the line width of crystalline Pt metal (FWHM = 1.8cm/s). Table 1 lists the values of the isomer shift, the quadrupole splitting, and the hyperfine splitting taken from the curves in Figure 2, including the results of Tsuei, et al.[2] for non-crystalline Pd-Fe-P, which will be included in the discussion as well.

The quadrupole interaction observed for the non-crystalline alloys investigated using the Fe 57 Mössbauer source suggests a non-isotropic environment of the Mössbauer atom. In the case of non-crystalline Pt-Ni-P no evidence for such an interaction was found. This suggests that, in this case, the environment of the Mössbauer atom is isotropic.

The model used for the interpretation of the experimental data for the alloys Pd(Fe)-Si, Fe-P-C, Pd-Fe-P, and Au(Fe)-Ge-Si is the Bernal-Polk model,[3] which assumes that the metal atoms form a dense random packing of hard spheres. In this structure, holes exist that are filled with the nonmetal atoms of the alloy (e.g., Ge, Si, P, C). Polk and Maitrepierre[4] also concluded that an atom in such an alloy has a neighborhood quite similar to the atomic arrangement in the crystalline intermetallic compound Pd_3Si, Fe_3P, and Pd_3P. These intermetallic compounds have a structure of noncubic symmetry similar to cementite (Fe_3C). Using the known structural data of these intermetallic compounds, one can calculate that a metal atom in the non-crystalline alloys has on the average 2 2/3 next nonmetal neighbours and approximately 10 next metal atom neighbours. The interatomic spacing between the nonmetal atoms and the center atom is smaller than the distance of the surrounding metal atoms. This number of the next nonmetal atom neighbours cannot be deduced from an x-ray measurement but can be verified by the Mössbauer experiements.

The experimental results were interpreted by applying the theoretical considerations of Sauer, et al.[5] and Brossard, et al.[6] Their experiments on iron solid solutions containing Ge and Si could best be interpreted by assuming the following relation between the measured isomer shift IS and the number N_{mn} of the next non-iron neighbours of an iron atom:

Table 1. The Experimental (e) and the Theoretical (t) Values of the Isomer Shift IS and the Number of the Next Nonmetal Neighbours N_{mn}, the Quadrupole Splitting QS, and the Magnetic Splitting H.

	IS(e) (mm/s)	IS(t) (mm/s)	N_{mn}(e)	N_{mn}(t)	QS (mm/s)	H (kG)	Ref.
$Fe_{80}P_{13}C_7$	0.18	0.18	2.4	2.7	----	255	[1]
$Pd_{79}(Fe_1)Si_{20}$	0.12	0.30	2.4	2.7	0.33	---	[1]
$Pd_{67}Fe_{13}P_{20}$	0.35	0.36	2.45	2.7	0.88	---	[2]
$Au_{76}(Fe_1)Ge_{13}Si_{10}$	0.23	-----	4.0	---	0.4	---	[1]
$Pt_{60}Ni_{15}P_{25}$	-0.26	---	---	---	----	---	[1]

The values of the isomer shifts are relative to α-Fe or Pt metal.

$$IS = IS_o + N_{mn} \cdot \Delta IS, \text{ where} \tag{1}$$

IS_o is the isomer shift at zero-concentration of the solute and ΔIS is the specific isomer shift depending on the kind of solute (e.g., Ge or Si). For the non-crystalline alloys, one can assume that IS_o is the isomer shift at a nonmetal concentration equal to zero and thus the shift has the known value of iron in the corresponding metal solvent, in this case Pd ($IS_o = 0.177$mm/s)[7] and Au ($IS_o = 0.637$mm/s).[8] IS_o is zero in the case of the non-crystalline Fe-P-C alloy because there is no isomer shift for iron in an iron matrix. The values of the specific isomer shift of Ge ($\Delta IS = 0.05$mm/s) and of Si ($\Delta IS = 0.05$mm/s) were taken from Brossard, et al.[6] and Sauer, et al.[5] The values of the specific isomer shift of C ($\Delta IS = 0.09$mm/s) and P ($\Delta IS = 0.07$mm/s) were obtained by analyzing crystalline Fe-C and Fe-P Mössbauer spectra.[1] Using these values of IS_o, ΔIS and N_{mn}, a value of IS can be calculated which should be equal to the measured one. N_{mn} is the average number of nonmetal neighbours of an arbitrary central metal atom. The calculations for the non-crystalline alloys were based on the corresponding intermetallic compounds Fe3C, Fe3P, Pd3Si, and Pd3P.

No calculations were made for the Au(Fe)-Ge-Si alloy since an intermetallic compound of the cementite type, e.g., Au_3Si is not known. However, the value to be expected should be greater than 0.637mm/s, i.e., the value of IS_o, assuming that Equation (1) is applicable. While there is good agreement between calculated and experimental value of the isomer shift for the alloys Pd-Fe-P and Fe-P-C, there is a large deviation for the alloys Au(Fe)-Ge-Si and Pd(Fe)-Si. The experimental values are in both cases smaller than expected. For the alloy Pd(Fe)-Si, the experimental value lies about 0.18mm/s below the calculated one (see Table 1, previous page).

The following is an attempt to explain this behaviour. In comparison with the alloys Pd-Fe-P and Fe-P-C, iron is not a structure supporting element in the alloys Pd(Fe)-Si and Au(Fe)-Ge-Si. One can assume that the iron atom in these two alloys is substituted for Pd or Au. The next neighbours of such a structural site should be the nonmetal atoms. It should therefore be possible to build the smaller iron atom (smaller compared to Pd and Au) into the structure in such a way that it would be totally screened off from its next metal neighbours by the nonmetal atoms. In this case, IS_o would be zero. However, these arguments are only valid for low iron concentrations, where the iron atoms do not greatly disturb the Bernal structure. The observed isomer shift can therefore be attributed entirely to the nonmetal neighbour atoms. Under this assumption the value of the isomer shift of the non-crystalline Pd(Fe)-Si alloy can be explained consistent with the Bernal-Polk model with 2 2/3 next Si neighbours. In the case of the non-crystalline Au(Fe)-Ge-Si alloy, the same considerations yield 4 next Ge or Si neighbours to a Au atom. Apparently, this latter value does not agree with the theoretical one.

X-RAY EXPERIMENTS

Taking the Mössbauer experiments' value for the number of nonmetal neighbours around a metal atom, $N_{mn} = 2.7$, then the number of neighboring metal atoms can be evaluated from the areas of the first peak of the radial distribution functions (RDF) obtained by x-ray diffraction measurements. However, many glassy metals are ternary alloys. In this case, 9 partial coordination numbers N_{ij} $(i,j = 1,2,3)$[9] contribute to the area N_1 under the first peak of the RDF:

347

TABLE 2. The Factors $c_i f_i f_j / \langle f \rangle^2$, the Experimental Areas N_1 and the Calculated Numbers of Nearest Metal Neighbours N_{mm} of Various Non-Crystalline Alloys.

	$\dfrac{c_m f_m^2}{\langle f \rangle^2}$	$\dfrac{c_n f_n^2}{\langle f \rangle^2}$	$\dfrac{2 c_m f_m\, f_n}{\langle f \rangle^2}$	N_1	REF	N_{mm}
$Au_{77}Ge_{13}Si_{10}$	1.09	0.03	0.67	9.65	[b]	7.2
$Fe_{80}P_{13}C_7$	1.01	0.05	0.92	13.1	[b]	10.5
$Fe_{32}Pd_{48}P_{20}$	1.04	0.04	0.82	13.2	[a]	10.6
$Mn_{75}P_{15}C_{10}$	1.00	0.07	0.92	12.2	[a]	9.7
$Ni_{32}Pd_{53}P_{15}$	1.03	0.03	0.79	12.7	[a]	10.3
$Ni_{45}Pt_{30}P_{25}$	1.09	0.04	0.68	11.8	[a]	9.2
$Pd_{84}Si_{16}$	1.06	0.02	0.65	12.5	[b]	10.2
$Pd_{80}Si_{20}$	1.08	0.03	0.66	11.6	[a]	9.1
$Pd_{77.5}Cu_6Si_{16.5}$	1.06	0.02	0.66	11.8	[b]	9.4
$Pd_{75}Fe_5Si_{20}$	1.08	0.03	0.67	11.7	[b]	9.2
$Pt_{60}Ni_{15}P_{25}$	1.16	0.02	0.51	10.3	[b]	7.7
$Au_{81}Si_{19}$ liquid	1.14	0.01	0.40	9.55	[b]	7.4
$Pd_{84}Si_{16}$ liquid	1.06	0.02	0.65	11.4	[b]	9.1

[a] acc. to ref.9   [b] measurements by the authors.

$$N_1 = \frac{1}{\langle f \rangle^2} \sum_{i,j=1}^{3} c_i\, f_i\, f_j\, N_{ij}, \tag{2}$$

where $\langle f \rangle$ is the mean atomic scattering factor averaged over the atomic concentrations c_i. This expression can be transferred in the quasi-binary case with three partial coordination numbers if the terms of the metal (1) atoms (indicated by m) and of the nonmetal atoms (2) and (3) (indicated by n) are factored:

$$N_1 = \frac{1}{\langle f \rangle^2} \{ c_m\, f_m^2\, N_{mm} + f_n^2\, N_{nn} + 2 c_m\, f_m\, f_n\, N_{mn} \} \tag{3}$$

$$f_n = \left(c_2 f_2 + c_3 f_3 \right) / \left(c_2 + c_3 \right), \quad c_n = c_2 + c_3, \quad c_m = c_1, \quad f_m = f_1$$

348

Furthermore, $c_m N_{mn} = c_n N_{nm}$, where (4)

N_{mm} is the number of surrounding metal atoms and N_{mn} the number of surrounding nonmetal atoms. In the derivation of Equation (3) it is assumed that the nonmetal atoms (2) and (3) are mutually distributed at random, i.e., the coordination numbers of the alloying nonmental elements are proportional to their concentrations:

$$N_{22}/c_2 = N_{33}/c_3 = N_{23}/c_3 = N_{32}/c_2 = N_{nn}/c_n$$

$$N_{31}/c_1 = N_{13}/c_3 = N_{12}/c_2 = N_{21}/c_1 = N_{nm}/c_m = N_{mn}/c_n$$

(5)

According to Equation (3), the measured area N_1 under the first peak of the RDF consists of a sum of three partial coordination numbers of the metal and of the nonmetal neighbours multiplied by weighting factors which depend on the atomic concentrations and on the scattering factors. In Table 2 (previous page), these factors are listed for a number of glassy alloys calculated from $f_i(K=0) = Z_i$. The values of $c_n f_n^2 / <f>^2$ are very small. Therefore the contributions of the non-metal pairs to the areas N_1 can be neglected. Furthermore, Table 2 shows the experimental values of the areas N and the number of nearest metal neighbours N_{mm} obtained from the solution of Equation (3) by assuming a value $N_{mn} = 2.7$. With the exception of Au-Ge-Si and Ni-Pt-P, all values of N_{mm} are between 9 and 11 neighbouring metal atoms. The average is 10. For Au-Ge-Si and Ni-Pt-P, a lower value of $N_{mm} = 7-8$ was found. Both alloys have different properties, not only regarding the Mössbauer spectra as discussed above, but also with regard to the x-ray interference and radial distribution functions. For instance, there is no shoulder on the second peak of the interference function, and the first maximum is not as high as that of other glassy metallic alloys.[9]

The Bernal-Polk model suggests that non-crystalline alloys have a liquid-like structure. A comparison of the reduced interference function (Figure 3) and the radial distribution functions (see Figure 1) of the glasses $Pd_{84}-Si_{16}$ and $Au_{77}-Ge_{13}-Si_{10}$ with those of molten $Pd_{84}-Si_{16}$ (T = 900C) and $Au_{81}-Si_{19}$ (T = 410C) indicates that there are many

similarities of the structure.[10] The positions of the maxima and minima differ only slightly in the amorphous and in the molten state. However, the curves obtained from the melts are more damped. The areas
N_1 under the first peak of the RDF are somewhat smaller (see Table 2)
in the liquids. The splitting of the second peak of the RDF which is
observed in these glasses (with the exception of Pt-Ni-P) is absent in
the melts.

CONCLUSIONS

The Mössbauer experiments and the x-ray experiments show that the
structure of the non-crystalline Fe-P-C, Pd(Fe)-Si, Pd-Fe-P can be
described by the Bernal-Polk model. A metal atom has 2.4 nonmetal
neighbours and 10 metal neighbours in agreement with this model. For
the alloys Au(Fe)-Ge-Si, Pt-Ni-P, a different experimental result is
obtained which implies a significant structural difference.

ACKNOWLEDGEMENTS

This work has been supported by the Deutsche Forschungsgemeinschaft.

REFERENCES

1. Glozbach, P., Dissertation TU Berlin (1975) D 83.
2. Tsuei, C.C., et al., Phys. Rev. B, 5, 1047 (1972).
3. Polk, D.E., Scripta Met. 4, 117 (1970).
4. Maitrepierre, P.L., J. Appl. Phys. 40, 4826 (1969).
5. Sauer, W.E., et al., ibid. 42, 1604 (1971).
6. Brossard, L., et al., ibid 42, 1306 (1971).
7. Mössbauer, Eff. Data Index, Plenum Press NY (1970).
8. Window, B., Phys. Rev. B, 6, 2013 (1972).
9. Wagner, C.N.J. in Liquid Metals (SZ Beer, ED), NY (1972).
10. Just, T., Dissertation TU Berlin D 83 (1973).

MÖSSBAUER SPECTROSCOPY OF AMORPHOUS AND CRYSTALLINE Zr$_x$Cu$_{.925-x}$Fe$_{.075}$ ALLOYS

B.C. Giessen and A.E. Attard[27]

Departments of Chemistry and
Mechanical Engineering
Northeastern University, Boston

R. Ray

Allied Chemical Corporation
Morristown, New Jersey

INTRODUCTION

Mössbauer spectroscopy is a sensitive local probe of the electromagnetic fields and electric charges at the site of the Mössbauer isotope; therefore, it has been applied to metastable alloys frequently during the last ten years.[15]

In many of these applications, ^{57}Fe was used as a probe of amorphous alloys; for these alloys, the magnetic hyperfine field and the electrical field gradient (EFG) have received most of the attention. The third major parameter, the magnitude of the s-electron density at the nucleus expressed by sign and magnitude of the isomer shift (IS) has not been treated comprehensively to date; in a paper[6] presented at this conference, an attempt has been made to calculate the IS using a relation[7] based on the number of nearest neighbors and the alloy concentration.

In the present paper, we report on an investigation of the effect of composition changes in amorphous Cu-Zr alloys to which a constant amount of Fe (7.5at.%) has been added as a Mössbauer probe. The existence of a previously unobserved crystalline phase with the approximate composition Zr$_{.50}$Cu$_{.425}$Fe$_{.075}$ is also reported.

Ray, Giessen, and Grant[8] first found that the Cu-Zr system contains alloy compositions at which an amorphous metal could be prepared readily by splat cooling[9]; since then, a number of detailed studies[10-12] of amorphous Cu-Zr alloys have yielded information on their structural, thermal, mechanical and magnetic properties. The possibility of preparing amorphous alloys over a fairly wide binary composition range (32.5 to 62.5at.%Cu in the present study) has made this sytem especicially suitable for systematic studies of the composition dependence of thermal properties[12] as well as others such as Mössbauer spectroscopy.

EXPERIMENTAL METHODS

Foils were produced by splat cooling, generally using an arc-furnace quenching method similar to that described in Reference 13 which was also used in a study of metastable refractory alloys reported at this conference.[14] The structures of all foils were monitored by x-ray diffraction.

Mössbauer measurements were made with an ASA drive system equipped with a Cu(Co) source of about 20mC; data were evaluated using the customary least-square program.

RESULTS

1. <u>Structural Data</u>: A ternary alloy of composition $Zr_{.50}Cu_{.425}Fe_{.075}$ [or $Zr(Cu_{.85}Fe_{.15})$] consisting in equilibrium of a mixture of ZrCu (of unknown structure[15]) and ZrNi (CrB type[16]) could be retained by splat cooling either as an amorphous phase or, alternatively, upon somewhat slower quenching, as a CsCl type phase with a lattice parameter $a_o = 3.241 \pm .003$Å. The diffraction lines of this phase were quite sharp and clearly resolved to at least $\theta = 55°$, indicating a well-crystallized, ordered phase. The alloy $Zr_{.45}Cu_{.475}Fe_{.075}$ was also retained as a CsCl type phase.

2. <u>Mössbauer Data on Metastable $Zr_{.925-x}Cu_xFe_{.075}$ Alloys</u>: The Mössbauer doublet spectrum given in Reference 14 for the single Fe-doped Nb-Ni alloy is characteristic of the spectra for the present series of amorphous Fe-doped Zr-Cu alloys; the crystalline Zr-Cu-Fe alloys yielded single-line spectra.

The spectra were evaluated assuming a symmetric or asymmetric doublet for amorphous alloys and a single line for crystalline phases; the results are given in the Table hence. We discuss briefly some of the main features:

a. The <u>isomer shifts</u> Δ of the $Zr_{.925-x}(Cu_xFe_{.075})$ spectra vary approximately linearly with composition (Figure 1a), regardless of

352

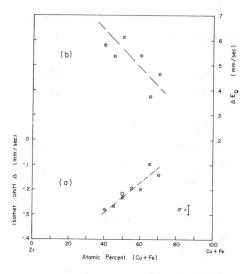

Figure 1. (a) Isomer shift Δ (mm/sec) and (b) quadrupole doublet splitting ΔE_Q (mm/sec) for amorphous (o) and crystalline ($\square$) $Zr_{.925-x}Cu_xFe_{.075}$ alloys.

Table I

Mossbauer Spectroscopic Data for Metastable $Zr_{.925-x}Cu_xFe_{.075}$ Alloys

Composition	Structure	Δ(mm/sec)*	ΔE_Q(mm/sec)	Γ(mm/sec) or Γ_1**	Γ_2**	Intensity Ratio** (areas)
$Zr_{.60}Cu_{.325}Fe_{.075}$	Amorph.	.280	.582	.291	.261	1.04 : 1
$Zr_{.55}Cu_{.375}Fe_{.075}$	Amorph.	.267	.539	.357	.264	1.12 : 1
$Zr_{.50}Cu_{.425}Fe_{.075}$	Amorph.	.230	.612	.184	.305	1 : 1.61
$Zr_{.50}Cu_{.425}Fe_{.075}$	CsCl	.215	——	.188		——
$Zr_{.45}Cu_{.475}Fe_{.075}$	CsCl	.193	——	.268		——
$Zr_{.40}Cu_{.525}Fe_{.075}$	Amorph.	.198	.538	.312	.223	1.15 : 1
$Zr_{.35}Cu_{.575}Fe_{.075}$	Amorph.	.096	.375	.264	.188	1.57 : 1
$Zr_{.30}Cu_{.625}Fe_{.075}$	Amorph.	.142	.465	.340	.199	1.53 : 1

* Relative to α-Fe,

** First peak listed corresponds to larger negative isomer shift.

of whether the alloy is amorphous or single phase crystalline (CsCl type). The IS thus depends only on the number of Zr and Cu atoms adjacent to an iron atom, not their relative orientations [in ordered crystalline $Zr(Cu_{.85}Fe_{.15})$, Fe atoms occupy Cu sites and an equivalent

353

statement probably is true also for the dense-random packed amorphous alloys]. Δ and its variations will be discussed further under d., below.

b. The quadrupole splitting of the amorphous alloys ranges from .38 to .60mm/sec, with a noticeable composition dependence (Figure 1b). These values agree well with E_Q = .4 to .55mm/sec typically observed in amorphous metals.[13] By contrast, the value of ΔE_Q = .30mm/sec for amorphous $Nb_{40}(Ni_{84}Fe_{.16})_{60}$[14] is rather small, indicating a smaller EFG in this alloy. On the other hand, the linewidths for the latter alloy of Γ_1 = .358 and Γ_2 = .411mm/sec are larger than those in the Zr-Cu-Fe alloys of Γ = .19 to .35mm/sec (see Table); this indicates a larger range of EFG values in the Nb-Ni-Fe alloy than in the Zr-Cu-Fe alloys, although with a smaller average EFG value.

c. The asymmetry of some of the Zr-Cu-Fe doublets is comparable in degree to that frequently found in amorphous alloys. While this a-symmetry is not likely to be a manifestation of the Goldanski effect,[17] it has been assumed in another case to be the result of many superim-posed doublets with correlation between EFG and IS.[1] This typically re-sults in a doublet with one narrow, straonger and one broader, weaker line, which was not found here. The recent observation of local pre-ferred orientation in amorphous foils,[18] and the strong effect of aniso-tropy on Mössbauer spectra suggest "texture" as another possible ex-planation for the peak asymmetry; this interpretation will be tested by addition measurements, e.g., on powdered foils.

d. To establish fundamental correlations of different non-ferromag-netic amorphous metals using Mössbauer spectra, the isomer shift ap-pears to be the most interesting property. Thus, for groups of related metals such as the inter-transition metal phases, one might hope to find a relationship between the ability of the amorphous metal to change the s-electron concentration of added Fe and properties such as its bond strength or stability. In the following, the IS values ob-served here are compared with those for Fe in several crystalline hosts (Figure 2).[19-21]

354

Strictly, in a comparison of this kind the IS determined at very small Fe solute concentrations or by extrapolation should be used rather than that at the 7.5at.% Fe level measured in this study; an indication of the effect of the Fe concentration on the IS can be obtained from the study[1] of amorphous $Pd_{.80-x}Fe_xSi_{.20}$ alloys, where $0.01 \leq x \leq 0.07$; it was found that Δ changes from -.047 for x = .07 to about -.105 mm/sec for x∿0.[1] If the Pd-Si and the Zr-Cu systems are somewhat similar in their effect on the Fe probe atoms, the negative Δ values of the present study should be decreased accordingly by $\leq$ .06 mm/sec; however, the value of the correction varies for each amorphous matrix composition and should be determined specifically, e.g., by a study of $Zr_{.5}(Cu_{.5-x}Fe_x)$ or $(Zr_aCu_{1-a})_{1-x}Fe_x$ alloys.

It is seen in Figure 2 that the isomer shifts roughly follow those of the 3d-metal solvent curve, although the Δ values are more negative; in fact, the magnitude of the observed negative IS is quite large by comparison to that of other Fe-containing transition metal alloys. According to the plot of IS versus charge density,[22] the largest negative Δ observed here corresponds to an Fe charge density $3d^{8-x}4s^x$ with x $\simeq$ 1.12. In any case, it is certain that the s-electron density at the Fe nucleus is substantially enhanced compared to elemental metallic Fe; in a rigid-band model, this is probably due to the transfer of Fe(3d)-electrons into Zr(4d)-orbitals of the amorphous Zr-Cu metal (which are already partly filled by Cu(4s)-electrons); this transfer will increase $|\psi_s(0)|^2_{Fe}$ and leads to the high negative value of Δ.

As additional Cu is added, the number of empty Zr bonding orbitals unfilled by Cu and available for Fe will decrease; less Fe(3d)-electrons can be transferred and $|\psi_s(0)|^2_{Fe}$ moves back towards its elemental value.

By contrast, the Nb-Ni system involves two transition metals. It is necessary to go beyond the rigid band model; the high concentration of Ni in the amorphous metal will affect the position of the Nb(d)-electron levels, moving them partly to higher energies. There is then no or almost no transfer of Fe(d)-electrons into empty d-electron levels of the amorphous matrix; some Fe(s)-electron transfer or, more likely,

355

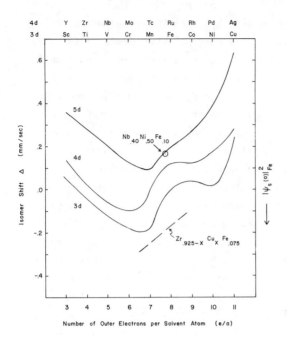

Figure 2. Isomer shifts
Δ for metastable
$Zr_{.925-x}Cu_xFe_{.075}$ alloys
(from Figure 1) and a-
morphous $Nb_{.40}Fe_{.10}Ni_{.50}$[14]
compared to values for
dilute ^{57}Fe dissolved in
3d, 4d, and 5d transi-
tion metals.[19][20]

s- and d-electron trans-
fer from Ni to Fe(d)-
levels will then account
for the observed positive
isomer shift. A study in-
cluding variations of the
matrix composition is re-
quired.

It may be noted that
the decrease of the abso-
lute value of the IS and
the increase of $|\psi_s(0)|^2_{Fe}$
in amorphous Zr–Cu–Fe alloys with increasing Cu content goes together
with an increase in the thermal stability of Zr–Cu metallic glass; the
temperatures of glass transition and crystallization increase while the
heat of crystallization decreases.[12]

The present results suggest the need for several follow-up studies,
especially measurements on alloys with changing matrix compositions,
to assess the IS and EFG variations through a wider range of d-metal
average group numbers, atomic sizes and potentials.

3. Mössbauer Spectroscopic Data for Other Metastable Alloys: Mössbauer
spectra have also been measured for other metastable phases such as:

Crystalline Ti–TiFe solid solutions;[23][24] in these alloys, there is a
phase transition from the W type structure of β-Ti to the CsCl type
structure of TiFe;

Concentrated non-substitutional terminal solid solutions or other metastable phases in systems such as La-Fe[4] and Gd-Fe;[4,25]

A metastable, off-stoichiometric compound $m-GeFe_{1-x}$[4,26] with a high concentration of Fe site vacancies.

These phases will be treated in detail elsewhere.

ACKNOWLEDGEMENT

The authors are grateful for support of this work under ONR contracts N14-68-A-207-3 and N14-75-C-713, as well as for equipment support under NASA Cooperative Agreement 22-011-071. They acknowledge the experimental assistance of Dr. M. Madhava and helpful discussions with Drs. A. Bansil and D.E. Polk.

REFERENCE

1. Sharon, T.E. and C.C. Tsuei, Solid State Comm. 9, 1923 (1971).
2. Sharon, T.E. and C.C. Tsuei, Phys. Rev. B 5, 1047 (1972).
3. Cochrane, R.W., R. Harris, M. Plischke, D. Zobin, and R. Zuckermann, Phys. Rev. B. 12, 1969 (1975).
4. Segnini, M., Ph.D. Thesis, Northeastern U., Boston (1972).
5. Giessen, B.C. and C.N.J. Wagner in Liquid Metals, Chemistry, and Physics, S.Z. Beer, Ed., Marcel Dekker, New York, p. 633 (1972).
6. Bartsch, G.E.A., P. Glotzbach, and T. Just, Proc. Second International Conference on Rapidly Quenched Metals, §I, N.J. Grant and B.C. Giessen, Eds., MIT Press, Cambridge MA (1976).
7. Sauer, W.E. and R.J. Reynik, J. Appl. Phys. 42, 1604 (1971).
8. Ray, R., B.C. Giessen, and N.J. Grant, Scripta Met. 2, 357 (1968).
9. Giessen, B.C. and R.H. Willens, in Phase Diagrams; Materials Science and Technology, III, A.M. Alper, Ed., Academic Press, New York, p. 103 (1970).
10. Revcolevschi, A. and N.J. Grant, Met. Trans. 3, 1545 (1972).
11. Vitek, J., J. Vander Sande and N.J. Grant, Acta Met. 23, 165 (1975).
12. Kerns, A., Ph.D. Thesis, Northeastern U., Boston (1974).
13. Ohring, M. and A. Haldipur, Rev. Sci. Instrum. 42, 530 (1971).
14. Giessen, B.C., M. Madhava, D.E. Polk, and J. Vander Sande, in Proc. Second International Conference on Rapidly Quenched Metals, §II, N. J. Grant and B.C. Giessen, Eds., Mat. Sci. Eng. (1976).
15. Bsenko, L., J. Less Comm. Met. 40, 365 (1975).
16. Kirkpatrick, M.E., D.M. Bailey, and J.F. Smith, Acta Cryst. 15, 252 (1962).
17. Goldanski, V.L., E.F. Makarov, I.P. Suzdalev, and I.A. Vinogradov, Sov. Phys. J.E.T.P. 27, 44 (1968).
18. Alben, R., C.S. Cargill III, and J. Wenzel, Phys. Rev. B 13, 835 (1976).

19. Weisman, I.D., L.J. Swartzendruber, and L.H. Bennett, in <u>Techniques of Metals Research</u>, Vol. VI, Part 2, R.F. Bunshah, Ed., Wiley-Interscience, New York, p. 401 (1973).
20. Qaim, S.M., <u>Proc. Phys. Soc.</u> <u>90</u>, 1065 (1967).
21. Wertheim, G.K., <u>Mossbauer Effect</u>: <u>Principles and Applications</u>, Academic Press, New York (1964).
22. Walker, L.R., G.K. Wertheim and V. Jaccarino, <u>Phys. Rev. Lett.</u> <u>6</u>, 98 (1961).
23. Ray, R., B.C. Giessen and N.J. Grant, <u>Met. Trans.</u> <u>3</u>, 627 (1972).
24. Ray, R., B.C. Giessen, and W. Reiff [to be published].
25. Ray, R., M. Segnini, and B.C. Giessen, <u>Solid State Comm.</u> <u>10</u>, 163 (1972).
26. Segnini, M., R. Ray, and B.C. Giessen [to be published].
27. Institute of Chemical Analysis, Applications and Forensic Science.

358

THE STRUCTURE OF AMORPHOUS Co-Gd FILMS PREPARED BY BIAS SPUTTER DEPOSITION FROM A Co$_3$Gd TARGET

C.N.J. Wagner

University of California
Los Angeles, California

Co-Gd alloy films with 77 at.%Co-23at.%Gd (no bias) and with 77 at.%Co-17%Gd-6%Ar (-100 V) were prepared by bias sputter deposition from a Co$_3$Gd target on 50μm thick Be substrates. The interference functions I(K) were determined with Mo-Kα radiation in reflection and transmission. Within experimental uncertainties, no difference between reflection and transmission data could be observed. The biased film with a Co-Gd ratio of 4.7 showed a shoulder below the first peak which is not present in the unbiased film of similar Co-Gd ratio, determined by Cargill. Assuming that the structure of the unbiased film can be described with the hard sphere model, the difference in I(K) of the biased and unbiased films has been interpreted as a consequence of short-range compositional order in the biased Co-Gd film.

INTRODUCTION

Recently, stable magnetic bubble domains have been observed in sputter-deposited transition metal-rare earth metal alloys.[1] The existence of these bubble domains in thin amorphous films depends on the presence of uniaxial anisotropy with the easy axis of magnetization perpendicular to the film plane. It was found by Chaudhari, et al.[1] that amorphous Co-Gd films prepared by sputter deposition with bias applied to the substrates possess such uniaxial anisotropy, which was thought to have structural origin.[2]

In order to test this hypothesis, Wagner, et al.[3] investigated the structure of Co-Gd alloy films with 77 at.%Co-23%Gd and 68 at.%Co-32%Gd (no bias) and with 77 at.%Co-17%Gd-6%Ar (-100 V bias on substrate) prepared by sputter deposition on 50μm thick Be substrates. Within experimental uncertainties, the interference functions I(K), obtained from

359

reflection and transmission of Mo-Kα radiation by a given sample, were
identical. A comparison of the reduced atomic distribution function
$G(r) = 4\pi r [\rho(r) - \rho_0]$ where $\rho(r)$ is the atomic distribution function
and ρ_0 is the average atomic density, for various compositions lead to
an identification of the resolved components of the first maximum as
Co-Co, Co-Gd, and Gd-Gd nearest neighbors at $r = 2.50$, 2.95, and 3.45Å,
respectively, in agreement with Cargill's finding in an amorphous 82
at.%Co-18%Gd film.[4]

However, the interference functions I(K) of the biased and unbiased
Co-Gd films showed differences in the shapes of the first and second
peaks. Attempts will be made in this paper to correlate these differences with short-range compositional order in the Co-Gd films prepared
by bias (-100 V) sputter deposition from a Co₃Gd target.

EXPERIMENTAL TECHNIQUES AND DATA ANALYSIS

Films of Co-Gd (20μm thick) were prepared by sputter depositions from
a Co₃Gd target in an Ar atmosphere at 25μm pressure with 0 and -100 V
bias applied to the Be substrate (50μm thick foil). The actual compositions were found to be 23.2 at.%Gd, 76.6%Co and 0.2%Ar for the unbiased film and 16.5 at.%Gd, 77.2%Co and 6.3%Ar for the biased film.[3]

The x-ray intensities were obtained with Mo-Kα radiation in reflection and in transmission using a graphite monochromator in the dif-
fracted beam. The total scattering was corrected for substrate scattering, absorption in the sample and substrate, inelastic scattering, and
multiple scattering, and then normalized to absolute units,[5] thus yielding the elastically scattered intensity per atom $I_a(K)$. From this quantity, the interference function I(K) was calculated, i.e.,

$$I(K) = \{ I_a(K) - [<f^2> - <f>^2]\}/<f>^2 \tag{1}$$

where $<f>$ and $<f^2>$ are the mean and mean-square scattering factors, respectively, of the alloy.

The reduced atomic distribution function G(r) was obtained by Fourier transformation of $F(K) = K[I(K) - 1]$, i.e.,

360

$$G(r) = 4\pi r[\rho(r) - \rho_o] = (2/\pi)\int_0^\infty F(K) \, e^{-\alpha K^2} \sin Kr \, dK \tag{2}$$

using a value of the artificial damping factor $\alpha = 0$. Large modulations were usually visible in $G(r)$ at small values of $r(r > 2\text{Å})$ due to slowly varying errors in $F(K)$, most likely arising from errors in the correction for inelastic scattering. Additionally, a short wavelength modulation of low amplitude was clearly visible at values $r > 6\text{Å}$. Both modulations were removed by the Kaplow procedure[6] yielding refined interference functions $I(K)$ and atomic distribution functions $G(r)$. Details of the experimental procedures and data analysis will be published elsewhere.

RESULTS AND DISCUSSION
___

The interference functions $I(K)$ of the Co-Gd alloys, sputter-deposited from a Co_3Gd target with zero and -100 V bias applied to the substrate are shown in Figure 1. The solid curves represent the transmission data and the solid dots, the reflection data. In order to illustrate the effect of the Kaplow refinement procedure,[6] the unrefined $I(K)$ (dashed curve) of the biased film measured in transmission is also plotted in Figure 1. The only effect of the refinement procedure was to raise the values of $I(K)$ between $K = 4$ and 10 A^{-1} so that they modulate uniformly about the value of one without changing the shape of the modulation. It is believed that the difference between the refined and unrefined data is the consequence of inadequate normalization at high K values because of errors in our corrections for inelastic scattering.

It is obvious from Figure 1 that the transmission and reflection data are practically identical (within experimental errors). This seems to indicate that the structures of the biased and unbiased films do not exhibit any noticeable preferred orientation. The observed $I(K)$ can be explained as the superposition of the partial interference functions $I_{ij}(K)$,[5] i.e.,

$$I(K) = W_{11}(K) \, I_{11}(K) + W_{22}(K)I_{22}(K) + 2W_{12}(K)I_{12}(K) \tag{3}$$

361

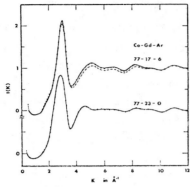

Figure 1.
Interference functions I(K) of sputter-deposited Co-Gd films. Solid curves represent refined transmission data; dots represent refined reflection data; dashed curve is the unrefined transmission data of biased film.

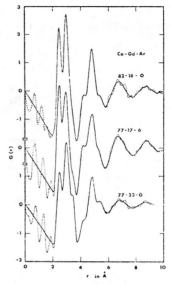

Figure 2.
Reduced atomic distribution functions $G(r) = 4\pi r \times [\rho(r) - \rho_0]$ of sputter-deposited Co-Gd films. Solid curves represent the average of the transmission and reflection data; dotted curves represent unrefined transmission data with $\alpha = 0$. The data for the 82at%Co-18%Gd film are due to Cargill (ref. [4] and private communication).

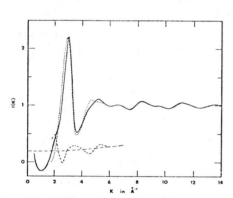

Figure 3.
Interference function I(K) of the biased 77at%Co-17%Gd-6%Ar film (solid curve) and the unbiased 82at%Co-18%.Gd film (dotted curve due to Cargill, ref. [4] and private communication). Dashed curve represents the difference of the biased and unbiased I(K) plotted about the Laue monotonic scattering $[<f^2> - <f>^2]/<f>^2$.

362

where

$$I_{ij}(K) = 1 + (4\pi\rho_o/K)\int r\left\{[\rho_{ij}(r)/(c_j\rho_o)] - 1\right\} \sin Kr \, dr \qquad (4)$$

and

$$W_{ij}(K) = c_i c_j f_i(K) f_j(K) / \langle f(K) \rangle^2 \qquad (5)$$

In these expressions, c_i and f_i are the atomic concentration and scattering factor, respectively, of element i, and $\rho_{ij}(r)$ is the number of j-type atoms per unit volume at the distance r from an i-type atom. It might happen that the partial functions $I_{ij}(K)$ possess maxima and minima at values of K such that the first peak of the total I(K) becomes rather broad and the second and third peaks are greatly reduced in height. Such behavior has been observed in liquid Cu-Sn alloys.[7]

The total distribution functions G(r), evaluated with Equation (2) from F(K) = K[I(K)-1], are shown in Figure 2. The solid curves represent the average of the refined transmission and reflection data, whereas the unrefined transmission G(r) are shown as the dotted curves. The most striking feature of the total distribution function G(r) is the splitting of the first peak. Similar observations were made by Cargill[4] on $Co_{82}Gd_{18}$ and $Fe_{64}Gd_{36}$ alloys and by Rhyne, et al.[8] on a $Fe_{67}Tb_{33}$ alloy.

It should be remembered that the original G(r) (dotted curves in Figure 2) contain short wavelength ripples which are clearly identifiable at r < 2Å and r > 6Å. Obviously, these ripples will also affect the exact shape of G(r) in the range 2 < r < 6Å, and indeed the transmission and reflection data showed slight differences in the heights of the split peaks. However, the amplitudes of the ripples decrease rapidly with increasing r. Thus, we conclude that these modulations cannot be solely responsible for the splitting of the first peak.

If we assume that $W_{ij}(k)$ is a slowly varying function of K, then we can approximate G(r) as the weighted sum of the partial $G_{ij}(r)$,[4,5]

$$G(r) = W_{11}(0)G_{11}(r) + W_{22}(0)G_{22}(r) + 2W_{12}(0)G_{12}(r) \qquad (6)$$

where $G_{ij}(r) = 4\pi r\rho_o[\rho_{ij}(r) / (c_j\rho_o) - 1]$ is the Fourier transform of $K[I_{ij}(K) - 1]$.

363

Thus, it is possible to associate the peaks at r_1 = 2.50Å and r_2 = 2.95Å with distances corresponding to the nearest neighbor separations between Co-Co and Co-Gd atoms. The separation between Gd-Gd is visible as a small hump or shoulder at r_3 = 3.45Å, a value which is slightly smaller than the atomic diameter of the Gd atom (3.60Å), whereas the Co-Co separation agrees well with the atomic diameter of the Co atom (2.50Å).[4]

The heights of the individual peaks between r = 2 and 4Å should be proportional to the weighting factors $W_{ij}(0)$ given in the Table below. However, when comparing the heights of the peaks at the first interatomic distances, particularly at the Co-Co distance, in the biased/unbiased films, it becomes apparent that the weighting factor $W_{CoCo}(0)$ must be higher in the biased film. This is a consequence of the increase in the Co-Gd atomic ratio in the biased film. Although the target material for both unbiased and biased Co-Gd film was identical, the ratio increased from a value of 3.30 in the unbiased film to 4.68 in the biased film because of the incorporation of 6.3 at.%Ar replacing Gd atoms. The latter ratio is quite close to the value of Cargill's Co-Gd sample[4] containing 18 at.%Gd, i.e., a Co-Gd ratio of 4.56, and prepared by sputter deposition without applying a bias voltage to the substrate. Thus, the biased film (77 at.%Co-17%Gd-6%Ar) should be compared with an unbiased 82 at.%Co-18%Gd film. This is supported by the fact that the G(r) curve of the 82 at.%Co-18%Gd film is quite similar to that of the 77 at.%Co-17%Gd-6%Ar sample as shown in Figure 2. Therefore, the interference functions of these films must also be similar, at least at large values of K. Indeed, this is the case as shown in Figure 3, where I(K) of the biased 77 at.%Co-17%Gd-6%Ar film and the unbiased 82 at.%Co-18%Gd film are plotted as the solid and dashed curves, respectively. As can be clearly observed, the interference functions I(K) of these two alloy films are practically identical at values of K larger than $6Å^{-1}$. Differences are noticeable in the neighborhood of the first and second peaks. Both peaks of the I(K) of the biased film are shifted slightly towards larger K values, and a shoulder is clearly visible below the first peak.

364

It is tempting to evaluate the difference in height between the bi-ased and unbiased films (Figure 3) with a similar Co-Gd atom ratio, i.e., ~4.6, at every value of K. This difference is also shown in Figure 3. We will attempt to correlate this difference with concentra-tion fluctuations in the Co-Gd alloys.

As shown by Bhatia and Thornton,[9] the interference function $I(K)$ can be written as

$$I(K) = S_{NN}(K) + [(\Delta f)^2/<f>^2]\left[S_{CC}(K) - c_1 c_2\right] + 2(|\Delta f|/<f>)S_{NC}(K)$$
(7)

where $\Delta f = f_1 - f_2$. The partial functions $S_{\alpha\beta}(K)$ are related to the par-tial interference functions $I_{ij}(K)$, i.e.,[9]

$$S_{NN}(K) = c_1^2 I_{11}(K) + c_2^2 I_{22}(K) + 2c_1 c_2 I_{12}(K)$$
(8)

$$S_{NC}(K) = c_1 c_2\left\{c_1[I_{11}(K) - I_{12}(K)] - c_2[I_{22}(K) - I_{12}(K)]\right\}$$
(9)

$$S_{CC}(K) = c_1 c_2\left\{1 + c_1 c_2[I_{11}(K) + I_{22}(K) - 2I_{12}(K)]\right\}$$
(10)

By adding and subtracting $S_{NN}(K)$ in Equation (10), we readily obtain:

$$S_{CC}(K) = c_1 c_2\left\{1 + [c_1 I_{11}(K) + c_2 I_{22}(K) - S_{NN}(K)]\right\}$$
(11)

which may be written on introducing $I_i(K) = c_1 I_{i1}(K) + c_2 I_{i2}(K)$ as:

$$S_{CC}(K) = c_1 c_2\{1 + [c_2 I_1(K) + c_1 I_2(K) - I_{12}(K)]\}$$
(12)

$$= c_1 c_2\{1 + (4\pi/K)\int r[c_2\rho_1(r) + c_1\rho_2(r) - \rho_{12}(r)/c_2]\sin Kr\ dr\}$$

$$= c_1 c_2\{1 + (4\pi/K)\int r\ \rho'(r)\ \alpha'(r)\sin Kr\ dr\}\quad \text{where}$$
(13)

$$\rho'(r) = c_2\rho_1(r) + c_1\rho_2(r) = c_2[\rho_{11}(r) + \rho_{12}(r)] + c_1[\rho_{22}(r) + \rho_{21}(r)]$$

and $\quad \alpha'(r) = 1 - \rho_{12}(r)/[c_2\rho'(r)]$
(14)

365

TABLE: Values of the weighting factors $W_{ij}(0) = c_i c_j \, f_i(0) f_j(0) / \langle f(0) \rangle^2$
for Co-Gd alloys of different Gd concentrations

Composition [at%]			$W_{GdGd}(0)$	$2W_{GdCo}(0)$	$W_{CoCo}(0)$
Co	Gd	Ar			
67.5	32.2	0.3	0.28	0.50	0.22
76.6	23.2	0.2	0.17	0.49	0.34
82	18	—	0.12	0.45	0.43
77.2	16.5	6.3	0.13	0.46	0.41

is the generalized Warren short-range order parameter.[5] In a crystalline solution, $\rho_1 (r) = \rho_2 (r) = \rho (r)$ which has also been asssumed to
hold in amorphous alloys by Keating.[10]

Bhatia and Thornton[9] have shown that for a hard sphere liquid[11] consisting of two kinds of atoms with different diameters, the function
$S_{CC}(K)$ deviates only very little from the value $c_1 c_2$. Thus, to a first
approximation, one might write the total interference function $I(K)$
for a hard-sphere liquid as

$$I^{HS}(K) \simeq S_{NN}(K) + 2 \left(|\Delta f| / \langle f \rangle \right) S_{NC}(K) \tag{15}$$

If we were to assume that the structure of the unbiased 82 at.%Co-
18%Gd film could be represented by a hard-sphere model,[*] i.e., $[I(K)]$
unbiased $= I^{HS}(K)$, we could correlate the difference shown in Figure 3,
between the interference functions of the biased film [Equation (7)]
and the unbiased film [Equation (17)], with the term $S_{CC}(K)$, i.e.,

$$[I(K)]_{biased} - [I(K)]_{unbiased} = \left[(\Delta f)^2 / \langle f \rangle^2 \right] [S_{CC}(K) - c_1 c_2] \tag{16}$$

We believe that such correlation is possible, at least qualitatively,
and conclude that there are concentration fluctuations in the biased
film.

That there might be a tendency towards local or short-range order
in the biased film similar to that in crystalline Co_5Gd has been suggested by Onton, et al.[12] Assuming that the structural coherence (or
"crystallite size") is of the order of 10Å, they calculated a diffraction
pattern which is in qualitative agreement with the observed pattern
of a biased $Co_{3.1}Gd$ film.   Both theoretical and experimental dif-
366

fraction patterns clearly showed a shoulder or subsidiary peak at the low-angle side of the first peak.

ACKNOWLEDGEMENT

The author would like to thank N. Heiman for the preparation of Co-Gd films, T.C. Huang and W. Parrish for their assistance in the x-ray analysis of the data, D.F. Kyser for electron microprobe analysis of the film compositions, and A. Onton for helpful discussions. Part of this work was carried out at the IBM San Jose Research Laboratory while the author was associated with the IBM/NSF Faculty Research Participation Program.

REFERENCES

1. Chaudhari, P., J.J. Cuomo, and R.J. Bambino, IBM J. Res. Devel. 17, 66 (1973).
2. Heiman, N., A. Onton, D.F. Kyser, K. Lee, and C.R. Guarnieri, Magnetism and Magnet Materials (1974), AIP Conference Proceedings No. 24; Graham, C.D., Jr., G.H. Lander, and J.J. Rhyne, Editors, American Institute of Phys., New York, 573 (1975).
3. Wagner, C.N.J., N. Heiman, T.C. Huang, A. Onton, and W. Parrish, Magnetism and Magnetic Materials (1975), AIP Conference Proceedings [in press].
4. Cargill, G.S., III, Solid State Physics 30; Seitz, F., D. Turnbull, and H. Ehrenreich, Editors, Academic Press, New York, 227 (1975).
5. Wagner, C.N.J., "Liquid Metals", Chemistry and Physics, Beer, S.Z., Editor, Dekker, New York, p. 257 (1972).
6. Kaplow, R., S.L. Strong, and B.L. Averbach, Phys. Rev. 138, A1336 (1965).
7. North, D.M. and C.N.J. Wagner, Phys. Chem. Liquid 2, 87 (1970).
8. Rhyne, J.J., S.J. Pickart, and H.A. Alperin, Magnetism and Magnetic Materials (1973), AIP Conference Proceedings, No. 18, Graham, C.D., Jr. and J.J. Rhyne, Editors; American Inst. of Phys., New York, 563 (1974).
9. Bhatia, A.B. and D.E. Thornton, Phys. Rev. B2, 3004 (1970).
10. Keating, D.T., J. Appl. Phys. 34, 923 (1963).
11. Ashcroft, N.W. and D.C. Langreth, Phys. Rev. 156, 685 (1967).
12. Onton, A., N. Heiman, J.C. Suits, and W. Parrish, IBM J. Res. Dev. [in press].

MECHANICS OF METALLIC GLASSES

Lance A. Davis

Materials Research Center
Allied Chemical Corporation
Morristown, New Jersey

At temperatures well below T_g, deformation of melt quenched metallic glasses occurs in the form of highly localized shear bands. In bending or compression multiple shear bands are observed; in tension, deformation occurs in a singularly intense shear band and failure occurs by shear rupture through this band. The occurrence of such localized plastic flow indicates an absence of work hardening, *i.e.* these materials behave as elastic-perfectly plastic solids. The strengths of metallic glasses are exceptional; alloys with yield strengths (σ_y) on the order of 370 Kg/mm^2 (525 Ksi) have been identified. The strengths (and hardnesses) of glassy alloys scale with their moduli; the ratio $\sigma_y/E \simeq 0.02$, *i.e.* the tensile strain to yield is $\simeq 2\%$. The fracture toughnesses of ferrous metallic glasses are consistent with those for steel when one allows for the higher σ_y and lower E of the glasses. Metallic glasses exhibit cycle dependent, *i.e.* not time dependent, fatigue. The rate of fatigue crack propagation is $\sim$ proportional to the stress intensity squared, as would be expected for crack extension by alternating plastic shear.

Introduction

The initial study of the mechanics of a metallic glass was conducted by Chen and Turnbull,[1] who examined the deformation of a Au$_{77}$Ge$_{13.6}$Si$_{9.4}$ alloy near its glass transition temperature. The initial reports[2-4] of "low temperature" ($T \ll T_g$) mechanical properties of metallic glasses, for alloys based primarily on Pd ($\sim$80 at.%) and Si, followed a few years later. These studies were the forerunners of a rapidly growing interest in the mechanics of glassy alloys. Numerous recent studies have issued, encouraged, it would appear, both by scientific curiosity and the growing expectation that metallic glasses will find technological utilization as structural materials. This expectation is founded, at least in part, on the knowledge that one can produce a considerable variety of glassy alloys based on inexpensive elements, *e.g.* Fe and Ni[5] and the demonstration that one can produce uniform cross section ribbons and wires rapidly ($\sim$1800 m/min) and continuously from the melt.[6]

Despite the brief history of metallic glasses in general, and study of their mechanical properties in particular, a number of features fundamental to their mechanics have been identified. The present paper will attempt to review the significant observations relevant to their deformation and (tensile) failure characteristics, their strength, elasticity fracture toughness and fatigue behavior.

Deformation Characteristics

Deformation Modes

When deformed above their glass transition temperatures (T_g), in the absence of crystallization, metallic alloys exhibit homogeneous (Newtonian) viscous flow, as demonstrated by Chen and Turnbull[1] for $Au_{77}Ge_{13.6}Si_{9.4}$. However, for $T<<T_g$ plastic flow in metallic glasses is extremely inhomogeneous, occurring in the form of highly localized shear deformation bands. Shear offsets may be readily observed on specimens deformed by rolling.[7,8] compression[7,9] or bending[4,8,10] (Fig. 1). In the first two cases shear bands lie in zones at ±45° to the rolling or compression direction; in the latter case they lie parallel to the bending axis.

The width of shear bands, such as those in Fig. 1, cannot be determined by optical microscopy. Using electron microscope replicas Masumoto and Maddin[3] found them to be ~200 Å thick in $Pd_{80}Si_{20}$. Corresponding shear offsets of ~2000 Å were observed; hence these materials exhibit enormous local shear ductility in the "low" temperature regime. Such ductility is notably absent in silicate glasses, precluded by their covalent bonding.

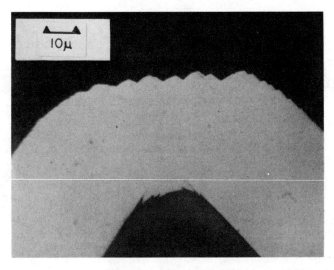

Fig. 1 Shear offset profile of a bent metallic glass strip (optical micrograph, from Chen and Polk[11]).

When a multiplicity of shear bands is activated, macroscopic ductility is observed. On loading in compression Pampillo and Chen[7] effected height reductions of about 40% in small cylinders of $Pd_{77.5}Cu_6Si_{16.5}$; 2 mm dia. rods of the same alloy may be cold rolled to strip $\sim$0.1 mm thick.[7] On rolling of Ni-Pd-P strips Takayama and Maddin[8] achieved thickness reductions up to 40%. As shown in Fig. 1, a thin glassy alloy strip may be sharply bent back on itself, *e.g.*, between the platens of a micrometer, without failure. Gilman[12] has noted that a strip of steel shim-stock of similar strength and dimensions will not sustain such a sharp bend. The greater ductility of the metallic glass arises, presumably, because it is a homogeneous elastic continuum, containing, in principle, no hard particles, voids or phase boundaries where local triaxial stresses can arise and promote crack initiation.

The occurrence of intense ($\varepsilon \simeq 10$ to 100) highly localized plastic shear bands is indicative of the absence of work hardening.[13] This is intuitively apparent; if the sheared material hardens, the band will expand laterally into undeformed material at low strain. Indeed, Argon suggests that strain localization requires work softening.[14] Direct evidence for a zero or negative rate of work hardening in a metallic glass was provided by Pampillo and Chen.[7] A specimen of $Pd_{77.5}Cu_6Si_{16.5}$ was first compressed to initiate deformation and then repolished to produce smooth lateral surfaces. Examination of the specimen before and after polishing indicated activity of the same shear bands on reloading. Hence deformation is favored in existing bands either due to work softening or an availability of deformation sources, in which case the rate of work hardening could be zero. It is also observed that shear bands produced by bending will reverse when the direction of bending is reversed.[15] This behavior is consistent with the absence of work hardening within the bands.[16]

Macroscopic evidence for the lack of work hardening in metallic glasses is provided by the curves of Fig. 2, which approximate the expected behavior for an elastic-perfectly plastic material. We attribute the roundness of the curves at low strain to extrinsic factors such as slight misalignment of the specimen ends.

Deformation Mechanism ($T \ll T_g$)

It was observed by Pampillo and Chen[7] that shear deformation bands terminate within compressed glassy alloy specimens. The elastic discontinuity which exists between the sheared and unsheared material is, by definition, a Volterra dislocation.[16] In principle, this macroscopic dislocation may result from a pileup of atomic size dislocations. According to Gilman[12,17-19] the slip vector of a microscopic dislocation in the glass will fluctuate in direction and magnitude along the dislocation line, but its mean value will be dictated by some structural parameter, such as the average nearest neighbor distance.

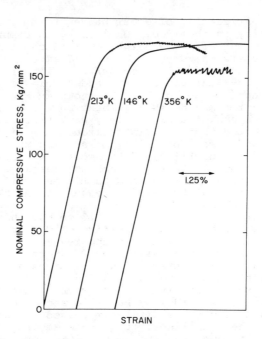

Fig. 2   Compression stress-strain curves for
$Pd_{77.5}Cu_6Si_{16.5}$ (from Pampillo and Chen[7]).

As the structural periodicity required for diffraction
contrast is absent in a metallic glass, microscopic dislocations are invisible in the electron microscope. It follows,
then, that evidence for their existence is necessarily indirect. Gilman[20] has noted, for example, that serrated
yielding (Fig. 2) and transient creep phenomena, observed in
$Pd_{80}Si_{20}$ by Maddin and Masumoto,[21] are readily explained in
terms of dislocation theory. Pampillo[15] has noted that the
observed temperature dependence of the fracture stress in Fe-
Ni metallic glasses[22] is similar to that in crystalline Fe or
Ni and, as such, suggests that deformation is produced by the
motion of microscopic line defects. Finally, the lack of
work hardening, *i.e.* strain localization, may be conveniently
explained based on a dislocation model; passage of a line defect through the material will destroy the local short range
order (Polk and Turnbull[23]), or generate dilatance (Gilman[20]),
when the dense random packed structure is disturbed. In
either event the local environment will be permanently altered and subsequent "dislocations" will be able to pass
through the same region more easily. Evidence for such a
local perturbation of the material was provided by Pampillo
and Chen,[7] who demonstrated that shear bands in compressed
$Pd_{77.5}Cu_6Si_{16.5}$ are susceptible to preferential etching.
Spaepen and Turnbull[24] have suggested that plastic flow
follows from a stress induced decrease of local viscosity,

i.e. deformation in the shear zone occurs by viscous flow. However, their model is constructed in terms of dilatance at the tip of a moving crack, and, as such, is a failure mechanism, rather than a flow mechanism.

Strength

Tensile/Yield Strength

Among the many unique physical properties of a metallic glass none is more striking, perhaps, than its strength at temperatures well below T_g. For alloys based on Ni and/or Fe tensile strengths ranging from $\sim$140 Kg/mm^2 (*e.g.*, for $Ni_{72}P_{18}B_7Al_3$[22]) to $\sim$310 Kg/mm^2 (*e.g.*, for $Fe_{80}P_{15}C_5$[25]) have been reported. While the author believes the latter value may err on the high side,[26] new glasses have been recently identified, *e.g.* $Fe_{80}B_{20}$, which do, indeed, exhibit remarkable strengths [$\sim$370 Kg/mm^2 (525 Ksi)[27,28]].

As metallic glasses do not work harden their tensile strengths must be equal to or less than their yield strengths. The former case obtains, of course, when yielding and failure occur simultaneously. One is able to uniquely identify this occurrence from the mode of failure of the sample. According to plasticity theory[29] a thin sheet will yield (following von Mises' criterion) in a zone whose normal makes an angle $\theta = 35.3°$ with the tensile axis and 90° with the thickness vector. According to Argon[30] θ will increase slightly if, as for metallic glasses, σ_y/E, (σ_y = yield stress, E = Young's modulus) is large; for $\sigma_y/E \approx 0.02$, typical for glassy alloys, the predicted angle θ is $\approx$37°. Accordingly, the included angle (90 - θ), measured on a wide face) at the fracture tip of a specimen which fails coincident with yielding should be $\sim$53° (see Fig. 3). This failure geometry has been reported for glassy $Ni_{49}Fe_{29}P_{14}B_6Si_2$ strips.[31] It was also observed for those alloys[28,32] whose tensile yield strengths are listed in Table I below (save for $Pd_{77.5}Cu_6Si_{16.5}$, where the tensile results are for wires[35]); σ_y values observed range from $\sim$147 Kg/mm^2 for $Pd_{77.5}Cu_6Si_{16.5}$ to $\sim$370 Kg/mm^2 for $Fe_{80}B_{20}$.

In order to observe the 53° mode of failure at $T \ll T_g$ one must test specimens with a reduced area gage section, with width to thickness (w/t) ratio of the order of 8:1. If smooth, uniform cross section ribbons are tested, failure is typically initiated at the grips and occurs by tearing (mode III, antiplane strain[34]). If a reduced section specimen with w/t>>8 is pulled, failure will occur by tearing across the gage section, but at a somewhat higher stress due to elimination of the grip constraint. One has, therefore, the sequence of failures shown schematically in Fig. 3 using data for Ni-Fe base metallic glasses.[35] Tensile strength data in the literature (save for those noted in Table I) invariably pertain to mode III failure on straight strips. If the specimen edges are ragged, one may expect, of course, even lower relative strength than indicated by Fig. 3. For composite reinforcement applications uniform cross section ribbons

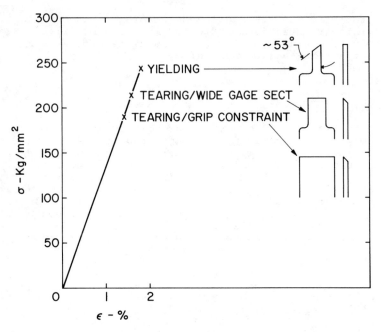

Fig. 3.   Tensile strength and failure mode for Ni-Fe
base metallic glasses.

(or wires) will be used.   In this case the lateral constraint
of the matrix may be sufficient to prevent mode III failure,
allowing the full intrinsic strength of ribbons to be realized.
        Fig. 3 is shown with the σ-ε curve linear to failure.
Curves are often reported in the literature which show considerable curvature (reduced slope) prior to separation.
This curvature is either an artifact associated with the nature of the specimen/grip/machine interaction[33] or, perhaps,
a nonlinear elastic response at high stress, as exhibited,
e.g. by Fe single crystal whiskers.[36]   Accordingly, "yield
stresses" predicted from the onset of nonlinearity, often
reported,[2-4,25] occasionally carefully labeled "apparent" σ_y,
are misleading.

Compression

        Yield stresses determined in compression are also listed
in Table I.   Palladium rich alloy specimens, with appropri-
ately large diameter (∼1 to 2 mm) for compression experiments, may be readily quenched to the glassy state.   Due
to the lack of work hardening σ_y is determined from the plateau of the σ-ε curve (Fig. 2).   Almost identical values
(157-158 Kg/mm²) are observed for $Pd_{77.5}Cu_6Si_{16.5}$ and
$Pd_{64}Ni_{16}P_{20}$.   It may also be noted that the yield stress in
compression (σ_{yc}) is slightly greater than that in tension

(σ_{yt}) for Pd-Cu-Si, indicating a small hydrostatic pressure or normal stress effect on yielding. Interrupted compression tests at ambient pressure and 6.2 Kb confirmed[33] the existence of this effect; the compressive flow stress is observed to increase by approximately 0.5% per kilobar of superposed pressure, *i.e.* $d\ln\sigma/dP \simeq 0.5 \times 10^{-3}/Kb$.

Theoretical Yield Stress

Theoretical estimates of the yield stress of a metallic glass have been provided by Li[37] and Gilman[12,20] based on the dislocation model for plastic flow. Li[37] modeled the glassy structure (two dimensional) as being produced by a lattice network of dislocations. The shear stress predicted for homogeneous deformation, requiring simultaneous motion of all the dislocations, is of the order 0.236 μ (μ = shear modulus) which of course far exceeds the observed shear yield stress ($\sigma_y/\sqrt{3}$, for a material which follows von Mises' criterion). The stress required to move a dislocation in the dislocation lattice is $\sim\mu/80$, which is within about a factor of 2 of the observed yield stress. Gilman[12,20] considers that dislocation drag develops due to a local dilatation at the dislocation core, generated when the dense random-packed structure is disturbed by shear displacement of the atoms. He derives the expression

$$\sigma_y = [8\pi\varepsilon^2/(1+\alpha)]B \tag{1}$$

where ε is the strain perpendicular to the direction of motion, B is the bulk modulus, and $\alpha = 3B/4\mu$. The macroscopic analogue of this model is the sliding behavior of soils, from whence it may be estimated that $\varepsilon \simeq 0.05$;[12] allowing for local compression of the atoms, $\varepsilon \simeq 0.04$.[12] One may then consider the case of $Pd_{77.5}Cu_6Si_{16.5}$ where the appropriate data for B,[38] μ[38] and σ_y (Table I) are available. Based on Eq. 1 with $\varepsilon \simeq 0.04$ σ_y is $\simeq 140$ Kg/mm^2, in rather good agreement with the observed value.

Hardness

An indication of the plastic strength of a material is provided by measurement of its microhardness.[39] It is typically observed, for example, that the ratio of hardness (Vickers/136° diamond pyramid) to yield stress (H/σ_y) for crystalline metals is ~ 3.[40] This value is predicted approximately (for the two dimensional case) by slip-line field analysis,[41] where a basic assumption is that the material underlying the indentation will act rigidly. Material adjacent to the indentation is expected to behave plastically, moving along slip lines and heaving up around the sides of the indenter to form a "coronet." Precisely such behavior has been observed for hardness indentations in metallic glasses.[4,26] Accordingly, the average H/σ_y observed (Table I) for metallic glasses (~ 3.2) is close to the theoretical value; the hardness of an ultrahigh-strength alloy such as $Fe_{80}B_{20}$ is then a remarkable 1100 Kg/mm^2. In principle, because the

375

hardness test is done under compression conditions, one should use the compression yield stress to calculate H/σ_y when σ_{yc} and σ_{yt} differ. If the results for $Pd_{77.5}Cu_6Si_{16.5}$ are typical (Table I), σ_{yc}/σ_{yt} for metallic glasses is of the order of 1.05. Hence H/σ_{yc} for glassy alloys is $\simeq 3$.

The indentation behavior of metallic glasses is distinctly different than that of nonmetallic glasses, *e.g.* silicate glasses, which, due to their covalent bonding, exhibit plastic-elastic, rather than plastic-rigid behavior, when indented. This results in a so-called "sinking in" impression.[42,43] Marsh[42,43] developed a theory for such behavior which predicts a decrease of H/σ_y with increasing σ_y/E and yields $H/\sigma_y \simeq 1.6$ for silicate glasses. This, then leads to a realistic estimate of σ_y, which is unmeasurable for such glasses; using the ratio $H/\sigma_y \simeq 3$ one makes the unreasonable prediction that σ_y is <u>less</u> than the observed <u>brittle</u> fracture strength. As noted above, an analogous situation obtains for metallic glasses, *i.e.* when the mode of tensile failure is brittle (tearing) it should be recognized that one <u>cannot</u> look to the σ-ε curve to define a yield stress.

TABLE I

MECHANICAL PROPERTIES OF METALLIC GLASSES

Alloy	H (Kg/mm²)	σ_y (Kg/mm²)	H/σ_y	E (10³ Kg/mm²)	E/σ_y
$Ni_{36}Fe_{32}Cr_{14}P_{12}B_6$ (METGLAS[R] 2826A)	880 (32)*	278 tension(32)	3.16	14.4 (32)	52
$Ni_{49}Fe_{29}P_{14}B_6Si_2$ (METGLAS[R] 2826B)	792 (31)	243 tension(31)	3.26	13.2 (44)	54
$Fe_{80}P_{16}C_3B_1$ (METGLAS[R] 2615)	835 (32)	249 tension(32)	3.35		
$Fe_{80}B_{20}$ (METGLAS[R] 2605)	1100 (28)	370 tension(28)	2.97	16.9 (28)	45
$Pd_{77.5}Cu_6Si_{16.5}$	498 (33)	157 comp. (7) 147 tension(33)	3.17 3.39	8.97 (44)	57
$Pd_{64}Ni_{16}P_{20}$	452 (46)	158 comp. (9)	2.86	9.37 (46)	59

*References cited below; H and σ_y have uncertainties of the order of ±5%.

The elastic properties of a variety of metallic glasses have been examined by several different methods. The shear, v_s, and longitudinal, v_ℓ, wave velocities of bulk specimens of Pd and Pt based glasses have been determined using Mhz pulse echo techniques.[38,45,46] These data and the density, ρ, allow calculation of the shear modulus, $\mu = \rho v_s^2$, the bulk modulus, $B = \rho v_\ell^2 - (4/3)\mu$, Poisson's ratio, ν, and E, only two of which, of course, are independent. Extensional wave velocities, v_e, have been determined by Khz pulse-echo techniques on ribbons[44,47-49] and by the method of resonance of vibrating reeds;[50-52] $E = \rho v_e^2$. The latter technique is more uncertain ($\sim\pm10\%$ of the modulus) than the pulse-echo techniques (~ 1-2%) because the square of the thickness of the reed enters as a factor. Pampillo and Polk[22] determined μ for two metallic glasses using an inverted torsion pendulum. Logan and Ashby[53] made load-deflection determinations of μ and E using helical springs. This particular static approach is viable because, in principle, one may obtain a reasonably large elastic deflection in the specimen relative to that in the loading machine. This is not typically true for tensile loading of straight ribbons and wires. Hence moduli determined from the load-elongation curves of such specimens are subject to considerable uncertainty and are not included here.

Tables IIA & B summarize most of the available elastic constant data for metallic glasses. We have not listed the value of $E/(1-\nu)$ [≈ 19] for Ni-P given by Jovanovic and Smith.[54] Also we have included only one entry each for the sets of data presented by Chen et al.[46,48] for $(PdNi)_{80}P_{20}$, $(PdFe)_{80}P_{20}$ and $(PdNi)_{83.5}Si_{16.5}$ alloys as a function of composition. For the ferromagnetic and magnetostrictive metallic glasses a stress induced incremental strain is provided by the motion of domain walls.[50,55] Accordingly, the apparent modulus of the material is reduced in the demagnetized state. The inherent modulus may be determined by one of the above techniques when the sample is magnetized to saturation; this is indicated in Table IIB by "s" in the "method" column. This variation of E with field is known as the ΔE effect[50,55] and can be of the order of 10-20% for metallic glasses. For the purposes of comparison with plastic properties (below) the inherent or "saturated" modulus is the appropriate parameter.

Inspection of Table IIA indicates rather similar data for the various Pd rich glasses which, if nothing else, indicates that metallic glasses can be synthesized in a reasonably reproducible fashion. In particular the elastic constants of the Pd-Si system appear to be insensitive to modest additions of a third element. The most studied glass is $Pd_{77.5}Cu_6Si_{16.5}$, where to facilitate comparison of the data, we have calculated the moduli for $\rho = 10.4$ gm/cm^3; reported room temperature values range from ~ 10.3 to 10.5 gm/cm^3. In detail the data appear to vary somewhat beyond the limits of precision expected for pulse-echo measurements. For most of the Pd free alloys data for only E or μ are known. This

TABLE IIA - ELASTIC PROPERTIES OF METALLIC GLASSES

Alloy	ρ	v_e	E	v_s	μ	v_ℓ	B	ν	method	ref.
$Pd_{77.5}Cu_6Si_{16.5}$	10.4	(0°K→)	8.88	1.74	3.15	4.48	16.7	0.41	1	38
	10.4		9.29	1.779	3.29	4.584	17.5	0.41	1	46
	10.4		9.36	1.797	3.44	4.60	18.0	0.41	1	45
	10.4	2.90	8.74						2	47
	10.4	3.015	9.45						2a	49
	10.4	2.92	8.88						2w	44
$Pd_{77.5}Ag_6Si_{16.5}$	10.4e		8.50	1.70	3.01	4.43	16.4	0.41	1	38
$Pd_{76}Au_6Si_{16.5}$			9.0						3	51
$Pd_{77.5}Ni_6Si_{16.5}$	10.4e		8.99	1.75	3.19	4.52	17.0	0.41	1	38
$Pd_{79.5}Ni_4Si_{16.5}$	10.5	2.92	8.99						2	48
$Pd_{81}Si_{19}$			8.0						3a	52
$Pd_{82}Si_{18}$	10.2		7.8						3	50
$Pd_{80}Si_{20}$	10.3	2.92	8.79						2	44
$Pd_{64}Ni_{16}P_{20}$	10.1		9.19	1.800	3.27	4.560	16.6	0.41	1	46
$Pd_{64}Fe_{16}P_{20}$	10.0		9.30	1.816	3.31	4.530	16.2	0.40	1	46
$Pt_{60}Ni_{15}P_{25}$	15.7		9.61	1.467	3.38	3.965	20.2	0.42	1	46

Units: $v-10^5$ cm/sec; E, μ, $B-10^{11}$ dyne/cm^2; $\rho-$gm/cm^3. Data for room temperature except as noted.

Method: 1-Pulse-echo/bulk sample; 2-Pulse-echo/ribbon; 3-Vibrating reed; a=sample annealed at $T<T_g$; e = estimate; w = wire.

TABLE IIB - ELASTIC PROPERTIES OF METALLIC GLASSES

Alloy	ρ	v_e	E	μ	method	ref.
$Ni_{76}P_{24}$			10.0		3ed	52
$Ni_{76}P_{24}$	7.8		9.5	3.5	4ed	53
$Ni_{74}P_{16}B_7Al_3$			8.6		3	52
$Ni_{75}P_{16}B_6Al_3$	7.71	4.065	12.7		2a	49
$Ni_{49}Fe_{29}P_{14}B_6Al_2$				5.4	5	22
$Ni_{49}Fe_{29}P_{14}B_6Si_2$	7.51	4.15	13.0		2	44
$Ni_{39}Fe_{38}P_{14}B_6Al_3$	7.6	4.12	12.9		2s	44
$Ni_{36}Fe_{32}Cr_{14}P_{12}B_6$	7.46	4.34	14.1		2	44
$Co_{85}P_{15}$	7.9		12.0	3.9	4 ed u	53
$Fe_{75}P_{15}C_{10}$	6.95		15.0		3s	50
$Fe_{75}P_{16}B_6Al_3$	7.1	4.27	12.9		2u	49
$Fe_{75}P_{16}C_5Al_3Si_2$			8.5		3u	52
$Fe_{74}P_{16}B_7Al_3$			9.9		3u	52
$Fe_{76}P_{16}C_4Al_2Si_2$				5.8	5u	52
$Fe_{80}B_{20}$	7.4	4.74	16.6		2s	44

Legend: as in Table IIA Method: 4-Spring; 5-Torsion
pendulum, ed = electrodeposited material; s = magnetic
saturation; u = unsaturated.

reflects the difficulty of producing bulk specimens of these
alloys. While the results of Logan and Ashby[53] include both
E and μ, they are insufficiently precise to calculate B re-
liably; for Co-P the calculated B is negative.
    For alloys dominated by a single metalloid, (e.g. P), E
appears to increase in the order Pd→Ni→Co→Fe. The results of
Chen et al.[46,48] also indicate a smooth increase of μ, B and/
or E as Pd is replaced by Fe or Ni in the Pd-P system[46] or by
Ni in the Pd-Si[48] system. However, their extrapolated results for $Ni_{80}P_{20}$ indicate an initial increase of E on addition of Pd. The importance of the nature of the metalloid is
suggested by comparison of E for $Fe_{80}B_{20}$ and $Fe_{75}P_{15}C_{10}$.
This latter effect may be related to the electronic structure
of the alloys. It is known from magnetic property measurements[56] that P is a more potent donor of electrons to the
transition metal d-band than B. In the case of the P alloy,
this may lead to a relative softening of the repulsive part
of the interatomic potential, reducing the curvature of the

379

potential well and, hence, the moduli. In general, compositional trends in the Tables are obscured by the multicomponent nature of the alloys.

A material which is bonded by centrally directed forces, and for which each atom is a center of symmetry, is known as a Cauchy solid.[57] For the case of a cubic material one then has only two independent elastic stiffness coefficients because $c_{12}=c_{44}$; for an isotropic solid $B=(5/3)\mu$. Inspection of Table IIA indicates that B/μ for metallic glasses is of the order of 5. This reflects the long range nature of their metallic bonding. The deviation from the Cauchy condition appears to be greater for metallic glasses than for the elemental crystalline metals. This results from the disordered small scale atomic rearrangements which presumably occur when an elastic pulse propagates in the glass. In their calculations of the elastic properties of amorphous metals Weaire et al.[58] found it necessary to allow internal displacements in order to account for the sharp reduction of μ (of the order of 30 to 50%) in the amorphous as compared to the crystalline state. (The bulk modulus differs by only some 5-10%; see e.g. ref. 38.) Similarly, the relatively high value of ν (~0.4) observed, intermediate between that typically exhibited by a crystalline transition metal (0.3) and that of a liquid (0.5), reflects internal displacements in the glass, and the intermediate structural stability of the glass compared to those of the crystal and liquid.

The anelastic properties of metallic glasses have been examined by Dutoit and Chen[38] using the pulse-echo method and by Berry and Pritchet[50] and Chen and coworkers[51,52] using the vibrating reed technique. In general, glassy alloys are characterized by low acoustic loss ($Q^{-1} \simeq 10^{-4}$ to 10^{-5}). A broad loss peak, of uncertain origin, exists around 40°K. From the low strength thermoelastic loss peak exhibited by flexing reeds one can obtain a measure of the thermal diffusivity and conductivity. Testardi et al.[47] have examined the anharmonic response of $Pd_{77.5}Cu_6Si_{16.5}$. They report a large value of $dE/d\sigma$ ($\sim7-12$) which would account for some non-linearity of the $\sigma-\varepsilon$ curve.

As indicated by the discussion above the yield stress (or hardness) levels of metallic glasses are expected to reflect their moduli. A comparison of E and H, exhibiting an increase of hardness with increasing E covering a range of H from ~500 to 1100 Kg/mm^2 for a number of glassy alloys,[44] is shown in Fig. 4. For the most part these alloys were only available in ribbon form for which only E is conveniently obtained. As the change in E reflects primarily the change in μ we expect that H does not vary linearly with μ through the compositions shown. Over a smaller range of hardness (~100 to 150 Kg/mm^2 near H $\simeq$ 500 Kg/mm^2) Chen et al.[46,48] observed nonlinear behavior of E vs. H for Pd-Ni-Si alloys and of μ vs. H for Pd-Ni-P and Pd-Fe-P alloys. The reduced hardness values (H/μ or H/E) apparently exhibit maxima (Pd-Ni-Si) or minima at intermediate compositions. Hence Fig. 4 may be a smoothed approximation to a series of short curvilinear segments. From the data of Chen et al.[46] one can show that σ

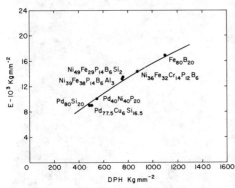

Fig. 4. Correlation of Young's modulus and Vicker's hardness for a variety of metallic glasses.

does not vary linearly with $B/[1+(4B/3\mu)]$, the appropriate elastic parameter for Gilman's model (Eq. 1).

Citing the nonlinear dependence of H on E or μ and the similarity between plots of H/μ (or H/E) and T_g vs. composition, Chen *et al.*[46,48] question the suitability of a defect, *i.e.* dislocation, model for plastic flow in metallic glasses. We note, however, that the dislocation models involve structural parameters as well as elastic parameters. In Li's model the parameter b/h enters, where h is a dislocation spacing parameter and b is the slip vector. In Gilman's model the strain, ε, perpendicular to the direction of motion, enters. One may expect these parameters to vary with structure as the composition varies and hence σ (or τ) will not vary linearly with modulus. For the whole range of data shown in Fig. 4 b/h need vary by only 30% to account for the lower ratio of E/H at high H. For change of temperature in a given alloy one would expect a linear dependence of H or σ_y on elastic modulus (if the deformation behavior is athermal). For temperatures from ambient to $\sim$450°K σ does appear to scale linearly with μ (or E) for a variety of glasses.[22] Appropriate data are not available to test the Gilman model in this case.

Tensile Failure

Axial Symmetry or Plane Stress

Tensile failure of glassy alloy wires and strips (plane stress) is accompanied by intense plastic shear deformation. This is true whether the macroscopic mode of failure is brittle (tearing; antiplane strain, mode III failure) or ductile (yielding). In the former case the shear deformation is localized at the tip of a crack, which propagates across a ribbon in the manner of a screw dislocation, *i.e.* on a plane of maximum shear stress in a direction perpendicular to the shear direction and the tensile axis. In the case of yielding the shear deformation zone develops over the entire cross section of the specimen (ribbon or wire) before, or coincident with, failure. In both cases, tearing and yielding, failure occurs by shear rupture through the intense shear zone and hence, in the microscopic sense, the failure mode is ductile and the fracture surface features observed are identical.

381

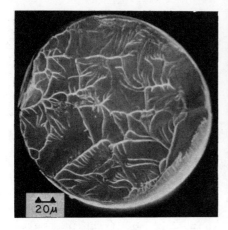 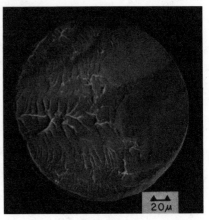

Figs. 5a (left) and 5b. Fracture surfaces of $Pd_{77.5}Cu_6Si_{16.5}$ wires fractured at 1 atm (a) and 6.9 Kb (b). The fracture surfaces are inclined at ~45°, sloping down from left to right.[33]

An example is shown in Fig. 5a for a $Pd_{77.5}Cu_6Si_{16.5}$ glassy alloy wire which failed coincident with (or close to) yielding[33] at ambient temperature and pressure. The fracture topography includes a featureless shear offset zone, which lies in an arc to the right of the fracture surface, (the brush markings in this region are an artifact associated with handling) and a ridge or vein pattern. The observation of this pattern was first noted by Leamy, Chen and Wang,[4] who used stereo pair scanning electron micrographs to demonstrate that the veins are hills, not steps, on a smooth background. They also found, on allowing for the relative shear of the specimen, that the opposing fracture surfaces are essentially mirror images of one another; this finding was confirmed by Pampillo and Reimschuessel.[59]

As separation occurs through the shear zone, it is apparent, on inspection of Fig. 5a, that failure is initiated by shear disk cracks, *i.e.* cracks with extended dimensions in the shear zone and minimal thickness ⊥ to the zone. These cracks expand like dislocation loops and their intersections produce the system of veins observed. As noted by Leamy *et al.*[4] the effects of adiabatic heating may be significant near the terminus of failure. This would lead to viscous necking of the material remaining between the closely approaching cracks, rather than cracking through, thereby producing the smoothly rounded veins observed.[59]

Fig. 5b shows the failure topography of a $Pd_{77.5}Cu_6Si_{16.5}$ wire fractured under a superposed hydrostatic pressure of 6.9 Kbar. One may note that the shear displacement prior to failure is a factor of two to three larger than for failure at 1 atm and that the vein pattern is no longer equiaxed. Rather, a system of fine veins lying more or less parallel

382

to the shear direction is observed. It is evident, therefore, that internal nucleation of shear cracks at isolated
imperfections, *e.g.* voids, hard particles *etc.* involves a
dilatational component which is opposed by superposed hydrostatic pressure. Consequently, at high pressure the shear
deformation proceeds until cracks are initiated at the periphery of the specimen. Short crack segments run from the
boundary into the specimen in the "shear" mode leaving behind
a system of closely spaced parallel veins which intersect near
the middle of the section.

While recognizing that yielding of a sheet is expected
to occur at 54.7° across its width, Pampillo[15] has suggested
that through thickness yielding on 45° planes can occur prior
to shear rupture in some cases. The evidence cited is the
observation of shear bands near and parallel to the fracture
(tearing) "plane." If Pampillo's suggestion is correct one
would have the peculiar situation of the tensile yield stress
depending on the geometry of failure. In our experience
(Fig. 3) specimens which fail in shear zones inclined through
the thickness always exhibit a lower tensile strength. We
also note that careful in situ observations of tensile deformation of glassy alloy sheets, either in the optical microscope[60] or the electron microscope,[61] have not detected uniform, *i.e.* over the entire cross section, yielding on 45°
planes prior to failure. It is the author's suggestion that
the shear bands observed near the fracture tip are generated
after failure due to the violent elastic recoil of the specimen. The rarefaction waves produced by this recoil are
known, for example, to lead to the pulverization of brittle
inorganic glass filaments on tensile failure.

Plane Strain

Plastic instability (yielding) and tearing are the favored modes of failure of nominally notch free glassy alloy
wires or ribbons at ambient temperature. At low temperature[15,22] or in the presence of a sharp crack in thick specimens at ambient temperature, crack propagation and failure
may occur under plane strain conditions. For example, by
means of electric discharge machining (EDM) one may place a
small hole ($\sim$0.1 mm dia.) at the center of a glassy alloy
"panel." When this panel is subjected to cyclic loading
fatigue cracks will nucleate at either side of the hole and
propagate until catastropic failure occurs.

Fig. 6 shows a section of a failure surface so produced.[35] The edge of the EDM hole lies at the extreme right
of the micrograph (the fracture features are symmetric about
the center of the hole). Adjacent to the hole the surface is
marked by a series of fatigue striations apparently associated with periodic oscillations of the direction of fatigue
crack propagation at ±45° from horizontal; near the center of
the micrograph a sharp boundary occurs, marking the onset of
catastrophic failure, to the left of which the surface marking
tend to approach horizontal. In larger sections it is

apparent that these markings have a "V" shaped conformation with the apex of the V pointing towards the origin of the crack. This pattern may be recognized as classical chevron markings of the type ob-

Fig. 6. A portion of the failure surface of a $Ni_{39}Fe_{39}P_{14}B_6Al_3$ strip.[35] The crack propagation direction is to the left (see text).

served in the failure of steel sheets.[62]

On a macroscopic scale the rapid fracture surface is oriented at 90° to the tensile axis, except for shear lips at its edges, *i.e.* a square fracture, typical of fully or partially plane strain crack propagation conditions, is observed. For medium thickness sheets such as that of Fig. 6 ($\sim$43 µm thick) a square to slant (mode III) transition occurs (outside the field of view to the left). For sufficiently thick specimens (>70 µ for alloys such as METGLAS[R] 2826B) fully plane strain conditions obtain and no transition to slant failure occurs. On examination of Fig. 7, which shows the chevron pattern for a different specimen at higher magnification, it is apparent that the chevrons have a sawtooth like configuration, *i.e.* the surfaces are inclined to the tensile axis. They are also marked by a fine scale, equiaxed vein pattern. It is apparent, therefore, that under macroscopic plane strain crack propagation conditions local failure still occurs by shear rupture.

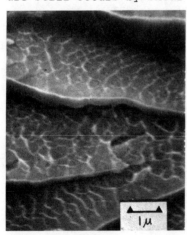

Fig. 7. A view of the chevron pattern surface for a $Ni_{48}Fe_{29}P_{14}B_6Al_3$ specimen. The crack propagation direction is to the left.[35]

Fracture Toughness

Using "center cracked panel" specimens of the type described above and specimens which failed with single edge notch (SEN) geometry (Ni_{48} alloy below) the author has measured the toughness of a series of Ni-Fe base metallic glasses.[35] Glassy alloy strips $\sim$25 µ ($Ni_{48}Fe_{29}P_{14}B_6Al_3$), $\sim$43 µ ($Ni_{39}Fe_{38}P_{14}B_6Al_3$) and $\sim$72 µ thick ($Ni_{49}Fe_{29}P_{14}B_6Si_2$) were tested. Fracture

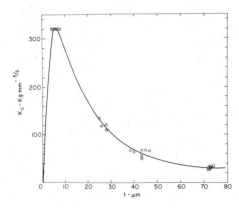

Fig. 8. Fracture toughness as a function of thickness for Ni-Fe base metallic glasses.[35]

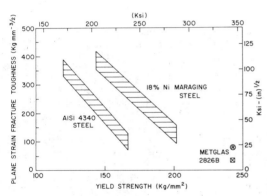

Fig. 9. Plane strain fracture toughness vs. yield stress for ferrous materials.[68]

toughnesses were calculated using the formula given by Irwin *et al.*[63] $K_C = \sigma\sqrt{\pi a}$, (with a suitable correction for the SEN geometry) where σ is the stress far removed from the crack and a is the crack half length. As a matter of convenience, most of the specimens were allowed to fail under cyclic loading conditions. For some specimens the "traditional" technique was used, *i.e.* cyclic loading was interrupted and the specimen was then pulled to failure at a constant deformation rate. The observed toughness values (for 43 μm thick samples) were identical within experimental scatter. Hence, at least for the specific conditions employed[35] for the Ni-Fe glasses, the cyclic failure tests yield valid toughness data.

As the above alloys have identical hardnesses, within experimental error, their different toughnesses are attributed to their different thicknesses; K_C is plotted vs. thickness in Fig. 8. The mode III fracture toughness for $Fe_{80}P_{13}C_7$, determined by Kimura and Masumoto[64] using tear tests, is plotted (shaded bar) at 2 thicknesses (4-8 μm), the shear lip thickness range for the Ni-Fe glasses. The shape of Fig. 8 is of the classical type observed, for example, for crystalline Al sheets.[65] The decrease of K_C from the peak value with increasing thickness reflects the transition from plane stress towards fully plane strain crack propagation conditions. According to the rule of thumb cited by Knott[66] the thickness t must be $\gtrsim 2.5$ $(K_{I_C}/\sigma_y)^2$ for a valid plane strain fracture toughness determination. For the Ni-Fe glasses σ_y is large (~ 243 Kg/mm²) and the measure K_C for the 72 μ thick sheets is modest (~ 30 Kg/mm³/²). Inserting these data into the above formula one finds $t \gtrsim 38$ μ. We conclude, therefore, that K_{I_C} for the Ni-Fe base glasses studied is $\simeq 30$ Kg/mm³/².

The fracture toughness of a Ni-Fe base metallic glass

(METGLAS[R] 2826; $Ni_{40}Fe_{40}P_{14}B_6$) has also been determined by
Ast and Krenitsky.[69] The toughness value observed for the
as-quenched material ($K_C \simeq 44$ Kg/mm$^{3/2}$) fits close to the
line (50 μm thick ribbons) in Fig. 8. These authors also examined the variation of toughness due to low temperature
annealing of the material. A decrease of toughness is observed (to ∿18 Kg/mm$^{3/2}$ after a 300°C anneal) which is attributed to the formation of a metastable crystalline phase.

Fig. 9, adapted by the author from a review by Zackay
et $al.$[67] compares the toughness behaviors of steels and
METGLAS[R] 2826B as a function of their yield strengths. One
concludes that toughness inevitably decreases as yield
strength increases. Hence the modest toughness values of the
exceptionally strong metallic glasses are as would be expected. The lower toughness of the metallic glass also reflects,
to some extent, its lower modulus (∿2/3 that of steel).

Fatigue

Crack Propagation Rate

Using the center cracked panel geometry one may conveniently monitor the advance of a propagating fatigue crack with
an optical microscope. One may then deduce da/dn, the cyclic
rate of crack advance, as a function of ΔK, the cyclic stress
intensity. The author has found[35] that da/dn (mm/cycle)
$\simeq 2 \times 10^{-8}$ ΔK$^{2.25}$, where ΔK has units of Kg/mm$^{3/2}$, for a
$Ni_{39}Fe_{38}P_{14}B_6Al_3$ metallic glass (43 μ thick). Making due
allowance for the difference in moduli, these data may be
compared with those for steels by plotting da/dn vs.
(ΔK/E); on such a plot the data for the Ni_{39} glass fall within the scatter band of data for various steels, as compiled
by Clark.[70]

Fatigue crack propagation rates in METGLAS[R] 2826 have
been determined by Ast & Krenitsky[69]; ΔK exponents (m) ranging
from 0 to 4 were observed. Ogura et $al.$[71] have determined
da/dn vs. ΔK for $Pd_{80}Si_{20}$; a ΔK exponent m≈4 was observed.
As noted by Knott[72] log da/dn vs. log ΔK plots typically have
three sections. There is a transient, high m, threshold
growth region at low ΔK and a similar high m region at large
ΔK where the crack begins to accelerate as K_{max} approaches
K_{I_C}. The intermediate ΔK region, where crack extension
occurs by reversed plastic opening at the crack tip, exhibits
m≈2. The average slope for all three regions typically is
∿4. The data for the Ni_{39} glass exhibit three regions but
the exponent given above (≈2) corresponds to intermediate ΔK
values. The high m (∿4) values observed by the above authors
may represent average values of the type just noted. The
source of low m (≈0) values is uncertain.

Lifetime

Fig. 10[73] shows a compilation of fatigue lifetime data
for Ni-Fe base metallic glasses. Cyclic loading frequencies
of the order of 0.5 to 30 hz were employed. As notched

samples will sustain static loads of, *e.g.*, 0.8 σ_y indefinitely (s = 0.8) it is apparent that the fatigue behavior of metallic glasses is cycle dependent, not time dependent as for hard, nonmetallic glasses.

As one would expect, sample lifetimes are minimized for notched samples (triangles), somewhat better for straight strips (open circles), which suffer interaction with the grips, and maximized for samples with a polished reduced area gage section (solid circles). The stress ratio for long life (>10^6 cycles) is also maximized for the latter samples. As a practical matter the endurance limit (σ_a for $\sigma_m = 0$) deduced for notched samples from Fig. 10 is comparable to those for notched steels.

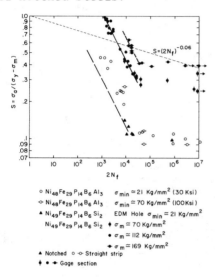

o $Ni_{48}Fe_{29}P_{14}B_6Al_3$ $\sigma_{min} \cong 21$ Kg/mm^2 (30 Ksi)
◇ $Ni_{48}Fe_{29}P_{14}B_6Al_3$ $\sigma_{min} \cong 70$ Kg/mm^2 (100 Ksi)
▲ $Ni_{49}Fe_{29}P_{14}B_6Si_2$ EDM Hole $\sigma_{min} \cong 21$ Kg/mm^2
$Ni_{49}Fe_{29}P_{14}B_6Si_2$ ♦ $\sigma_m \cong 70$ Kg/mm^2
● $\sigma_m \cong 112$ Kg/mm^2
◆ $\sigma_m \cong 169$ Kg/mm^2
▲ Notched   o ◇ Straight strip
♦ ● ◆ Gage section

Fig. 10. Stress vs. reversals to failure ($2N_f$) for Ni-Fe metallic glasses.[73] Stress is given in terms of the parameter S = [σ_a/($\sigma_y - \sigma_m$)] where σ_a = stress amplitude (1/2 the difference between max and min stress), σ_m = mean stress and σ_y = yield stress. The solid line for the Ni_{49} alloy ($\sigma_m \lesssim 140$ Kg/mm^2) is given by S = $16.9(2N_f)^{-0.4}$. The dotted line given by S = $(2N_f)^{-0.06}$ is the upper limit of fatigue lifetimes observed for high strength steels.[74]

An intriguing observation in Fig. 10 is the finding that reduced section samples of Ni-Fe glasses will sustain ∿600 stress cycles when loaded within ∿1% of their yield (and fracture σ_f) stress. By comparison high strength steels [S=$(2N_f)^{-0.06}$][74] will sustain only a few cycles when loaded close to their fracture stress, which, in fact, is about the same, *i.e.* ∿245 Kg/mm^2, as for the metallic glasses. This comparison reflects the high elastic limit of the metallic glasses. At stresses close to σ_f glassy alloys behave elastically and hence crack initiation is difficult. Steels cyclically loaded to the same absolute stress level experience gross plastic deformation because their yield stresses are well below σ_f. For lower stress ratios, *e.g.* of the order of 0.4, the lifetime for steels is longer. We expect that in this regime metallic glasses suffer because no mechanism for work hardening exists to effect crack tip blunting. Qualitatively similar behavior was observed for $Pd_{77.5}Cu_6Si_{16.5}$ wires.[73] These data were also in reasonable agreement with the low stress fatigue data presented by Ogura *et al.*[75] for $Pd_{80}Si_{20}$.

Conclusion

Deformation of metallic glasses ($T << T_g$) is highly inhomogeneous, occurring in the form of localized shear bands. Local shear strains on the order of 10 to 100 are observed in bands formed by bending, rolling, indentation, compression or tension. It has been observed that shear bands terminate within deformed glassy alloy specimens. By definition the boundary between the deformed and undeformed regions is a Volterra dislocation; one supposes that this macroscopic dislocation is formed by a pileup of atomic size dislocations, *i.e.* on a microscopic scale deformation is controlled by the motion of line defects. Dislocation models have been proposed which yield reasonable estimates of the yield strengths of metallic glasses.

Properly prepared glassy alloy ribbons exhibit tensile yielding; catastropic failure occurs coincident with yielding due to the absence of work hardening. The more common mode of failure is antiplane strain tearing, for which the tensile stress is, of course, less than the yield strength. Metallic glasses are exceptionally strong and hard; the binary alloy $Fe_{80}B_{20}$ exhibits a tensile yield strength (σ_y) of ~ 370 Kg/mm^2 (525 Ksi) and hardness (H) of ~ 1100 Kg/mm^2. The ratio H/σ_y for metallic glasses is typically ~ 3.2.

Glassy alloys are characterized by shear moduli which are ~ 30 to 50% lower than for the crystalline state. This reflects the random internal displacements of the atoms which occur when an elastic pulse propagates in the glass. Their bulk moduli are within about 10% of those for the crystalline state. The observed yield strengths are large fractions of the Young's moduli for metallic glasses; $E/\sigma_y \simeq 50$. The B glasses (*e.g.* $Fe_{80}B_{20}$) apparently exhibit the highest Young's moduli (16.5×10^3 Kg/mm^2; 24 Mpsi) ever observed for metallic glasses.

Glassy alloys exhibit classical fracture toughness behavior, *i.e.* toughness decreases with thickness as crack propagation conditions change from plane stress to plane strain. Due to their high strength and modest toughness ($\sigma_y \simeq 243$ Kg/mm^2, $K_{I_c} \simeq 30$ Kg/mm$^{3/2}$) fully plane strain crack propagation conditions obtain for sheets ~ 70 μm thick of $Ni_{49}Fe_{29}P_{14}B_6Si_2$.

Fatigue of metallic glasses is cycle dependent not time dependent. It occurs, therefore, due to the generation of deformation induced cracks, not due to the propagation of pre-existing flaws. Due to the high elastic limits, *i.e.* the coincidence of yield and fracture, metallic glasses will sustain an exceptional number of stress cycles (600 or more) when stressed with a percent or so of their fracture strength. As a practical matter the endurance limits for notched Ni-Fe base glassy alloy specimens are similar to those for notched steels.

388

References

1. H. S. Chen and D. Turnbull, J. Chem. Phys. **48**, 2560 (1968)
2. H. S. Chen and T. T. Wang, J. Appl. Phys. **41**, 5338 (1970)
3. T. Masumoto and R. Maddin, Acta Met. **19**, 725 (1971)
4. H. J. Leamy, H. S. Chen and T. T. Wang, Met. Trans. **3**, 699 (1972)
5. D. E. Polk and H. S. Chen, J. Non-Cryst. Solids **15**, 165 (1974)
6. Chem. Eng. News **19**, 24 (1973)
7. C. A. Pampillo and H. S. Chen, Mat. Sci. Eng. **13**, 181 (1974)
8. S. Takayama and R. Maddin, Acta Met. **23**, 943 (1975)
9. H. S. Chen, Scripta Met. **7**, 931 (1973)
10. H. S. Chen, H. J. Leamy and M. J. O'Brien, Scripta Met. **7**, 415 (1973)
11. H. S. Chen and D. E. Polk, J. Non-Cryst. Solids **15**, 174 (1974)
12. J. J. Gilman, J. Appl. Phys. **46**, 1625 (1975)
13. A. H. Cottrell, The Mechanical Properties of Matter, John Wiley, NY (1964) p. 322
14. A. S. Argon in Polymer Materials, ASM, Ohio (1975) p. 411
15. C. A. Pampillo, J. Mat. Sci. **10**, 1194 (1975)
16. See A. E. H. Love, The Mathematical Theory of Elasticity, Dover (1944) p. 221
17. J. J. Gilman, in Dislocation Dynamics (A. R. Rosenfield, G. T. Hahn, A. L. Bement, Jr. and R. I. Jaffee, Eds.) McGraw-Hill, NY (1968) p. 3
18. J. J. Gilman, in Physics of Strength and Plasticity (A. S. Argon, Ed.) MIT (1969) p. 3
19. J. J. Gilman, J. Appl. Phys. **44**, 675 (1973)
20. J. J. Gilman, Physics Today, **28** May 1975, p. 46
21. R. Maddin and T. Masumoto, Mat. Sci. Eng. **9**, 153 (1972)
22. C. A. Pampillo and D. E. Polk, Acta Met. **22**, 741 (1974)
23. D. E. Polk and D. Turnbull, Acta Met. **20**, 493 (1972)
24. F. Spaepen and D. Turnbull, Scripta Met. **8**, 563 (1974)
25. T. Masumoto and R. Maddin, Mat. Sci. Eng. **19**, 1 (1975)
26. L. A. Davis, Scripta Met. **9**, 431 (1975)
27. Materials Research Report, Nov. 1975, Allied Chemical Corporation
28. R. Ray, S. Kavesh, L. A. Davis, P. Chou and L. E. Tanner, to be published
29. R. Hill, The Mathematical Theory of Plasticity, Osford Univ. Press, London (1967) p. 300
30. A. S. Argon, in The Inhomogeneity of Plastic Deformation, ASM, Ohio (1973) p. 161
31. L. A. Davis, Scripta Met. **9**, 339 (1975)
32. S. Takayama and L. A. Davis, to be published
33. L. A. Davis and S. Kavesh, J. Mat. Sci. **10**, 453 (1975)
34. A. S. Tetelman and A. J. McEvily, Jr., Fracture of Structural Materials, John Wiley, NY (1966) p. 94
35. L. A. Davis, J. Mat. Sci. **10**, 1557 (1975)

36. S. S. Brenner, J. Appl. Phys. 27, 1484 (1956)
37. J. C. M. Li, to be published in Distinguished Lectures in Materials Science, Marcel-Dekker
38. M. Dutoit and H. S. Chen, Appl. Phys. Lett. 23, 357 (1973)
39. J. J. Gilman, in The Science of Hardness Testing and its Research Applications (J. H. Westbrook and H. Conrad, Eds.) ASM, Metals Park, Ohio (1973) p. 51
40. D. Tabor, The Hardness of Metals, Oxford Univ. Press, London (1951)
41. R. Hill, The Mathematical Theory of Plasticity, Oxford Univ. Press, London (1967) p. 213
42. D. M. Marsh, Proc. Roy. Soc. A279, 420 (1964)
43. D. M. Marsh, Proc. Roy. Soc. A282, 33 (1964)
44. L. A. Davis, P. Chou, L. E. Tanner, R. Ray and S. Kavesh, to be published
45. B. Golding, B. G. Bagley and F. S. L. Hsu, Phys. Rev. Lett. 29, 58 (1975)
46. H. S. Chen, J. T. Krause and E. Coleman, J. Non-Cryst. Solids 18, 157 (1975)
47. L. R. Testardi, J. T. Krause and H. S. Chen, Phys. Rev. B 8, 4464 (1973)
48. H. S. Chen, J. T. Krause and E. Coleman, Scripta Met. 9, 787 (1975)
49. H. S. Chen, Scripta Met. 9, 411 (1975)
50. B. S. Berry and W. C. Pritchet, J. Appl. Phys. 44, 3122 (1973)
51. H. S. Chen, H. J. Leamy and M. Barmatz, J. Non-Cryst. Solids 5, 444 (1971)
52. M. Barmatz and H. S. Chen, Phys. Rev. B 9, 4073 (1974)
53. J. Logan and M. F. Ashby, Acta Met. 22, 1047 (1974)
54. S. Jovanovic and C. S. Smith, J. Appl. Phys. 32, 121 (1961)
55. B. S. Berry and W. C. Pritchet, Phys. Rev. Lett. 34, 1022 (1975)
56. R. C. O'Handley, D. S. Boudreaux and R. Hasegawa, to be published
57. C. Kittel, Introduction to Solid State Physics, J. Wiley, NY (1956) p. 95
58. D. Weaire, M. F. Ashby, J. Logan and M. J. Weins, Acta Met. 19, 779 (1971)
59. C. A. Pampillo and A. C. Reimschuessel, J. Mat. Sci. 9, 718 (1974)
60. S. Takayama and R. Maddin, Met. Trans, in press
61. S. Takayama and R. Maddin, this conference
62. A. H. Cottrell, The Mechanical Properties of Matter, John Wiley, NY (1964) p. 366
63. G. R. Irwin, J. A. Kies and H. L. Smith, Proc. Am. Sco. Test. Mat. 58, 640 (1958)
64. H. Kimura and T. Masumoto, Scripta Met. 9, 211 (1975)
65. J. M. Krafft, A. M. Sullivan and R. W. Boyle, Proc. Symp. Crack Propagation, Cranfield (1961) p. 8
66. J. F. Knott, Fundamentals of Fracture Mechanics, Butterworths, London (1973) p. 137
67. V. F. Zackay, E. R. Parker, J. W. Morris, Jr. and

G. Thomas, Mat. Sci. Eng. 16, 201 (1974)

68. L. A. Davis, Battelle Colloquium on Fundamental Aspects of Structural Alloy Design, Harrison Hot Springs, B.C., Sept. 1975, proceedings to be published
69. D. G. Ast and D. Krenitsky, this conference
70. W. G. Clark, Jr., ASM Metals Eng. Quart., Aug. (1974) 16
71. T. Ogura, K. Fukushima and T. Masumoto, Scripta Met., in press
72. J. F. Knott, Fundamentals of Fracture Mechanics, Butterworths, London (1973) p. 246
73. L. A. Davis, J. Mat. Sci., in press
74. R. W. Landgraf in Achievement of High Fatigue Resistance in Metals and Alloys, STP467 ASTM, Philadelphia (1970) p. 3
75. T. Ogura, T. Masumoto and H. Fukushima, Scripta Met. 9, 109 (1975)

DEFORMATION, FRACTURE, AND CORROSION BEHAVIOR OF AN AMORPHOUS ALLOY

D. Lee and T.M. Devine

General Electric Corporate Research and Development
Schenectady, New York

INTRODUCTION

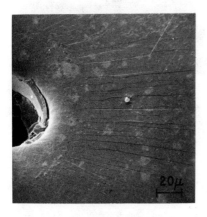

Figure 1. SEM photograph of notched Metglas 2826A pulled in tension and unloaded prior to fracture.

Among the variety of interesting characteristics exhibited by amorphous alloys, one of the unique behavior patterns is related to the onset of surface deformation markings, fracture mechanisms, and corrosion behavior. While considerable attention has been given to the mechanical behavior of amorphous alloys [1,2], relatively little work has been reported on their corrosion characteristics. The main purpose of the present investigation is to identify the key phenomenology of deformation and corrosion behavior of an amorphous alloy so that the underlying mechanisms can be examined. In order to accomplish this goal, experiments have been carried out with an amorphous alloy, Metglas 2826A, in parallel with 304 stainless steel fabricated to the same specimen geometry. Test results include those obtained from mechanical testing at different temperatures with and without notches and from anodic polarization measurements.

EXPERIMENTAL METHODS

The amorphous alloy was tested in the form of ribbon samples, 0.0069cm in thickness and 0.36cm in width, from Allied Chemical Corp. The alloy was designated as Metglas[TM] 2826A with nominal composition of 34 Ni,

393

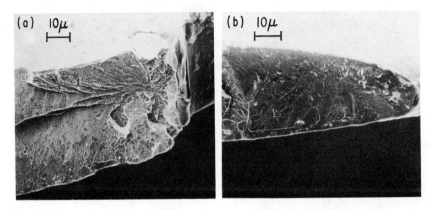

Figure 2. SEM photographs of fractured surfaces in the amorphous alloy:
(a) notched specimen, and (b) unnotched specimen.

29 Fe, and 15 Cr by weight with the remainder consisting of P and B
described as corrosion resistant ferrous alloy.

In order to make a valid comparison of behavior between the
amorphous metal and a reference crystalline metal, a 304 stainless
steel sheet was cold rolled to the same thickness of 0.0069cm. Starting
with an annealed sheet of 0.15cm in thickness, a total of 35 rolling
passes was required to produce the final thickness. The total reduction
in thickness was 95.3%.

Mechanical properties were examined with an Instron testing machine at both room temperature and 200°C. An Instron box furnace was
used for the elevated temperature tests. Simple tension tests were
made with strip specimens that had a reduced gage section of 0.33cm in
width and 3.12cm in length. In other cases, edge notches were machined
on the strip; dimensions of 60° V-notches were 0.064cm deep and 0.0013
cm root radius. Corrosion specimens approximately 7cm in length were
cut from the spool of material, soldered at one end to an electrical
test lead, ultrasonically cleaned in acetone and rinsed in distilled
water. The soldered joint was then masked off from the remainder of
the sample using Glyptal paint. This produced an effective sample size
394

of 4cm in length by 0.36cm in width by 0.0066cm in thickness. The
experimental apparatus used to determine the anodic polarization
characteristics of the samples consists of an electro-chemical cell
containing the deaerated test solution 0.17M NaCl, the test electrode,
two platinum-mesh counter-electrodes, and the standard calomel reference electrode ($\varepsilon° = 0.268V$). During testing, the anode potential is
increased with the aid of a potentiostat, starting from its free corrosion potential, in 20mV steps every 20 seconds and the cell current
which is equivalent to the corrosion rate is simultaneously recorded.

The cold worked amorphous metal ribbon used for the corrosion
test was obtained by the multi-pass rolling method with the total reduction of thickness of 32%.

RESULTS AND DISCUSSION

Deformation and Fracture Behavior

It has been shown that glassy metals exhibit deformation markings on
the surface prior to fracture reflecting the ability of the material
to undergo intensive plastic deformation on a local scale.[1] While the
exact nature of deformation markings as shown by discrete steps on
the surface is not known at the present time, they could be used to
identify conditions governing the initiation of the bands. For example,
when a thin strip specimen is notched and pulled in tension, deformation markings are expected to appear near the root of the notch in a
form which reflects a particular state of stress favoring the initiation of the deformation process.

The surface appearance of some of the notched specimens pulled
at room temperature and unloaded prior to fracture is shown in Figure
1. The surface markings appear initially away from the root of notch
where the sum of axial and transverse stress is the greatest. Since
there was no evidence of shear stress controlling the initiation of deformation markings, the result may be interpreted as an indication that
a biaxial stress state favors the nucleation of surface markings in the
direction normal to the largest normal stress, similar to crazing in

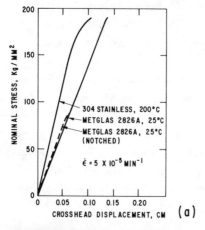

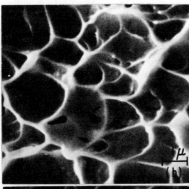

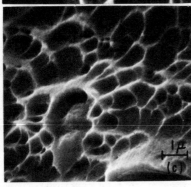

glassy polymers.[3] The role of hydrostatic stress in the initiation of deformation markings is not possible to evaluate because of the absence of through thickness stress on the surface.

When the notched specimen was pulled continuously to fracture, the fractured surface was not flat but showed ridges spreading radially from the site where cracks appeared to have initiated (Figure 2(a). Some evidence of vein-like structure has been observed at higher magnification, but the fracture surface topography is markedly different from that of unnotched specimen (Figure 2(b). Implications are that crack propagation in the notched specimen is a combination of two competing processes starting either from the surface offsets or internal defects adjacent to the root of a notch.

In order to gain more insight into the process of fracture in

Figure 3. (a) Nominal stress and displacement relationships under different testing conditions, (b) SEM photograph of fracture in the amorphous alloy near notch and (c) SEM photograph of fracture in cold worked 304 stainless steel.

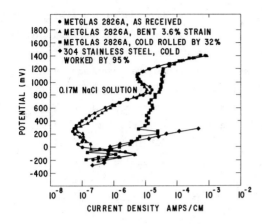

Figure 4. The anodic polarization characteristics of Metglass 2826A and 304 stainless steel under different conditions in 0.17M NaCl solution.

the amorphous alloy, heavily cold worked 304 stainless steel specimens were tested under different test conditions and the mode of fracture was compared with that of Metglas 2826A. It was observed that the nominal fracture stress of cold worked 304 stainless steel was 188 Kg/mm^2 when pulled at 200°C and at the strain rate of 5×10^{-5} min^{-1} which was nearly identical to the fracture stress of the alloy 2826A, as shown in Figure 3a. Furthermore, the fracture morphology of 304 stainless steel was quite similar to that of the amorphous metal in two respects: the fractured surface in 304 stainless steel was flat with little gross plastic deformation and the details of fracture were quite similar to that of notched Metglas 2826A fractured at room temperature. The details of fracture in the two materials are shown in Figure 3b and 3c.

The fractograph shown in Figure 3b indicates that a vein-like structure may assume a regular equiaxed morphology as a result of void nucleation and coalescence mechanism in the presence of hydrostatic stress. Microvoids within the stretched ligaments as seen in Figure 3b provide additional support for a ductile fracture process governed by the void nucleation mechanism. Cuplets are not necessarily elongated, suggesting that the plane of fracture was nearly normal to the direction of principle stress. This, in turn, is consistent with the grease model explaining the morphology of vein-like structure in glassy metals.[4,5]

397

The remarkably similar fracture morphology observed in 304 stainless steel with that of the amorphous metal provides additional insight into the fracture process of metallic materials with non-equilibrium structure. A related phenomenon known as work softening has been observed in crystalline metals when a metal is deformed plastically at a high temperature after being heavily deformed at a lower temperature.[6] The softening process was explained by the breakdown of the low temperature dislocation substructure during the subsequent application of load at the higher temperature. Highly localized deformation and fracture processes that took place in 304 stainless steel are consistent with above observations. The similarity in fracture morphology combined with the high localized deformation and little macroscopic ductility in both materials suggests that amorphous alloys may also undergo work softening process due to the non-equilibrium nature of the structure. The fact that others have attributed the fracture morphology to viscous behavior of deformed layer[4,5] is also consistent with the work softening hypothesis.

Corrosion Behavior — The anodic polarization characteristics of as-received amorphous alloy and the effect of cold work on its corrosion behavior are shown in Figure 4. Also indicated for the sake of comparison is the polarization behavior of cold rolled 304 stainless steel ribbon. The latter exhibits a passive corrosion current density of 7×10^{-7} amps/cm^2 and undergoes localized pitting attack at potentials as high as +1400mV and exhibits a minimum passive corrosion current density of 5×10^{-8} amps/cm^2. Evidently, the passive film formed on the amorphous alloy is highly protective and strongly resistant to localized corrosion attack. Bending the as received amorphous ribbon to an outer fiber strain of 3.6% results in no major change in the polarization behavior. The corrosion rate of the amorphous alloy ribbon is markedly increased following a 32% reduction in thickness by cold rolling. However, since pure bending produced no increase in corrosion, the nearly two orders of magnitude increase in passive corrosion rate is probably primarily due to localized crevice corrosion attack at the fine cracks introduced in the sample during the heavy cold reduction and not due to deformation enhanced dissolution.

398

The amorphous alloy Metglas 2826A is highly passive and strongly resistant to localized passive film breakdown which results in pitting in less passive alloys such as 304 stainless steel. Plastic deformation does not appear to impair the passivity of the amorphous alloy.

CONCLUSIONS

Some of the concluding remarks on deformation, fracture and corrosion behavior of an amorphous alloy, Metglas 2826A, are outlined in the following:

1. Initiation of deformation markings on the surface occurs normal to the direction of maximum normal stress and is aided by the transverse stress component.

2. Fracture initiates by void nucleation and coalescence mechanisms in the presence of hydrostatic stress; in this case the vein structure on the fracture surface is less pronounced.

3. Similar features in fracture morphology as well as the highly localized mode of deformation observed in the amorphous alloy and 304 stainless steel suggest that the fracture process in amorphous alloys is governed by the work softening mechanism.

4. Metglas 2826A is highly passive and strongly resistant to pitting corrosion in 0.17M NaCl solution as compared to cold worked 304 stainless steel.

5. Corrosion behavior of the amorphous alloy was relatively unaffected by plastic deformation.

ACKNOWLEDGEMENTS

The authors are grateful to W.T. Catlin for his assistance in mechanical testing and to M. Gill for SEM work.

REFERENCES

1. Gilman, J.J., Physics Today, May (1975), 46.
2. Masumoto, T. and R. Maddin, Mat. Sci. & Eng. 19, 1 (1975).
3. Lee, D., J. Mat. Sci., 10, 661 (1975).
4. Pampillo, C.A. and A.C. Reimschuessel, J. Mat. Sci. 9, 718 (1974).
5. Takayama, S., Ph.D. Thesis, U. Pennsylvania (1974).
6. Cottrell, A.H. and R.J. Stokes, Proc. Roy. Soc., 233A, 17 (1955).

CRYSTALLINITY AND MECHANICAL PROPERTIES OF METALLIC GLASSES

J. Megusar, J.B. Vander Sande, and N.J. Grant
Center for Material Science and Engineering
Massachusetts Institute of Technology

1.0.0 INTRODUCTION

On annealing, metallic glasses transform progressively from the amorphous to a fully crystalline structure. This transformation may proceed, depending on the particular alloy system, through several intermediate metastable phases.[1,2] As a result, a spectrum of microstructures may be obtained ranging from microcrystallites within the amorphous matrix to various amorphous-crystalline structures with different degrees of crystallinity.

A study of the properties of these alloys is interesting for two reasons: from a practical point of view it is interesting to know to what extent varying degrees of crystallinity alter the properties of metallic glasses; from a theoretical point of view a study of the underlying mechanisms of deformation and fracture in these alloys is most challenging. This work addresses these questions. The observations presented are based on research currently in progress. The Pd-Si and Cu-Zr systems were chosen to study tensile properties and fracture morphology in the amorphous, micro-crystalline, and mixed amorphous-crystalline states.

2.0.0 THE STRUCTURE OF ANNEALED $Pd_{80}Si_{20}$ AND $Cu_{60}Zr_{40}$

2.1.0 $Pd_{80}Si_{20}$

Masumoto and Maddin identified the transformation sequence in initially amorphous $Pd_{80}Si_{20}$ as consisting of the following four successive stages:

2.1.1 an incipient stage of crystallization during which some degree of ordering in the atomic arrangements occurs in the amorphous structure;

2.1.2 the appearance of Pd crystallites with an fcc structure within the amorphous matrix (MS-I) phase;

2.1.3 the formation of an ordered metastable phase (MS-II) over the entire matrix with dispersed MS-I phase;

2.1.4 formation of the stable phases, Pd solid solution and Pd_3Si.

Annealing temperatures of 200° and 300°C were selected for initially amorphous $Pd_{80}Si_{20}$. At 200°C, annealing gradually changes the amorphous matrix to a single fcc phase with the same composition. Annealing times of 60, 300, and 6000 minutes were chosen at 200°C. These annealing treatments resulted, according to Masumoto and Maddin, in the growth of microcrystals within the amorphous matrix to approximate sizes of 20, 40, and 80Å. At 300°C, crystallization proceeds through two metastable phases: the MS-I phase at shorter annealing times and the MS-II phase with dispersed MS-I phase at longer annealing times. Based on the temperature-time transformation diagram of amorphous $Pd_{80}Si_{20}$, annealing times from 100 to 1000 minutes were selected to transform the amorphous matrix progressively to the MS-I phase.[1]

2.2.0 $Cu_{60}Zr_{40}$

Vitek, Vander Sande, and Grant[2] identified the transformation sequence in an amorphous $Cu_{60}Zr_{40}$ alloy as consisting of two consecutive stages. In Stage 1, the original amorphous matrix transforms to a transformed amorphous matrix, with a simultaneous precipitation of Cu-rich crystallites. In Stage II, the transformed amorphous matrix and Cu-rich precipitates transform to the equilibrium structure. Giessen, et al.[3] identified the equilibrium structure as $Cu_{10}Zr_7$. Typically, the microstructure after annealing 1 hour at 425°C consisted of approximately 50% original amorphous matrix, 20% transformed amorphous matrix and 30% equilibrium phase $Cu_{10}Zr_7$.

An annealing temperature of 425°C and aging times of 15, 60, and 150 minutes were selected. These treatments resulted in approximately 30, 50, and 80% transformation to the equilibrium phase based on the transmission electron microscopy, calorimetry, and microhardness data of Vitek, et al.[2]

3.0.0 TENSILE PROPERTIES OF AMORPHOUS AND ANNEALED $Pd_{80}Si_{20}$ AND $Cu_{60}Zr_{40}$ ALLOYS

Amorphous $Pd_{80}Si_{20}$ was prepared by the Pond-Maddin technique. Amorphous $Cu_{60}Zr_{40}$ was prepared by the piston and anvil technique. Alloys

402

were judged to be amorphous by x-ray examination. Tensile specimens with a 10 mm gage length and a cross section of 0.4 x 0.035mm were cut from the ribbons of amorphous $Pd_{80}Si_{20}$. Similarly, tensile specimens with a 10mm gage length and with a cross section of 0.4 x 0.06mm were cut from amorphous areas of $Cu_{60}Zr_{40}$ foils. Tensile specimens of amorphous and annealed $Cu_{60}Zr_{40}$ were polished with 600 grit paper after which tensile testing was performed at room temperature on an Instron machine. Fracture surfaces were examined by scanning electron microscopy.

3.1.0 $Pd_{80}Si_{20}$

The results of the tensile tests with the amorphous and annealed $Pd_{80}Si_{20}$ specimens are summarized in Table 1.

It is shown in Table 1A that the fracture strength increases with increasing annealing time at 200°C. The shapes of the stress-strain curves of the amorphous and annealed $Pd_{80}Si_{20}$ specimens were observed to be similar and showed an "apparent yield strength".[1] These results suggest that ordering during the incipient stage of crystal-lization leads to an increase in tensile strength.

Table 1B shows the room temperature test results for $Pd_{80}Si_{20}$ specimens annealed at 300°C. The test data show an increase in fracture strength at shorter annealing times, followed by a decrease in fracture strength with progressive transformation from the amorphous to the MS–I phase.

The results shown in Table 1B are in agreement with the data of Masumoto and Maddin on the tensile strength of $Pd_{80}Si_{20}$ annealed at 290°C.[4] An increase in fracture strength from 127 kg/mm^2 for amorphous $Pd_{80}Si_{20}$ to 138 kg/mm^2 was observed after annealing for 100 minutes at 300°C, thus extending the observed[4] initial increase in fracture strength to shorter annealing times.

The observed tensile properties of amorphous and annealed $Pd_{80}Si_{20}$ (as shown in Table 1) are correlated, in research presently in progress,[5] with the structure of amorphous and annealed $Pd_{80}Si_{20}$. High-resolution electron microscopy is being used to provide the experimental evidence for the underlying mechanisms of deformation and fracture in amorphous, microcrystalline, and amorphous-crystalline $Pd_{80}Si_{20}$.

403

TABLE 1. TENSILE PROPERTIES OF AMORPHOUS AND ANNEALED $Pd_{80}Si_{20}$

A. Annealed at 200°C

Annealing time / crystal size min / Å		Amorph.	60/20	300/40	6000/80
Fracture strength	kg/mm^2*	127	131	135	136
Fracture strength	ksi	180	186	191	193

B. Annealed at 300°C

Annealing time / % transf. to MS-1 Phase: min / %		100/25	300/50	600/80	1000/100
Fracture strength	kg/mm^2*	138	135	132	125
Fracture strength	ksi	196	191	187	177

* Values are based on a minimum of four tests.

TABLE 2. TENSILE PROPERTIES OF AMORPHOUS AND ANNEALED $Cu_{60}Zr_{40}$

A. Annealed at 425°C

Annealing time / % trans. to equil. phase: min / %		Amorph.	15/30	60/50	150/80
Fracture strength	kg/mm^2*	125	140–150	140–150	85–95
Fracture strength	ksi	178	200–213	200–213	121–135
Apparent yield strength as fraction of σ_F [Figure 1]		$0.5\sigma_F$	$0.6\sigma_F$	$0.7\sigma_F$	$1.0\sigma_F$

B. Amorphous $Cu_{60}Zr_{40}$

Selected fraction of fracture strength σ_F	0.87	0.91	0.96	0.98
Strain (%) measured by:				
Instron machine	0.1	0.12	0.17	0.22
Optical method	<0.015	<0.015	<0.015	<0.015

* Values are based on a minimum of four tests.

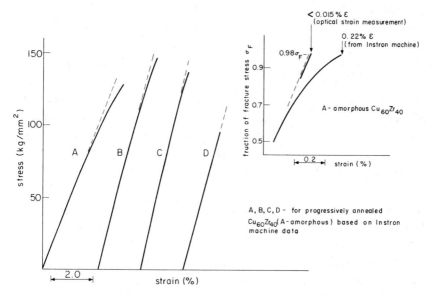

Figure 1. Stress-strain curves of amorphous and partially crystallized Cu60Zr40 alloys based on Instron machine data. Curve A is for amorphous material, Curves B-D for progressively longer anneals (See Table 2B).

3.2.0 Cu60Zr40

Results of mechanical testing of the amorphous and annealed Cu60Zr40 alloys are shown in Table 2 and Figure 1. It can be observed in Table 2A that the fracture strength increases after 15- and 60-minute annealing at 425°C. The range of 140-150 kg/mm^2 is given for the fracture strength rather than a single value because of lower reproducibility of data on partially crystalline specimens as compared to the amorphous specimens in this alloy. Annealing for 150 minutes at 425°C resulted in a decrease of fracture strength, below the value for amorphous Cu60Zr40.

Results on the fracture strength of Cu60Zr40 annealed at 425°C can be compared qualitatively with the results on the fracture strength of Pd80Si20 annealed at 300°C. The fracture strength in both systems increases initially with increasing crystallinity and then decreases as crystallization progresses.

405

Examination of the stress-strain curves of amorphous and mixed amorphous-crystalline $Cu_{60}Zr_{40}$ specimens allowed a further conclusion, namely that the apparent yield strength moves closer to the fracture strength with increasing crystallinity. Table 2A and Figure 1 show that the apparent yield strength of amorphous $Cu_{60}Zr_{40}$, measured as the deviation from straight line behavior on the stress-strain curve, corresponds to approximately 50% of the fracture strength. As a result of annealing 150 minutes at 425°C, the stress strain curve shows no apparent yield strength. Similar observations on the apparent yield strength as a function of degree of crystallinity were reported by Chou and Spaepen[6] after annealing amorphous $Pd_{74}Au_8Si_{18}$ at 392°C.

The stress-strain curve, taken directly from the Instron machine, showed an approximate 0.2% deviation from straight-line behavior at the fracture point. To separate anelastic and plastic behavior, the elongation was also measured by a sensitive optical method after a series of loading and unloading cycles. The results of the measurements are shown in Table 2B and Figure 1. The stress was progressively increased from about 87 to 98% of the fracture stress, with measurements made after each loading plus unloading cycle. The stress-strain curve taken directly from the Instron machine showed an 0.22% deviation from linear behavior at 98% of the fracture stress. The optical method showed no macroscopic ductility after a loading and unloading cycle at 98% of the fracture stress (within the limits of the accuracy of the measurement of 0.015%).

4.0.0 FRACTURE MORPHOLOGY OF AMORPHOUS AND ANNEALED $Pd_{80}Si_{20}$ AND $Cu_{60}Zr_{40}$ ALLOYS

4.1.0 $Pd_{80}Si_{20}$

Figure 2 shows the fracture morphology of amorphous $Pd_{80}Si_{20}$ as observed in a SEM at high magnification. The main vein runs parallel to the edge of the fracture surface, the tributary veins originate from the main vein and extend across the fracture surface. Fracture originated at one side of the specimen and extended in the direction

Figure 2. Fracture surface of amorphous $Pd_{80}Si_{20}$; 5,700X.

of the tributary veins. Fracture occurred, in both the amorphous and partially transformed states, at an angle of 90°C to the tensile axis. The fracture surface was flat and smooth.

The fracture morphology of $Pd_{80}Si_{20}$ was essentially unchanged as a result of annealing at 200° and 300°C. Figure 3 shows a fracture surface of $Pd_{80}Si_{20}$ annealed 1000 minutes at 300°C. One may similarly observe the main vein, and the tributary veins originating from the main vein.

Experimental observations suggest that the ordering during the incipient stage of crystallization (annealing at 200°C) and the appearance of Pd crystallites (annealing at 300°C) does not lead to a change in fracture morphology in $Pd_{80}Si_{20}$.

4.2.0 $Cu_{60}Zr_{40}$

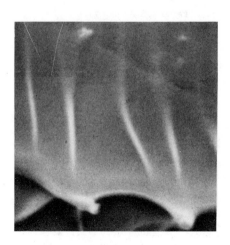

The fracture surface of amorphous $Cu_{60}Zr_{40}$ was characterized by a featureless zone associated with the initial crack displacement and the vein pattern zone formed during final rupture. Fracture occurred at an angle of 90° to the tensile axis.

Figure 3. Fracture surface of $Pd_{80}Si_{20}$ annealed 1000 min at 300°C; 11,800X

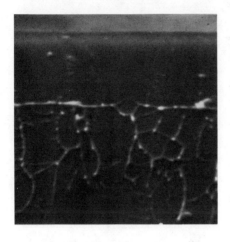

Figure 4. Fracture surface of $Cu_{60}Zr_{40}$ annealed 15 min at 425°C; view at edge of sample; 12,000X.

Annealing of $Cu_{60}Zr_{40}$ at 425°C resulted in a distinct change in fracture morphology with time. Some areas of the fracture surface exhibited features characteristic of the amorphous state, i.e., a featureless zone and a vein pattern zone. This can be observed in Figure 4 which shows the fracture surface of $Cu_{60}Zr_{40}$ annealed 15 minutes at 425°C. Other areas of the fracture surface of the same specimen exhibited a different morphology, described as veining on a chevron pattern, as can be observed in Figure 5. Fracture occurred macroscopically at 90° to the tensile axis. On a finer scale, the chevrons exhibited a sawtooth structure. The smooth surfaces show a fine scale equiaxed vein pattern.

4.3.0 Fracture Morphology in Fe-base Glasses

Pampillo and Polk reported[7] that an $Fe_{76}P_{16}C_4Al_3B_1$ glass fractured in tension showed a pseudo-cleavage mode of fracture (featureless zone, vein pattern zone, 45° angle with the tensile axis and the thickness vector) only when tested above room temperature. Between room temperature and 200°K they observed a "mixed fracture mode" consisting of the

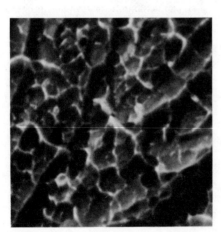

Figure 5. Fracture surface of $Cu_{60}Zr_{40}$ annealed 15 min at 425°C; view away from edge of sample; 12,000X.

Figure 6. Fracture surface of amorphous Metglas 2826; view at edge of sample; 1,185X.

pseudo-cleavage mode and of the final fracture along the surface normal to the tensile axis. Final fracture exhibited a chevron pattern with a fine scale equiaxed vein structure. They reported similar observations in some other Fe-based glasses such as $Fe_{77}P_{16}C_3Al_3B_1$, $Fe_{37}Ni_{37}P_{14}B_6Al_3Si_3$, and $Fe_{38.5}Ni_{38.5}P_{14}B_3Al_3Si_3$.

They further observed that the metallic glass systems exhibiting a "mixed fracture mode" also show an embrittlement effect. They observed a catastrophic decrease in fracture stress in $Fe_{76}P_{16}C_4Si_2Al_2$ and $Ni_{37}Fe_{37}P_{14}B_6Al_3Si_3$ when tested at 100°K and 175°K, respectively.

Davis[8] reported that fracture surfaces of Fe-Ni glasses when tested in tension exhibited a featureless shear deformation zone and a zone

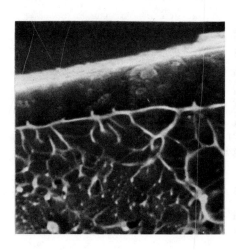

marked by a vein pattern, produced by shear rupture. When tensile failure occurred in the presence of a sharp fatigue crack, the same strips exhibited varying degrees of plane strain fracture, depending on sample thickness. The plane strain zone, macroscopically at 90° to the tensile axis,

Figure 7. Fracture surface of Metglas 2826 annealed 8,600 min at 300°C; view at edge of sample; 11,300X.

409

Figure 8. Fracture surface of Metglas 2826 annealed 8,600 min at 300°C; view away from edge of sample; 11,300X.

was marked by a classical chevron pattern. Fracture toughness decreased from 120 to 30 kg/mm^2 with an increase in sample thickness from 25 to 72μm. The "mixed fracture morphology" as observed by Pampillo and Polk[7] and by Davis[8] is similar to the fracture morphology observed in the annealed $Cu_{60}Zr_{40}$ alloy shown in Figures 4 and 5.

This observation prompted an extension of this study of fracture morphology to include amorphous and annealed Metglas 2826. When tested at room temperature in the as-received condition Metglas 2826 showed a fracture morphology consisting of a featureless zone and a vein pattern zone. This is shown in Figure 6. After annealing 8600 minutes at 300°C, a "mixed fracture morphology" was observed. Figure 7 shows the edge area of the fracture surface with the featureless zone and the vein pattern zone. The formation of a cellular pattern may be observed within the vein pattern zone. Figures 8 and 9 show at low and high magnification, respectively, the area of

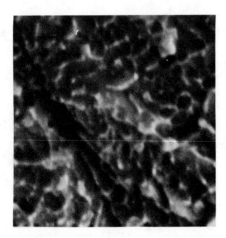

Figure 9. Fracture surface of Metglas 2826 annealed 8,600 min at 300°C; view away from edge of sample; 22,800X.

the fracture surface with the chevron pattern. The sawtooth structure and fine scale equiaxed vein pattern are observed.

Our observations on the fracture morphology in the amorphous and amorphous-crystalline $Cu_{60}Zr_{40}$ and Metglas 2826 and the noted similarity with the fracture morphology of some Fe-base glasses[7] suggests that the "mixed fracture mode" may not be a distinct fracture mode characteristic of amorphous alloys. It may rather reflect the fact that the respective metallic glass systems exhibited ab initio some degree of crystallinity. Since the embrittlement effect occurred only in the glass systems characterized by a "mixed fracture mode" a further deduction that the embrittlement phenomenon in metallic glasses, as observed and described by Pampillo and Polk,[7] may not be an inherent property of the amorphous state is suggested.

Davis observed a "mixed fracture mode" in the rapid fracture section only after tensile failure in the presence of a sharp fatigue crack. This suggests the possibility that a phase transformation may occur in amorphous alloys during fatigue testing. Ogura, Masumoto, and Fukushima, on the other hand,[9] observed no changes in structure on the fatigue fracture surface of amorphous $Pd_{80}Si_{20}$ using a micro-focus x-ray technique.

REFERENCES

1. Masumoto, T. and R. Maddin: Mat. Sci. & Eng'g 19, 1-24 (1975).
2. Vitek, J.M., J.B. Vander Sande, N.J. Grant: Acta Met. 23, 165-176 (1975).
3. Giessen, B.C. [private communications].
4. Masumoto, T. and R. Maddin: Acta Met. 19, 725-741 (1971).
5. Megusar, J., J.B. Vander Sande, and N.J. Grant [unpublished data].
6. Chou, C-P.P. and F. Spaepen: Acta Met. 3, 609-613 (1975).
7. Pampillo, C.A.: J. Mat. Sci. 10, 1194-1227 (1975).
8. Davis, L.A.: J. Mat. Sci. 10, 1557-1564 (1975).
9. Ogura, T., T. Masumoto, and K. Fukushima: Scripta Met. 9, 109-114 (1975).

STRUCTURE RELAXATION AND HARDNESS OF METALLIC GLASSES

H.S. Chen

Bell Laboratories
Murray Hill, New Jersey

C.C. Lo

Bell Laboratories
Columbus, Ohio

INTRODUCTION

In a preceeding paper,[1] we reported the structural relaxation in
metallic glasses of composition $(Pd_{.6}Ni_{.4})_{.80}P_{.20}$ and $Pd_{.775}Cu_{.06}$
$Si_{.165}$. It was suggested that the structural relaxation in metallic
glasses involves two distinct mechanisms:

o   local atomic regroupings, and

o   cooperative structural rearrangements.

Therefore, a different dependence of physical properties on the annealing temperature is predicted. We report here the effect of heat
treatment on the Vicker's hardness of several metallic glasses,
namely $Pd_{.775}Cu_{.06}Si_{.165}$, and $M_{.75}P_{.16}B_{.06}Al_{.03}$ with $M = Fe$, Co, and
Ni. Results are discussed in terms of structural relaxation.

EXPERIMENTAL

Pd-Cu-Si glass samples in cylindrical shape were obtained by quenching melts in a quartz capillary into water. Glassy samples of M-P-B-
Al Alloys were obtained in ribbon form using a centrifugal spinning
method.[2] The cylindrical samples were about 3mm in diameter and the
ribbons were about 25 μ thick and 1mm wide.

The glassy samples were vacuum sealed in a fused quartz tube and
annealed at various temperatures. The annealing time for Pd-Cu-Si
and M-P-B-Al glasses is, respectively, 10 minutes and 20 hours at
each temperature.

413

Glassy metal ribbons about 0.5cm long were glued on a flat metal surface with the free surface (surface not in touch with the rotating drum during rapid quenching) up. The surface was then lightly polished with 0.5μ chromium oxide paper to remove any excess glue. After the polish, a few scratch marks could be found on the sample surface. The hardness was measured on a Zeiss Ultraphot microhardness tester with a Vickers diamond indenter, using a 50gm load and 1600 magnification. Due to local surface waviness, some indentations did not make perfect impressions. These were not measured. The hardness values reported here are the average of eight measurements. The indentation depth was less than 2μ whereas the ribbon thickness was ~ 25μ. Therefore there should have been no anvil effect. With constant apertures and identical wavelength setting, the standard deviation of the measurements was estimated to be 0.1 to 0.2μ.

The relaxation spectrum of $Ni_{.75}P_{.16}B_{.06}Al_{.03}$ glass was obtained with a differential scanning calorimeter. Detailed procedures have been described in the preceding paper.[1]

RESULTS AND DISCUSSION

The Vickers hardness H_V of Pd-Cu-Si glassy samples is plotted against annealing temperature T_a in Figure 1. Also shown in the figure is the previously reported relaxation spectrum of the bulk glassy sample.[1] The scanning rate for the relaxation spectrum measurements was $20°K/min$. With increasing T_a, H_V of the glass

initially decreases then increases and peaks at $T_a \approx 600^\circ K$. On further increase in T_a, H_V goes through another minimum just above the
glass transition temperature T_g and then increases rapidly at higher annealing temperatures. H_V attains 610 k g/mm^2 at $T_a = 673^\circ K$
(not shown in the figure). The first minimum (T_1') and maximum
(T_2') in the H_V versus T_a curve coincide roughly with peak temperatures, T_1 and T_2, in the relaxation spectrum.

The apparent specific heat C_p of the Ni-P-B-Al glass in the as-
quenched and annealed states using a scanning rate of $20^\circ K / min$
is exemplified in Figure 2. The heat evolution of the structure relaxation $(- \frac{d \Delta H}{dt})$ is plotted against temperature on the top of the
figure. The relaxation spectrum of Ni-P-B-Al shows two peaks as does
the Pd-Cu-Si glass. Figure 3 illustrates the annealing effect on
the hardness H_V of M-P-B-Al glasses. The glassy samples remained
glassy except for the Fe-P-B-Al and Co-P-B-Al samples annealed at
$350^\circ C$ which were partially crystalline. The Co-P-B-Al glass annealed at $300^\circ C$ showed a trace of crystallinity.

All the H_V versus T_a curves exhibit dips and peaks very similar
to those of the Pd-Cu-Si glass. This feature is most pronounced in
binary (Fe, Co)-P-B-Al glasses. The first minimum and plateau positions $(T_1'$ and $T_2')$ in the H_V versus T_a curves of the M-P-B-Al glasses are lower by $\sim 130^\circ K$ than the corresponding peak temperatures $(T_1$
and $T_2)$ in the activation energy spectra in Figure 2. The second minimum or plateau (T_g') in the H_V versus T_a curves is also shifted by
$\sim -150^\circ K$ with respect to the thermally manifested glass transition

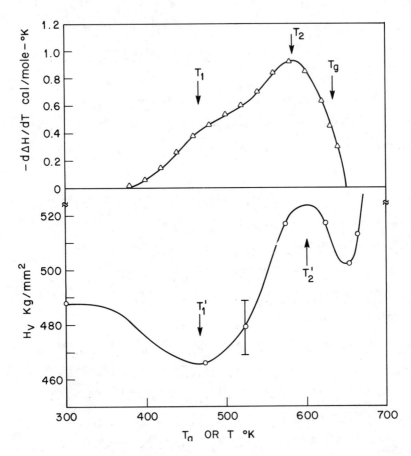

Figure 1. Vicker's hardness H_V versus annealing temperature (T_a) curve for a bulk $Pd_{.775}Cu_{.06}Si_{.165}$ glass. Annealing time is 10 min. Also shown in the figure is the tempering curve $-d\Delta H/dT$ versus T of the glass. Scanning rate = 20°K/min.

temperature T_g. This shift simply reflects a time-temperature scaling factor, and can be estimated as $\Delta T \approx -RT \cdot T' \ln(h) / \varepsilon$. Where T and T' are the temperatures at which atomic ensembles with activation energy ε undergo structural relaxation when the scanning rates are C and C' respectively. Here, $h = C/C'$, and R is the gas constant. Taking $C = 20°K/min$, $C' = 5 \times 10^{-2.0} K/min$, which roughly corresponds to 20 hours isothermal annealing, and $\varepsilon = aRT$ with a ≈ 25, [1,3] we obtain $\Delta T \approx -T'/4 \approx -(100 \sim 150°K)$.

The present results clearly indicate that the low temperature structure relaxation lowers H_v whereas the high temperature relaxation raises H_v of metallic glasses. A plausible explanation is that the local structure regroupings at low temperatures would enhance atomic structural inhomogeneity. Upon applying stress, this leads to stress concentrations and lowers the hardness of glasses. The high temperature equilibration, in contrast, raises the flow stress via cooperative structural stabilization.[1] In what follows, the effect of structure relaxation on the hardness of metallic glasses will be elaborated phenomenologically using the activated flow model based on configurational entropy.[4,5]

We will assume a single activated flow process with shear activation volume V* and activation energy ΔE for viscous flow. According to the entropy model for viscous flow, $\Delta E = R\Delta\mu*/S_c$.[4] When the local flow stress $\tau V* \gg RT$, then the applied shear stress for flow rate $\dot{\varepsilon}$ is given by:[5]

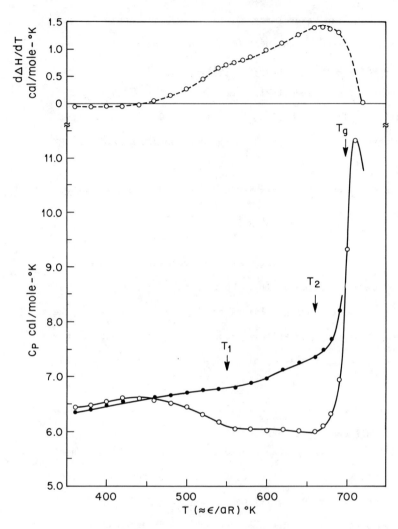

Figure 2. The apparent specific heat C_p of $Ni_{.75}P_{.16}B_{.06}Al_{.03}$ glass. o = as quenched, ● = annealed. The difference between the apparent specific heat in the as-quenched and annealed states ($-d\Delta H/dT$) is also shown.

$$\tau_y = \frac{RT_o}{KV^*} \left[\Delta\mu * / TS_c + \ln (2\dot{\varepsilon} \eta_o V^* / RT) \right] \tag{1}$$

where $\Delta\mu*$ is the minimum activation energy of a rearrangeable region, η_o a constant of the order 10^{-2}, R the gas constant, K the stress concentration factor, and S_c the configurational entropy.

Differentiating Equation (1) with respect to T_a, assuming $T = T_o$, the room temperature, $\dot{\varepsilon}$, and V^* to be constant

$$\left. \frac{d\tau_y}{dT_a} \right|_{\dot{\varepsilon}, V^*} = - \tau_y \frac{d \ln K}{dT_a} - \frac{R}{KV^*} \frac{\Delta\mu*}{S_c} \frac{d \ln S_c}{d T_a} \tag{2}$$

It is assumed that the local structure rearrangements at low temperature cause little change in cooperative structural ordering so that $\left| d \ln S_c / dT_a \right| < \left| d \ln K/d T_a \right|$. Equation (2) reduces to $d\tau_y / dT_a \approx - \tau_y d \ln K / dT_a < 0$. The flow stress τ_y, and therefore H_V of metallic glasses decrease with increasing annealing temperature T_a. A 10 to 20% change in both τ_y and K is probable as observed in low temperature annealing.

At $T_a \approx T'_2$, the cooperative structure equilibration dominates the flow properties. As $\left| d \ln S_c / dT_a \right| > \left| d \ln K / dT_a \right|$, $d\tau_y / dT_a \approx - \frac{R}{KV^*} \frac{\Delta\mu*}{S_c} \frac{d \ln S_c}{d T_a}$. Taking K = 2, $V^* \approx 15$ cm^3/mole, $d \ln S_c / dT_a \approx - T_g^{-1}$, and $\Delta\mu* / T_g S_c \approx 20$,[1,5] we obtain $d\tau_y / dT_a \approx 3 \times 10^7$ dynes/cm^2-$^\circ$K.

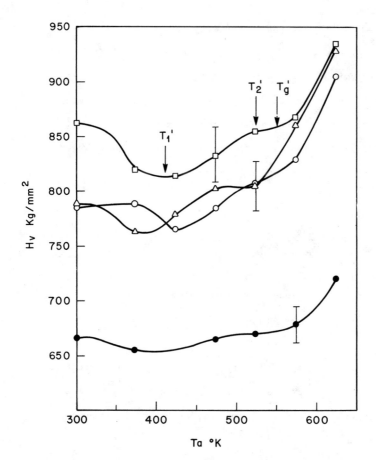

Figure 3. The Vicker's hardness H_V versus the annealing temperature T_a of $M_{.75}P_{.16}B_{.06}Al_{.03}$ glasses. Annealing time = 20 hrs. ● = $Ni_{.75}P_{.16}B_{.06}Al_{.03}$, o = $Co_{.75}P_{.16}B_{.06}Al_{.03}$, △ = $Fe_{.75}P_{.16}B_{.06}Al_{.03}$, □ = $(Fe_{.4}Co_{.6})_{.75}P_{.16}B_{.06}Al_{.04}$.

Many metallic glasses tend to phase separate into two glassy phases. Evidence of phase separation in Pd-Si based glasses near T_g has been observed using electron transmission microscopy[6] and small angle x-ray scattering.[7] Preliminary observations on the magnetic properties of Fe, Co, and Ni based glasses seem to suggest that these glasses also tend to phase separate into two glassy phases. The compositional segregation would cause stress concentrations, so that $d \ln K / dT_a > 0$. Concurrently, however, the configuration entropy S_c decreases, i.e., $d \ln S_c / dT_a < 0$. The flow strength τ_y, therefore, may decrease or increase depending on whether or not phase separation occurs.

On the other hand, the structure equilibration above the pseudo-glass transition temperature T_g' results in more structural disorder and less compositional and structural segregation so that $d \ln K / dT_a < 0$. Since S_c is primarily determined by the quenching rate $\dot{T}$ of the sample, $d \ln S_c / dT_a \simeq 0$, when $\dot{T} = $ constant. H_V increases above T_g'. H_V therefore decreases and increases respectively as T_a passes through T_g'. Consequently the H_V versus T_a curves near T_g' are S-shaped.

The present results tend to suggest that all the glasses studied exhibit a strong tendency to phase separate, so that the flow stress τ_y of these glasses is dominated by the stress concentration factor K. τ_y as well as H_V of the glasses therefore initially decrease and then increase near the pseudo-glass temperature T_g'. The initial decrease

421

Table 1

Structure Relaxation and Flow Strength in Metallic Glasses

T_c is the crystallization temperature, τ_y in dynes/cm^2

T_a Structure	T_1' Local	T_2' Cooperative	$\lessgtr T_g'$ Phase-Separation	$> T_g'$ Equilibration	T_c
$d \ln K/dT_a$	> 0	≈ 0	> 0	< 0	≈ 0
$d \ln S_c/dT_a$	$<_? 0$	< 0	< 0	$>_? 0$	< 0
$d\,\tau_y/dT_a$ cal	< 0	$\sim 3 \times 10^7$	≥ 0	> 0	> 0
$d\,\tau_y/dT_a$ exp	< 0	$(1-3) \times 10^7$	$<_? 0$	> 0	> 0

422

in H_V of Pd-Au-Si glass upon annealing near T_g, which has been observed,[8] simply reflects an increase in the stress concentration.

During the course of crystallization, the configurational entropy of the glassy phase (S_c) decreases, so the flow stress of most metallic glasses has been observed to increase drastically.[9,10]

Table 1 summarizes the predicted and observed hardness of metallic glasses, upon heat treatment. The agreement is fairly good.

CONCLUSIONS

The effect of annealing on the hardness of metallic glasses may be interpreted in terms of various structural relaxation mechanisms using an activated flow model. Low temperature local atomic regroupings lower the hardness of metallic glasses whereas high temperature cooperative structural relaxation increases the hardness.

All the glasses studied tend to phase separate into two glassy phases. This leads to stress concentrations which dominate the flow stress. At $T_a < T'_g$, the pseudo-glass temperature, $d \ln K / dT_a > 0$ and the hardness decreases. Above T'_g, structure equilibration leads to less compositional segregation, so that $d \ln K / dT_a < 0$. The H_V of metallic glasses increases with increasing T_a.

Partial crystallization in general results in lower configurational entropy of the glassy phase so that the flow stress τ as well as H_V of the glasses increases rapidly.

REFERENCES

1. Chen, H.S. and E. Coleman, Appl. Phys. Letters, March (1976).
2. Chen, H.S. and C.E. Miller, Mat. Res. Bull., January (1976).
3. Primak, W., Phys. Rev. 100, 1677 (1955).
4. Adam, G. and J.H. Gibbs, J. Chem. Phys. 43, 139 (1965).
5. Chen, H.S. [unpublished].
6. Chen, H.S. and D. Turnbull, Acta Met. 17, 1021 (1969).
7. Chou, C.P. Peter and D. Turnbull, J. Non-Cryst. Solids 17, 169 (1975).
8. Chou, C.P. Peter and Frans Spaepen, Acta Met. 23, 609 (1975).
9. Leamy, H.J., H.S. Chen, and T.T. Wang, Met. Trans. 3, 699 (1972).
10. Vitek, J.M., J.B. Vander Sande, and N.J. Grant, Acta Met. 23, 165 (1975).

UNIAXIAL STRESS DEPENDENCE OF YOUNG'S MODULUS
BY AN ULTRASONIC TECHNIQUE
Data for Magnetic and Non-Magnetic Metallic Glasses

J.T. Krause and H.S. Chen

Bell Telephone Laboratories
Murray Hill, New Jersey

INTRODUCTION

A previous study[1] revealed the large anharmonicity of a $Pd_{.775}Si_{.165}Cu_{.06}$ alloy. The ultrasonic technique utilized to obtain these results was only briefly described and since then several refinements in its use have been made. It is the purpose of this report to describe this technique in more detail and to offer by way of illustration data for the metallic-glass system $(Fe_{1-x}Ni_x)_{.75}P_{.16}B_{.06}Al_{.03}$. This system was chosen since, in addition to having the unique mechanical property of an amorphous metal, it is magnetic for the iron-rich compositions; thus ΔE effects are also revealed in this present data.

EXPERIMENTAL

Figure 1 illustrates diagramatically the apparatus for measuring the uniaxial stress dependence of Young's modulus of samples in fiber or ribbon form. The samples were prepared by the method of centrifugal spinning[2] and were typically $25\mu m$ x $200\mu m$ x $25\mu m$ in dimensions. These were gripped directly in opposing smooth-faced grips of pneumatically operated jaw assemblies of the tensile tester. Stress was uniformly applied at predetermined rates using a model 1101 slow-speed table model Instron machine. The output of the load cell in terms of grams force was fed as one parameter to an x-y recorder. The other parameter, the Young's modulus sound velocity, was obtained using a Panametrics[3] 5010 Panatherm ^{R} Pulser-receiver together with a Remendur-delay line operated at 100KHz. The Panatherm measures digitally the time interval in μsec. between any preselected echo pairs and provides an analog signal to an x-y recorder. The 100KHz sound wave was coupled into the sample by impedence matching

425

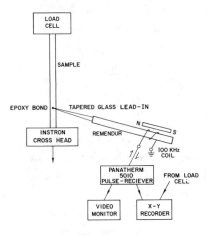

EXPERIMENTAL APPARATUS

LOAD CELL

SAMPLE

EPOXY BOND

TAPERED GLASS LEAD-IN

INSTRON CROSS HEAD

REMENDUR

N

S

100 KHz COIL

PANATHERM 5010 PULSE-RECIEVER

FROM LOAD CELL

VIDEO MONITOR

X-Y RECORDER

Figure 1. Experimental arrangement. A Young's modulus wave from the Remendur delay line is coupled into the sample by the tapered glass lead-in. The transit time of the echo pair between epoxy bond and load cell is measured as a function of uniaxial tension applied by the Instron cross head.

the Remendur to the sample through the use of a tapered glass lead-in. This was epoxy-bonded to the sample and the Remendur (Fig. 1). Since each joint or impedance discontinuity in the entire delay line gives rise to an echo, the correct echo pair must be identified. This is easily done by touching the delay line assembly at various places, thereby giving rise to identifying echoes which move in relation to the points being touched. The correct echo pair corresponds to the impedance discontinuities at the sample lead-in and the sample upper-grip junctions. The linear distance between these junctions is the sample length (ℓ). The delay time (τ) measured for this echo pair is the round trip delay time; therefore, in calculating the delay time/unit length, the round trip length (2ℓ) must be used. The Young's modulus velocity V_E and Young's modulus E are given by:

$$V_E = 2\ell/\tau, \qquad E = \rho\, V_E^2$$

where ρ is the sample density. By straining the sample in tension the differential delay is obtained as a continuous function of applied load (this can be plotted directly as stress using an area compensator). The stress dependence of Young's modulus is then given by

$$\frac{dE}{d\sigma} = 1 + 2\upsilon - \frac{2E_o}{\tau}\,\frac{d\tau}{d\sigma}$$

where E_o is the modulus at zero stress, σ is the stress, and υ is Poisson's ratio. The term $1 + 2\upsilon$ corrects for changes in length and cross-

426

sectional area of the sample under stress.[1] If υ is not known, as in the present study, then $\Delta E/\Delta\sigma$ can be evaluated making corrections for length and crossectional area using the stressed modulus calculated by reiteration.

The arrangement of Figure 1 is highly precise with respect to relative values. The analog output of the Panatherm is in steps of 0.1μsec. which allows one to either count the steps in the recorder trace or to use them as a standard of calibration. Retrace shows no shift or hysteresis and the relative accuracy increases with total delay time or sample length. Where $dE/d\sigma$ is large and/or where total delay time is large, reliable data can be obtained at 10^{-4} strain levels. Inaccuracy in absolute values stems from two main causes: (1) The inability to obtain true sample length due to the relatively large area of the sample lead-in junction, and (2) Ambiguity in the selection of corresponding wave cycles of the echo pair (often, in addition, these are distorted or out of phase). Absolute values of transit time and length were obtained separately using a precision rule graduated in 100ths of an inch. The sample was mounted on the ruled scale and sound propagated through the sample. Discrete echoes were initiated by pressing a razor blade against the fiber and noting the scale position and associated delay time. By a series of such successive measurements, reliable absolute unit delay time values to within 0.2% were obtained. These were then used to correct all subsequent relative measurements.

RESULTS AND DISCUSSION

Figure 2 shows the uniaxial stress dependence of delay time for the system $(Fe_{1-x}Ni_x)_{.75}P_{.16}B_{.06}Al_{.03}$ for compositions ranging in x from 0 to 1. For those compositions rich in Fe, there is shown a ΔE effect superimposed on the anharmonic behavior. Extrapolation of the linear portion of these curves to zero stress provides a measure of the magnitude of the ΔE effect.

Figure 2 shows, in addition to the stress levels required for saturation of the ΔE effect, some other unique details. For composition x = 0, there is a slight peak at 3.5kg/mm^2. From this point on, there is

427

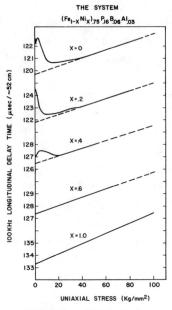

THE SYSTEM
$(Fe_{1-x}Ni_x)_{.75} P_{.16} B_{.06} Al_{.03}$

100KHz LONGITUDINAL DELAY TIME (μsec/~52cm)

X=0

X=.2

X=.4

X=.6

X=1.0

UNIAXIAL STRESS (Kg/mm²)

Figure 2. Delay time of a 100KHz Young's modulus wave as a function of uniaxial stress for the amorphous system $(Fe_{1-x}Ni_x)_{.75} P_{.16} B_{.06} Al_{.03}$.

stiffening up to about $10kg/mm^2$. Above this stress softening occurs which is governed largely by the anharmonic behavior. The ΔE effect disappears at $45kg/mm^2$ for this composition. As Ni is substituted for Fe, the ΔE effect decreases in magnitude and the stress level required for saturation also decreases. At $x = .6$ and above, no ΔE effect is observed, the compositions being nonmagnetic at room temperatures.[4]

These details in behavior of the ΔE effect correspond to some extent with changes in domain structure as shown by the work of Obi, Fujimori, and Saito[5] on the system $Fe_{80}P_{13}C_7$. They used a magnetic powder technique to identify domain patterns in stressed ribbons of this material. Their results indicate a maze pattern whose area rapidly disappears with tension up to $9.4kg/mm^2$. At this stress 180° wall domains appear directed parallel to the stress direction. At a stress of $18.8kg/mm^2$ the maze pattern has disappeared and the 180° wall domains predominate. The effects are reversible. They don't, however, show data at higher stresses. With reference to curve $x = 0$ in Figure 2, their data suggest that the rapid decrease in modulus up to $10kg/mm^2$ represents a rapid decrease in area of the maze type domain pattern. The abrupt change in slope above this stress signals the beginning of 180° domain wall alignment together with a slow decrease in maze pattern disappearing at $45kg/mm^2$.

Figure 3 is a comparison of the actual zero stress values of Young's modulus with those obtained by extrapolation of the data of Figure 2 to zero stress. These extrapolated values agree well with those obtained in a saturated magnetic field ($\sim$130 gauss). The solid curve in Figure 3 indicates a positive deviation of the modulus from linearity which may be

428

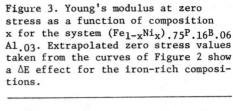

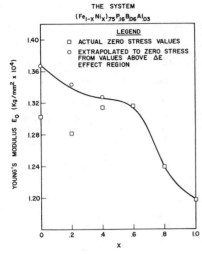

THE SYSTEM
$(Fe_{1-x}Ni_x)_{.75}P_{.16}B_{.06}Al_{.03}$

LEGEND

□ ACTUAL ZERO STRESS VALUES

○ EXTRAPOLATED TO ZERO STRESS FROM VALUES ABOVE ΔE EFFECT REGION

Figure 3. Young's modulus at zero stress as a function of composition x for the system $(Fe_{1-x}Ni_x)_{.75}P_{.16}B_{.06}$ $Al_{.03}$. Extrapolated zero stress values taken from the curves of Figure 2 show a ΔE effect for the iron-rich compositions.

partially attributed to a denser packing of the alloy upon mixing as has been observed for Pd-Ni-P glasses.[6]

Figure 4 shows the stress dependence $\left(\frac{dE}{d\sigma}\right)$ of Young's modulus as a function of compositional change. These values, evaluated at stress levels of $100 kg/mm^2$ (strain $\sim 8 \times 10^{-3}$), indicate the large anharmonicity for these materials and are similar in magnitude to that obtained for a $Pd_{.775}Si_{.465}Cu_{.06}$ metallic glass. The positive deviation in Figure 4 indicates a less anharmonic structure for the mixed alloys than for the end members.

SUMMARY

An ultrasonic technique is described for measuring the uniaxial stress dependence of Young's modulus. Sufficient sensitivity is available to detect

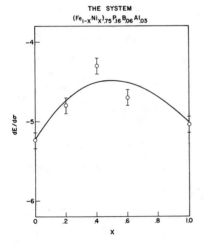

THE SYSTEM
$(Fe_{1-x}Ni_x)_{.75}P_{.16}B_{.06}Al_{.03}$

Figure 4. The uniaxial stress dependence of Young's modulus as a function of composition x for the system $(Fe_{1-x}Ni_x)_{.75}-P_{.16}B_{.06}Al_{.03}$. A positive deviation toward a less anharmonic structure is indicated.

429

anharmonicity at 10^{-4} strains. Illustrative data are presented for the metallic glass series $(Fe_{1-x}Ni_x)_{.75}P_{.16}B_{.06}Al_{.03}$ showing, in addition to the large anharmonic behavior, a ΔE effect for the iron-rich compositions. Some fine structure in the ΔE dependence is also indicated and appears to represent changes in the magnetic domain structure.

TABLE				SYSTEM $(Fe_{1-x}Ni_x)_{.75}P_{.16}B_{.06}Al_{.03}$				
	X	0	.2	.4	.6	.8	1.0
	ρ in g/cm^3	7.10	7.22	7.34	7.46	7.59	7.71
V_E Km/sec.	At Zero Stress	4.24	4.17	4.19	4.16	4.00	3.90
	Extrap.	4.34	4.27	4.21	—	—	—
	At 130 Gauss	4.35	4.26	4.21	4.16	4.00	3.90
E_o Kg/mm^2	At Zero Stress	13019	12806	13144	13168	12387	11961
	Extrap.	13666	13427	13270	—	—	—
	At 130 Gauss	13704	13365	13270	13168	12387	11961
	dE/dσ	−5.24	−4.80	−4.31	−4.70	—	−5.04

REFERENCES

1. Testardi, L.R., J.T. Krause, and H.S. Chen, "Large Anharmonicity of Amorphous and Crystalline Phases of a Pd-Si Alloy", Phys. Rev. B 8, 4464-4469 (1973).
2. Chen, H.S. and C.E. Miller, "Centrifugal Spinning of Metallic Glass Filaments", Mat. Res. Bull., 11, 49 (1976).
3. Panametrics, a subsidiary of Esterline Corporation, Waltham, MA USA.
4. Sherwood, R.C., E.M. Gyorgy, H.S. Chen, S.D. Ferris, G. Norman, and H.J. Leamy, "Ferromagnetic Behavior of Metallic Glasses", AIP Proceedings, 20th Magnetism and Magnetic Materials Conference, San Francisco CA USA (1974).
5. Obi, Y, H. Fijimori, and H. Saito, "Magnetic Domain Structure of an Amorphous Fe-P-C Alloy," J.J. Appl. Phys. [in press].
6. Chen, H.S., J.T. Krause, and E. Coleman, "Elastic Constants, Hardness and Their Implications to Flow Properties of Metallic Glasses", J. Non. Cryst. Solids 18, 157-171 (1975).

EFFECTS OF COLD ROLLING
ON LOW TEMPERATURE ANELASTIC PROPERTIES
OF METALLIC GLASSES

M. Barmatz, K.W. Wyatt*, and H.S. Chen

Bell Laboratories
Murray Hill, New Jersey

INTRODUCTION

Recent studies of the elastic behavior in metallic glasses revealed a broad absorption peak[1,2] and velocity dispersion[1] at low temperatures, which are similar to previous observations in non-metallic glasses.[3] In the case of the non-metallic glasses, the low temperature anelastic behavior was attributed to structural rearrangements,[4] however, at present there is no completely satisfactory microscopic theory. New experimental studies are needed to further clarify the nature of these low temperature anomalies.

In the present investigation, we have studied the effect of cold rolling on the low temperature anelastic properties of the metallic glass alloy $Fe_{0.74}P_{0.16}C_{0.05}A\ell_{0.03}Si_{0.02}$. This alloy is ferromagnetic with a Curie temperature of 598K. Previous measurements[1] on initially quenched samples of this material showed a rather large low temperature anelastic anomaly. Recently, Chen[5] measured the effect of cold rolling on several metallic glasses from room temperature to the glass transition. He found that cold rolling induces two distinct structural rearrangements and leads to a decrease in Young's modulus. The present study is intended to complement Chen's work by investigating the effect of cold rolling on the low temperature properties of metallic glasses.

Measurements of the Young's modulus and internal friction were carried out in the temperature range 1.3 - 300K using a vibrating reed technique.[1,6] Cold rolling substantially increased the magnitude of the low temperature damping and dispersion, however the measured tempera-

---

* Present address: School of Engineering and Technology, Dept. of Electrical Engineering, Tennessee State University, Nashville.

ture dependence was similar to the results of previous investigations of initially quenched samples.[1] Measurements at several flexural resonant frequencies revealed a new and unusual frequency dependence of the damping and dispersion in the range 0.1 - 5 kHz. This paper is intended to provide a preliminary analysis of these new measurements.

EXPERIMENTAL MEASUREMENTS
_____

A roller quenching technique[7] was used to prepare initially quenched amorphous ribbons ∿20 cm long, 0.2 cm wide, and 0.005 cm thick. X-ray analysis confirmed the amorphous nature of the ribbons. Initially quenched specimens, ∿2 cm long, were cut from one half of a ribbon while the other half was cold rolled to produce a 9% reduction in sample thickness.

Elastic measurements were carried out for both as quenched and cold rolled samples. Each specimen was clamped at one end between two copper plates. Flexural vibrations were excited and detected at the free end by electrostatic means.[1,6] The Young's modulus, E, is related to the flexural resonant frequency, f_n, of the nth mode of a clamped-free rectangular reed by the expression

$$f_n = A_n V_E , \tag{1}$$

where A_n is a geometric factor.[1] The Young's modulus velocity $V_E = (E/\rho)^{1/2}$, where E is the modulus along the reed and ρ is the density. The temperature dependence of the modulus is thus determined from the temperature dependence of the resonant frequency; and the internal friction, given by the inverse of the quality factor Q, may be obtained from the half power points of the resonance. A detailed description of the apparatus and electronics has been published previously.[1,6]

Non-ideal sample geometry may lead to large uncertainties in the absolute magnitude of the Young's modulus. However relative modulus measurements have an accuracy better than 0.2%. In this paper, we present all modulus measurements in the dimensionless form $\Delta E/E_o = E(T)/E(300) - 1$.

Measurements were initially carried out on an as quenched specimen. The observed internal friction had the same magnitude and temperature

432

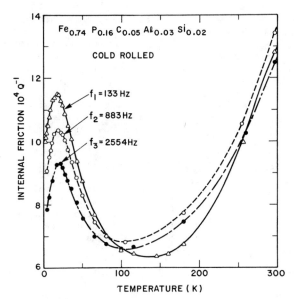

Figure 1. Temperature dependence of the internal friction in cold rolled amorphous $Fe_{0.74}P_{0.16}Co_{0.05}A\ell_{0.03}Si_{0.02}$. All the flexural modes exhibit a broad low temperature maximum which decreases with increasing frequency. The frequences given are room temperature values.

dependence as previously published measurements.[1] The maximum in the internal friction occurred at $T_M = 20.0 \pm 0.5K$. The Young's modulus also had the same temperature dependence as previous measurements, however the present data gave a larger absolute value for the modulus $E_0(300K) = 1.15 \times 10^{12}$ dynes cm^{-2}.

In Figure 1, we present internal friction measurements for the first three flexural modes of a cold rolled sample. All the modes exhibit a broad low temperature anomaly in the damping with maxima at 17.5K, 19.3K and 21.5K for flexural resonant frequencies f_1, f_2, and f_3 respectively.

The maximum internal friction for the fundamental resonance after cold rolling was ∿3.6 times its value in the as quenched sample (a constant background clamp loss was assumed in both cases). The most striking feature of the low temperature loss is the increase in the internal friction maximum for decreasing frequency. This dependence on frequency is not expected for simple relaxation processes and is rarely observed. Between 100-300K, other anelastic processes become important. In this

433

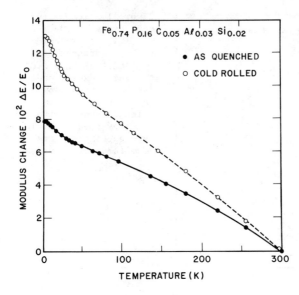

Figure 2. Temperature dependence of the Young's modulus for the fundamental resonance in as quenched (f_1=159Hz) and cold rolled (f_1=133Hz) amorphous samples. The moduli of the as quenched and cold rolled measurements were normalized to their room temperature values of E_0= 11.5 x 10^{11} dyn cm^{-2} and 7.3 x 10^{11} dyn cm^{-2} respectively. Cold rolling increases the low temperature dispersion and temperature dependence of the modulus.

temperature range, the damping in previously measured initially quenched samples of $Fe_{0.74}P_{0.16}C_{0.05}A\ell_{0.03}Si_{0.02}$ was attributed primarily to thermoelastic relaxation.[1]

Young's modulus measurements in the as quenched and cold rolled samples are shown in Figure 2 for the temperature range 1.3-300K. Both sets of data correspond to the fundamental resonance and were normalized to their respective modulus values at room temperature, E_0(300K). Cold rolling enhances the temperature dependence of the modulus and leads to a 13% change in modulus between liquid helium and room temperature. At low temperatures ($\lesssim$80K), both the cold rolled and as quenched modulus experience dispersion effects. As with the internal friction measurements, the magnitude of the dispersion in the cold rolled sample is greater than in the as quenched sample by at least three-fold. As the resonant frequency is increased, the modulus temperature dependence and low temperature dispersion are reduced. Values for the internal friction maximum temperature, T_M, the Young's modulus, E(300K), the modulus temperature dependence at 300 K, $(1/E)(\partial E/\partial T)$, and the

434

Table   Young's modulus and its temperature dependence
for various flexural modes in the amorphous metal
$Fe_{0.74}P_{0.16}C_{0.05}Al_{0.03}Si_{0.02}$. T_M is the
temperature of the internal friction maximum.
$(1/E)(\partial E/\partial T)$ is evaluated at 300K and the
expression $\Delta E/E = [E(4.2K) - E(300K)]/E(300K)$.

	$E_O(300K)$ 10^{11} dyn cm^{-2}	$(1/E)(\partial E/\partial T)$ 10^{-4} K^{-1}	$\Delta E/E_O$ %	$T_M{}^a$ K
AS QUENCHED				
f_1(159Hz)	11.5 ± 3.0	−3.2	7.8	20.0
COLD ROLLED				
f_1(133Hz)	7.3 ± 2.0	−3.9	13.0	17.5
f_2(883Hz)	8.2^b	−3.7	11.7	19.3
f_3(2554Hz)	8.8^b	−3.6	11.0	21.5
f_4(5100Hz)	9.2^b	∿−3.1	10.4	

a   All temperatures of the internal friction maxima are
uncertain to within ±0.5K.

b   Error in modulus relative to f_1 is ∿0.5%.

modulus change, $E(4.2K)/E(300K) - 1$ are given in the Table for the as
quenched and cold rolled measurements. For the fundamental resonance, we
see that cold rolling decreases the absolute magnitude of the room temperature modulus and decreases the temperature of the internal friction
maximum. It is important to note that at all temperatures between 0K
and 300K the absolute value of the modulus is larger at higher frequencies.

The difference in the temperature dependence of the Young's modulus
between the fundamental and the first three higher overtones is shown
in Figure 3 for a cold rolled sample. The relative dispersion is presented in a dimensionless form since the absolute magnitude of the modulus is not determined with high precision. All the measurements show
an anomalous decrease in the relative dispersion in the same low temperature region associated with the internal friction maxima. The unusu-

435

Figure 3. Relative dispersion between the fundamental (n=1) and higher flexural modes of cold rolled amorphous Fe alloy.

The modulus measurements at each frequency were normalized to their room temperature value $E_0(300K)$.

The low temperature relative dispersion is inconsistent with a simple relaxation process having a constant relaxation strength.

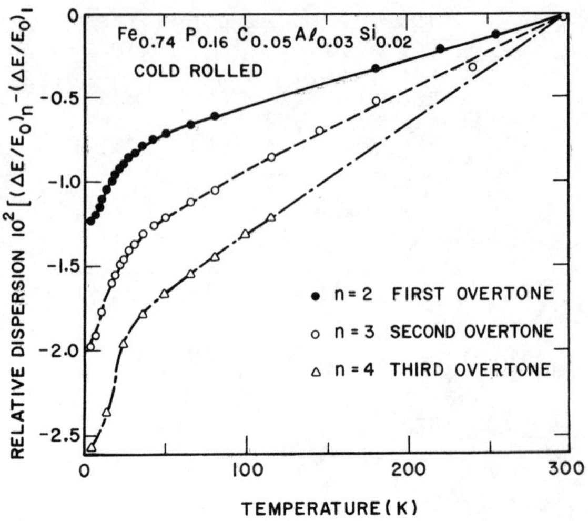

al frequency dependence of both the dispersion and the damping suggest that the coupling between the low temperature anelastic process and the acoustic phonons decreases with increasing frequency.

DISCUSSION

In discussing these new measurements we initially assume that the low temperature anomalies may be described by a relaxation process with a broad distribution of relaxation times associated with a particular defect distribution. The internal friction and Young's modulus dispersion may be represented by the general relaxation expressions

$$Q^{-1}(\omega,T) \int \psi(\xi) \frac{\omega\tau(T,\xi)}{1+\omega^2\tau^2(T,\xi)} d\xi \qquad (2)$$

$$\frac{E(\omega,T) - E(0,T)}{E(0,T)} = \int \psi(\xi) \frac{\omega^2\tau^2(T,\xi)}{1+\omega^2\tau^2(T,\xi)} d\xi \qquad (3)$$

436

where $\psi(\xi)$ is the relaxation spectrum and ξ corresponds to one or more defect parameters.

For a thermally activated process the relaxation time τ_{TA} has the Arrhenius form

$$\tau_{TA}(T,\xi) = \tau_o(\xi)e^{H(\xi)/kT} \quad , \qquad (4)$$

where $H(\xi)$ is the activation energy and k is Boltzmann's constant. The most probable value of the activation energy may be estimated from a plot of $\ln\omega$ versus T_M^{-1}. However, a quantitative analysis of the temperature dependence of the thermally activated process requires a knowledge for the relaxation spectrum which is not generally known for glasses.

Recent high frequency ultrasonic measurements by Dutoit[2] in Pd-Si based metallic glasses revealed a broad low temperature attenuation peak similar to the measurements presented here. The temperature shift with frequency of the attenuation maxima measured by Dutoit obeyed the Arrhenius expression [Equation (4)], with an average activation energy of 130 cal mole^{-1} (5.7 meV). These measurements were fit by Dutoit to the empirical relation

$$\log\,(Q^{-1}/Q_M^{-1}) = C^2[\log\,(T/T_M)]^2 \quad , \qquad (5)$$

where C is a frequency dependent parameter, and Q_M^{-1} and T_M are the internal friction and temperature at the maximum loss. We have also compared the present measurements to Equation 5 and find a reasonably good fit for $T > T_M$ for both the as quenched and cold rolled samples ($C \sim 1.5$-3). However, Equation (5) is not satisfied for $T < T_M$ in the measured frequency range (100-2500 Hz).

A similar Arrhenius plot for the data at the three frequencies of the cold rolled sample yields an average activation energy of $\bar{H}_{CR} = 590 \pm 150$ cal mole^{-1}. Measurements in the as quenched amorphous Fe alloy gave $\bar{H}_{AQ} \sim 850$ cal mole^{-1}. While these values of the activation energy are typical for low temperature loss processes in both metallic and non-metallic glasses, the present measurements also suggest that the internal friction maxima may not be associated with a thermally activated re-

437

laxation process. At a temperature where the condition $\omega\tau < 1$ is satisfied, the internal friction at a given frequency (e.g., f_1) must be lower than that of all higher frequencies (e.g., f_2 and f_3). This statement is valid even for a complicated relaxation process with a broad distribution of relaxation times and a non-constant relaxation strength. The measurements in Figure 1 show that for $T > T_M$, where $\omega\tau$ would be less than one for a thermally activated relaxation process, the internal friction at f_1 is larger, <u>not smaller</u>, than at f_2 or f_3. This situation is only possible in a relaxation process for $\omega\tau > 1$. Thus, it appears that this low temperature damping and dispersion is due to 1) a relaxation process where $\omega\tau > 1$ and the relaxation rate and/or the relaxation strength go through a maximum near 20K or 2) a process, other than relaxation, possibility associated with the magnetic properties of the material.

A theoretical model of glasses was proposed by Anderson, Halperin, and Varma[8] and Phillips[8] which, at least qualitatively, explained anomalies in thermal conductivity, specific heat and ultrasonic measurements below 1K. This model assumes that glasses contain a large number of two level systems with a broad distribution of level splittings. At low temperatures, a relaxation process associated with these two-level systems may take place by quantum-mechanical tunneling as well as thermal activation. Recently, this tunneling approach was used to explain high frequency ultrasonic attenuation[9] and dispersion[10] measurements in vitreous silica. The present measurements show anomalies at higher temperatures where excited states of the two-level systems may become populated and multi-phonon processes also must be considered. In contrast to the relatively open structure of non-metallic glasses, amorphous metals are closely packed systems and may have a considerably different distribution of two-level states. Furthermore, the acoustic frequencies in this study are $\sim 10^5$ times smaller than the ultrasonic frequencies used in other investigations where the tunneling model has been applied. At present, there is insufficient information concerning the distribution and other parameters of the two-level systems in metallic glasses to permit a quantitative comparison between theory and experiment.

CONCLUSION

Cold rolling substantially increases the low temperature damping and dispersion and decreases the temperature of the internal friction maximum in the amorphous metal alloy $Fe_{0.74}P_{0.16}C_{0.05}Al_{0.03}Si_{0.02}$. However, cold rolling did not change the general features of this anelastic process previously observed in as quenched samples.[1] It appears that cold rolling essentially increases the number of defects which may interact with the sound wave. The frequency dependence of the damping and dispersion found in this acoustic investigation is considerably different from the results of higher frequency ultrasonic studies in both metallic and non-metallic glasses. Further insight into this low temperature anelastic process will come from new investigations particularly at lower frequencies and lower temperatures.

ACKNOWLEDGMENT

We wish to thank L.R. Testardi, P.C. Hohenberg, B. Golding, and B.I. Halperin for helpful discussions and G.F. Brennert for technical assistance.

REFERENCES

1. Barmatz, M. and H.S. Chen, Phys. Rev. B9, 4073 (1974).
2. Dutoit, M., Phys. Lett. A50 (1974).
3. See, for example: Anderson, O.L. and H.E. Bömmel, J. Ceramic Soc. 38, 125 (1955) and Maynell, C.A., G.A. Saunders, and S. Scholes, J. Non-Cryst. Solids 12, 271 (1973).
4. See: Vukcevich, M.R., J. Non-Cryst. Solids 11, 25 (1972) and references therein.
5. Chen, H.S., Scripta Met. 9, 411 (1975).
6. Barmatz, M., L.R. Testardi, and F.J. DiSalvo, Phys. Rev. B12, 4367 (1975).
7. Chen, H.S. and C.E. Miller, Rev. Sci. Instr. 41, 1237 (1970).
8. Anderson, P.W., B.I. Halperin, and C.M. Varma, Phil. Mag. 25, 1, (1972); and W.A. Phillips, J. Low. Temp. Phys. 7, 351 (1972).
9. Golding, B.,J. Graebner, B.I. Halperin, and R.J. Schutz, Phys. Rev. Letters 30, 223 (1973); J. Jäckle, L. Piché, W. Arnold, and S. Hunklinger [to be published].
10. Piché, L., R. Maynard, S. Hunklinger, and J. Jäckle, Phys. Rev. Lett. 32, 1426 (1974).

ELECTRICAL AND MAGNETIC PROPERTIES
OF RAPIDLY QUENCHED METALS

C.C. Tsuei

IBM Thomas J. Watson Research Center
Yorktown Heights, New York

1.0.0 INTRODUCTION

A variety of metastable crystalline and amorphous metallic phases can
be formed with the tehniques of rapid-quenching such as splat-cooling
or vapor deposition, etc. The purpose of this article is to review
some of the electrical and magnetic properties of materials thus obtained. The emphasis will be on those phenomena associated with the
metastable nature of these metals and alloys.

The types of alloys resulting from rapid quenching can be summarized as follows:[1,2]

1] Supersaturated solid solutions:

> If the rate of quenching is sufficiently high to retard the nucle-
> ation and growth of a second solid phase, solute atoms may be re-
> tained in a parent lattice in excess of the equilibrium concentra-
> tion. The effect of rapid quenching is thus to extend the terminal
> solid solubility limit or the range of homogeneity of an intermedi-
> ate solid phase. For example, the solubility of magnesium in alum-
> inum is increased from 18.9 to 37at.% by liquid quenching, and
> that of aluminum in magnesium from 11.6 to 23at.%.[2] Of particular
> interest, it has been demonstrated that by liquid quenching, com-
> plete solubility can be achieved in certain alloy systems such as
> Cu-Ag, GaSb-Ge, etc.[2] Another interesting example is the extension
> of the homogeneity range of the A-15 compound Nb_3Ge by sputtering.[3]
> More examples of alloys with extended solute solubility can be
> found in several review articles.[1,2]

2] Non-equilibrium crystalline phases:

> In this case, the effect of rapid quenching is to form a crystalline
> phase which does not exist under equilibrium conditions. The simple
> cubic alloys (such as Au-Te,[4] Pd-Sb[5] alloys) obtained by liquid
> quenching are examples of this type of alloy.

3] Amorphous phases:

> The crystallization process can be by-passed by rapid quenching from
> a melt, vapor, or solution, providing that certain structural and

441

kinetic conditions[6] are satisfied. The resulting metastable phases are characterized by the absence of any long-range structural order in the atomic arrangement beyond nearest neighbors. An amorphous state has been achieved in pure elements as well as well as in alloys by using an appropriate technique of rapid quenching.[2,7] The occurrence and structure of amorphous phases have been well documented in the literature.

2.0.0 METASTABLE SIMPLE CUBIC ALLOYS

One of the most novel non-equilibrium crystalline phases obtained by rapid quenching is probably the metastable simple cubic phase formed in splat-cooled alloys. The simple cubic crystal structure (space group Pm3m, one atom/unit cell), although used frequently to demonstrate calculations involved in various solid-state physics problems, is rare in equilibrium. With the exception of one allotropic form of the element Po, this structure does not occur in any known equilibrium elements or alloys. Metastable simple cubic phases have been obtained in a number of alloy systems by liquid quenching.[2] Attempts to prepare simple cubic alloys by the techniques of vapor deposition or sputtering have not been successful.

Metastable simple cubic phases were first found in liquid-quenched Te-Au and Te-Ag binary alloys by Luo and Klement.[4] It was also found that several Sb-base alloy systems tended to form this phase. The simple cubic alloys have been studied with the Mössbauer effect and by measuring the lattice parameter, superconducting transition temperature, thermoelectric power and magneto-resistance. The results of these studies can be summarized as the following:

1] All simple cubic alloys exhibit metallic conduction. In fact, they are also <u>all</u> superconducting with transition temperatures ranging from $1.3°$ to $\sim 7°K$. [See Table, following page.]

2] Anomalies in the variation of lattice parameter, thermoelectric power, and superconducting transition temperature with concentration in the simple cubic Te_xAu_{1-x} alloys (x = 60 to 85at.%) can be qualitatively explained in terms of a Fermi-surface-Brillouin-zone interaction.[12,17]

Those who are interested in the result [2] are referred to the original references. The significance of superconductivity in simple cubic alloys will be discussed in some detail.

TABLE: Critical Temperature and Lattice Parameter
of Simple Cubic Alloy Phases

COMPOSITION (at. %)	T_c (°K)	Lattice Parameter (Å)
$Au_{25}Sb_{75}$	6.7[8]	2.93[9]
$Pd_{16.5}Sb_{83.5}$	4.9[8]	3.00[9]
$In_{25}Sb_{75}$	4.1[8]	3.06[10]
$Au_{16.7}Pd_{16.6}Te_{66.7}$	4.6[8]	2.93[8]
Au_xTe_{100-x} (x=15 to 45)	1.6 to 3.0[11,12]	3.10-2.93[4]
$Ag_{25}Te_{75}$	2.6[11]	3.06[4]
$In_{50}Te_{50}$	2.19[13]	3.07[14]
P	4.7[15] (110 Kb)	----
$Sn_{25}Te_{75}$	1.7[16]	3.15[16]

First of all, the fact that metallic conduction prevails in simple
cubic alloys containing as much as ∿90at.% Te requires some explanation. This can be qualitatively interpreted as a result of the dilution
of the covalent bonding that characterizes the semiconducting hexagonal
tellurium. The tellurium atom has six outer electrons (s^2p^4) beyond the
closed shells. Two of these electrons are paired in the s-orbital, two
in one of the three p-orbitals, while the remaining two are available
for covalent bonding in half-filled p-orbitals. In the hexagonal tellurium, the atoms form spiral chains with three atoms per turn and the
axes of these chains are located at the corner of a hexagon. Each tellurium atom forms two covalent bonds with its two nearest neighbors.
The chain structure, however, does not exist in the simple cubic lattice.
The coordination number is six instead of two, giving approximately 0.67
electrons per bond. Therefore, it is believed that an important part of
the binding energy of the simple cubic alloy results from the mutual
sharing of the unpaired electrons. As a consequence of this unsaturated
covalent bonding, electrons can move with ease from one atom to another and hence give rise to the metallic conduction.

To provide more insight to this problem, augmented plane-wave energy-band calculations[18] have been performed for Te having a hypothetical

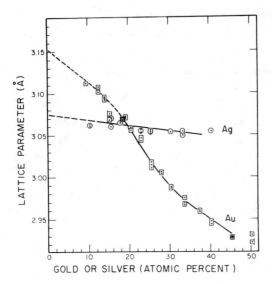

Figure 1. Lattice parameter as a function of composition for simple cubic Au-Te and Ag-Te alloys.[4]

simple cubic crystal structure with a lattice parameter based on extrapolation of the Te-Au experimental data as shown in Figure 1. The energy bands along the five segments joining the symmetry points of the first Brillouin zone are shown in Figure 2 for a simple cubic tellurium with a lattice parameter of 3.18Å. From these energy-band calculations, one concludes that the Fermi energy (E_F) of the simple cubic Te probably falls in the region $0.40 < E_F < 0.60$ Ry and it should be a metallic conductor. Furthermore, it is estimated that the density of states at the Fermi energy $N(E_F)$ is of the order of 12-15 electron states/atom/Ry. This value is about three or four times larger than the $N(E_F)$ for a metal like Cu, and about 50 to 60% as large as that for most of the A-15 type superconducting compounds. It should be pointed out that it has been proposed that the three-orthogonal-chain structure in the A-15 type is partially responsible for the attractive electron pairing[19] and the relatively high $N(E_F)$[20] found in this type of material. The simple cubic structure also has three orthogonal chains of atoms, although the interatomic distance in the chain is not as short as those found in the A-15 compounds. This analogy does suggest, however, that the simple cubic structure favors the occurrence of superconductivity in a similar

444

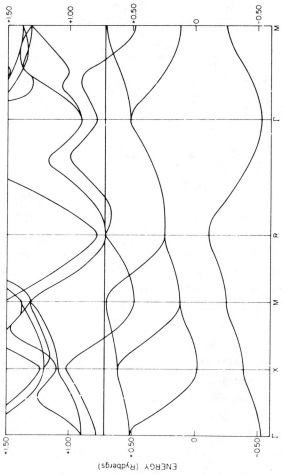

Figure 2. APW energy bands for hypothetical simple cubic tellurium with a lattice constant of 3.18Å.[18]

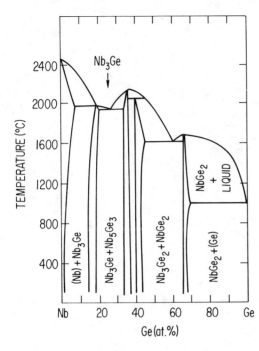

Figure 3. Nb–Ge phase diagram.[23]

manner as the A-15 structure,
and is consistent with the
fact that all the simple cubic
materials are superconductors.

3.0.0 THE SUPERCONDUCTING TRANSITION TEMPERATURE OF Nb_3Ge

As mentioned in the Introduction, one of the effects of
rapid-quenching is to produce
supersaturated solid solutions
and intermetallic compounds
with extended homogeneity
ranges. There has been considerable effort devoted to the study of the
electrical and magnetic properties of these metastable alloys. Although
the crystal structure of these alloys is not altered by rapid quenching,
an extension of the composition range and/or an alteration in morphology
can sometimes lead to some interesting properties which do not exist in
the equilibrium alloys. For instance, superconductivity has been observed as a bulk property of some eutectic alloys[21] (e.g., Al-Si and Al-Ge).
Transition temperatures (T_C) well in excess of those for the pure metals
were found. Quenching the alloys from the liquid state was found to increase the superconducting transition temperature and critical field
(H_C), and decrease the characteristic size of the metal and semiconduc-
tor domains. A correlation between the Fermi energy of the metal and
the enhancement of T_C was observed. On the other hand, in metal-metal
eutectic systems (e.g., Al-Al$_2$Cu) with comparable microstructure no enhancement of T_C was found. This suggests that a metal-semiconductor
interface is necessary for the T_C enhancement. These experimental find-

446

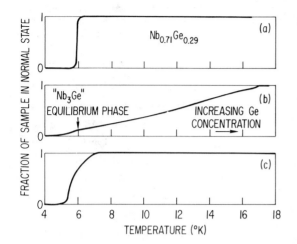

Figure 4. Superconducting transitions of an alloy $Nb_{0.79}Ge_{0.21}$ (a) as prepared in the arc, (b) after liquid quenching and subsequent annealing up to 1000°C, (c) after annealing at 1100°C for three days. [25]

ings were tentatively attributed to the exciton mechanism discussed by Ginsburg and by Allender, Bray and Bardeen although there is some uncertainty and controversy about this mechanism.

Another example is the high T_c phase found in Nb-Ge alloys. Recently, Gevaler[3] and Testardi[22] showed that T_c values as high as 23°K can be achieved by low energy sputtering onto hot substrates (∿700-900°C). It should be pointed out that liquid quenching has played a significant role in the development of achieving Nb_3Ge with high T_c. As indicated in the Nb-Ge equilibrium phase diagram[23] (Figure 3), the A-15 phase normally exists over a non-stoichiometric composition range (∿14-18at.% Ge). The non-stoichiometry of the equilibrium Nb_3Ge phase was, in fact, pointed out by Geller[24] based on the lattice parameter data. The T_c of the equilibrium A-15 compound is ∿6°K (Figure 4a). Realizing that rapid quenching can extend the homogeneity range of an intermetallic compound, Matthias, et al.,[25] liquid-quenched Nb-Ge alloys with nearly stoichiometric composition and obtained a broad superconducting transition with an onset at 17°K (Figure 4b). The equilibrium value of T_c of this metastable alloy was restored by annealing at 1100°C for three days (Figure 4c). Experimental evidence was also presented to show that this unusual increase in T_c can be attributed to an optimization of the Nb:Ge stoichi-

447

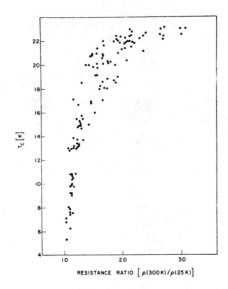

RESISTANCE RATIO $[\rho(300\,K)/\rho(25\,K)]$

Figure 5. Correlation between T_c and resistance ratio $[\rho(300°K)/\rho(25°K)]$ for Nb-Ge films prepared under various conditions.[22]

ometry achieved by liquid quenching. It was also recognized that there is a detrimental effect on T_c and its transition width due to the disorder in the atomic arrangement, an unavoidable by-product of rapid-quenching from the melt. It was concluded[25] that "it is likely that stoichiometric-ordered Nb_3Ge would have a transition of at least 18°K," a prediction that was borne out by Gavaler's discovery. The importance of the exact 3:1 Nb:Ge ratio on the highest possible T_c is, however, controversial[3,22,26] at the present time. On the other hand, the disorder along a chain in the A-15 structure is probably quite crucial in controlling T_c. Testardi, et al.[22] showed there is a simple correlation of T_c and electrical resistance ratio (Figure 5, supra). This correlation is essentially independent of all sputtering conditions and composition. The estimated electronic mean free path for films with resistance ratios between 2 to 3 ($T_c > 21°K$) is about 50Å which is about the same order of magnitude as a typical coherence length of most A-15 compounds. This observation suggests that order in the atomic arrangement (especially the chain integrity) that extends over more than one coherence length is rather important in achieving high T_c, regardless of how this condition is satisfied. A mean free path much longer than the coherence length probably will lead to a T_c not much higher than 23°K, as implied by Figure 5.

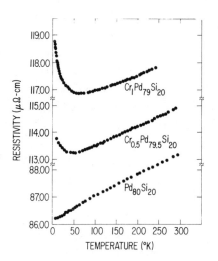

Figure 6. Resistivity as a function of temperature for amorphous Pd-Si alloys containing a small amount of Cr (0 to 1%).[28]

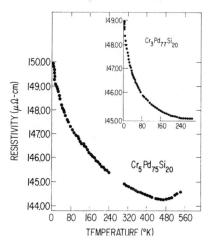

Figure 7. Resistivity as a function of temperature for amorphous Pd-Si alloys containing a small amount of Cr (3 to 5%).[28]

4.0.0 AMORPHOUS METALLIC PHASES

In recent years considerable effort has been devoted to the study of the electrical and magnetic properties of non-crystalline metallic alloys. From these studies, two interesting features emerge characterizing the amorphous metallic state.

4.1.0 The Atomic-Like Characteristics Play a More Prominent Role in the Amorphous Phase than in Its Crystalline Counterpart

There is no first-principle theory as yet that predicts that the above statement should be true. The fact that the electron mean free path is of the same order of magnitude as the interatomic distance suggests that the electrical and magnetic properties of amorphous metallic alloys preferentially probe the details of the atomic environment over other long-range properties. This effect also leads to a relatively weak correlation between spins or electrons in the amorphous solids, some examples of which follow:

4.1.1 Kondo-type resistance minimum: A well-defined minimum in the electrical resistivity versus temperature curve is found for a number

449

of amorphous alloys (e.g., Pd–Si alloys) containing small amounts of magnetic elements such as Fe, Cr, etc. (Figures 6 and 7).[28] Below the resistivity-minimum temperature, the resistivity difference $\Delta\rho$ varies as $-\ln T$ as predicted by the Kondo theory[29] for exchange scattering between conduction electrons and localized moments. At lower temperature, $\Delta\rho$ tends to level off, suggesting the formation of the spin-compensated state (Figure 8). The resistivity-minimum anomaly is always accompanied by a negative magneto-resistivity approximately proportional to the square of the magnetization and a susceptibility obeying the Curie-Weiss law for a wide temperature range. It should be mentioned that this Kondo-type effect has been observed in amorphous alloys containing as much as 7at.% Cr. This is quite a contrast to the case of most crystalline Kondo systems[30] in which the resistivity-minimum is completely suppressed due to the spin-spin correlation, when the magnetic impurity content exceeds ∿0.1at.% or less. From the point of view of the single-impurity based Kondo theory, the most unexpected result is probably the observation that the Kondo-type resistance minimum can co-exist with ferromagnetism in amorphous alloys. The existence of a Kondo-type effect

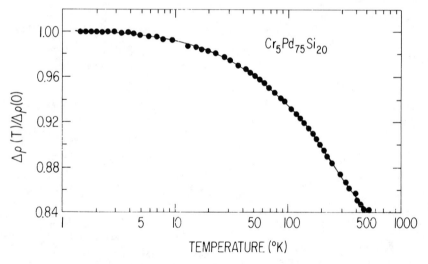

Figure 8. $\Delta\rho(T)/\Delta\rho(o)$ versus $\ln T$ plot for amorphous $Cr_5Pd_{75}Si_{20}$.[28]

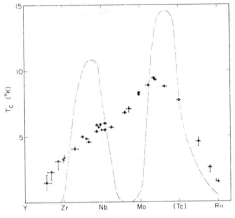

Figure 9. The superconducting transition temperature for amorphous 4d metals and their alloys.[37]

in a ferromagnet was first observed in the Co-Pd-Si alloy system.[31] Similar results were also reported in several amorphous ferromagnetic alloys with relatively high Curie temperatures.[32,33] Based on the experimental findings just described, one can conclude that the experimental results obtained so far suggest strongly that the effect of the d-d spin correlation in reducing the s-d exchange effect is relatively weak in amorphous alloys.

4.1.2 The systematics of T_c for amorphous transition-metal superconductors: Over two decades ago, it was demonstrated that a number of amorphous superconductors can be prepared by vapor deposition onto a substrate held at liquid helium temperature. These metastable films are superconducting with a T_c which is usually higher than the equilibrium value. Most of these films are stabilized by impurities and transform to one or more crystalline phases below liquid nitrogen temperature. For example, amorphous Bi films with a T_c ∿6°K are found to crystallize above 20°K,[35] and the amorphous phase of Ga films (T_c = 8.4°K) transforms into a crystalline state (β-Ga with T_c ∿6.4°K) at 15°K. Further annealing at ∿60°K results in the room-temperature stable phase, α-Ga with a T_c = 1.08°K. More recently, amorphous superconducting alloys have been reported; some 4d and 5d transition metal alloy films are found to be superconducting to ∿8 °K and are stable at room temperature.[37] Other examples of relatively stable amorphous superconductors are: Nb$_3$Ge (T_c ∿3.3°K) prepared by sputtering onto a room-temperature substrate[38] and La-Au alloys (T_c ∿3.8°K, in bulk form) obtained by liquid quenching.[39]

451

As discussed in Anderson's theory of dirty superconductors, the mechanism for electron-electron attraction in an amorphous superconductor is based on pairing each one-electron state in the presence of structural and compositional disorder scattering with its exact time reverse, a generalization of the $\vec{k}$ up, $-\vec{k}$ down pairing of the BCS theory which is independent of such scattering. In general, then, amorphous superconductors are not expected to behave in a fundamentally different way from their crystalline counterparts. There are, however, some general characteristics which can be associated with the amorphous nature of the material. One of such features is the trend of T_c in amorphous transition-metal (4d and 5d series) superconducting metals and alloys: Recent work by Collver and Hammond[37] shows that the variation of T_c as a function of the valence electron per atom ratio (z) has a sharp triangular peak in the middle of the transition-metal series (Figure 9). For the 4d transition metals and their alloys, the peak is located at $\sim z = 6.4$ which corresponds to a valence electron per atom ratio at which the d-shell of the atom is half-filled with five electrons with parallel spins. Therefore, the systematics of T_c for the amorphous superconductors is believed to be a manifestation of the dominance of atomic-like parameters in determining the superconducting properties of these amorphous materials. There are two attempts to explain the T_c dependence of z in the amorphous state.[41,42] Both conclude that the observed trend in T_c reflects the dependence of a local atomic parameter $\eta \approx N(o)<I^2>$ on z, where $N(o)$ is the electronic density of states at the Fermi level, and $<I^2>$ the average squared electron-phonon interaction strength. This type of atomic parameter was first used by Hopfield for transition-metal superconductors.[43] To carry this concept one step further, one can assume that $N(o)$ varies smoothly or even is constant for the most part of the transition-metal series as a result of simple smearing or averaging of the electronic density of states or other effects as discussed by Collver.[44] The T_c of the amorphous superconductors is then characterized essentially by the strength of electron-phonon coupling, an atomic-like parameter.

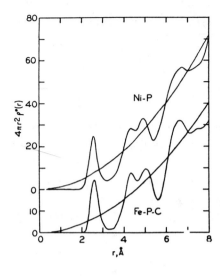

Figure 10. Radial distribution functions for $Ni_{77}P_{23}$ and $Fe_{80}P_{13}C_7$ (taken from Reference 46).

4.2.0 Many Electrical and Magnetic Properties of Amorphous Metallic Solids can be Described in Terms of a Distribution Function for Certain Physical Parameters.

In amorphous metallic alloys, there are at least two kinds of disorder which are responsible for the broadening effect on many characteristic properties: (1) Compositional disorder; (2) structural disorder. The former exists in both the crystalline and amorphous alloys while the latter is unique for the non-crystalline state. The structure of amorphous metals and alloys has been studied extensively by x-ray and electron diffraction.[45,46] The radial distribution functions (RDF) of the amorphous metallic alloys are found to be quite similar if the differences in the sizes are taken into consideration. The RDF's for two amorphous metallic alloys are shown in Figure 10. In general, it is found that there are about 12-13 nearest neighbors for a given metal atom. The interatomic distances are on the average of ~ 2.6Å (for the amorphous Fe-P-C alloys) and with a spread of ~ 0.5Å. Undoubtedly, such a distribution in the atomic arrangement will have a significant effect on many physical properties. Results of magnetization and Mössbauer measurements on a typical amorphous ferromagnet $Fe_{75}P_{15}C_{10}$ prepared by splat-cooling illustrate this point:

4.2.1 Magnetization and fluctuations in the exchange interaction: The spontaneous magnetization as a function of temperature up to the Curie temperature T_c (=596.5°K) has been measured for the amorphous alloy $Fe_{75}P_{15}C_{10}$.[47] The spontaneous magnetization at 0°K for this alloy is found to be 151 emu/g (or $1.8\mu_B$ per Fe atom). The reduced magnetization $[M_s(T) / M_s(o)]$ versus reduced temperature (T/T_c) is shown in Figure 11.

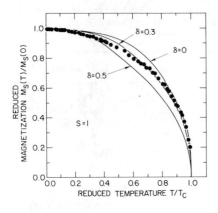

Figure 11. Magnetization as a function of temperature for amorphous $Fe_{75}P_{15}C_{10}$ alloy as compared with the Handrich prediction for the cases of S=1; δ=0, 0.3 and 0.5.[47]

It is found that the corresponding states curve is considerably lower than that of Fe_3P (the closest crystalline counterpart of the amorphous alloy) or crystalline Fe. This is apparently a general property common to many amorphous ferromagnets.[48,49]

To gain a better understanding of the mechanism that is responsible for the temperature dependence of magnetization, the experimental data are compared with Handrich's theory for non-crystalline ferromagnets. This theory is based on a molecular-field approach and assumes that the compositional and structural disorder can lead to fluctuations in the exchange-interaction strength (J_{ij}) between the neighboring spins. The effect of these fluctuations on the reduced magnetization can be expressed by the following formula:

$$\sigma(T) \equiv M_s(T) \ / \ M_s(0) \ = \ \frac{1}{2} \ \{B_S[(1+\delta)\chi] + B_S[(1-\delta)\chi]\}$$

where $\chi = 3S\sigma T_c \ / \ (S+1)T$, and δ is a measure of the degree of disorder and is defined as the root mean square of deviation from an average exchange integral between two nearest-neighbor spins: $\delta^2 = <\Delta J^2> \ / \ <J>^2$.

As shown in Figure 11, the theoretical curve with S = 1, δ = 0.3 gives the best overall fit to the experimental data. In terms of the Handrich theory,[50] the experimental results (δ ≃ 0.3) for amorphous $Fe_{75}P_{15}C_{10}$ suggest that the average fluctuation in J can be the same order of magnitude as J itself.

4.2.2 <u>Mössbauer spectrum and hyperfine field distribution</u>: The room temperature Mössbauer spectrum for the amorphous alloy $Fe_{75}P_{15}C_{10}$ is shown in Figure 12. For comparison, the Mössbauer spectrum for crystalline

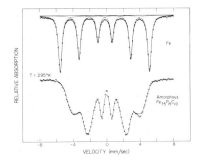

Figure 12. Mössbauer spectra for crystalline Fe and amorphous $Fe_{75}P_{15}C_{10}$
alloys.

iron at room temperature is also presented in Figure 12. The well-resolved spectrum for the amorphous ferromagnet is
characterized by sharp inner peaks and
broad outer ones. These characteristics
are indicative of a distribution of hy-
perfine fields which reflects all possible local environments for Fe
atoms. As discussed previously[49] the apparent absence of quadrupole interactions below T_c is probably due to a directional range of electric
field gradients, with respect to the hyperfine field, which gives a line
broadening rather than a line shift. The Mössbauer spectrum can therefore be fitted in terms of a distribution of hyperfine fields P(h), an
average isomer shift and a line width which allows for the effect of
quadrupole broadening. A reasonably good fit to the experimental data
can be obtained by using the following model[49] for P(h):

$$P(h) = \begin{cases} A[(h-h_o)^2 + (\Delta_o/2)^2]^{-1} & \text{for } 0 \le h \le h_o, \\ \dfrac{4A}{\Delta_o^2} \exp[-(h-h_o)^2/2\Delta_1^2] & \text{for } h > h_o \end{cases}$$

where A is a normalization constant determined by the condition:

$$\int_0^\infty P(h)\,dh = 1$$

The calculated spectrum based on this
model is shown as the solid curve in Figure 12. The fitting parameters are:

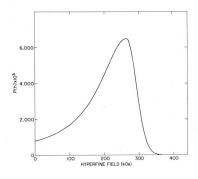

Figure 13. The hyperfine field distribution function P(h). [47]

455

$h_0 = 264.8$, $\Delta_0 = 195.6$, and $\Delta_1 = 27.6$ (all in units of kOe). The $P(h)$ curve obtained from the fitting is shown in Figure 13. Using these numbers, one can calculate the following quantities:

$$\bar{h} = \int^{\infty} P(h)h\,dh \qquad \text{and} \qquad \Delta^2 = \int_0^{\infty} (h-\bar{h})^2 P(h)\,dh$$

It is found that the ratio $\Delta/\bar{h}$ for the amorphous $Fe_{75}P_{15}C_{10}$ is equal to 0.36. If the magnetic hyperfine field seen by a given Fe^{57} atom is proportional to its local magnetization, the quantity $\Delta/\bar{h}$ is then a measure of the ratio of root mean square deviation to the average of the magnetization. The value for $\Delta/\bar{h} = 0.36$ is consistent with the estimate of $\delta \sim 0.3$ from the magnetization data.

5.0.0 SUMMARY

In summary, three types of metastable metals and alloys have been obtained with various techniques of rapid quenching; supersaturated solid solutions, non-equilibrium crystalline phases, and amorphous metallic phases. Some of the electrical and magnetic properties of several metastable phases selected from the above categories were discussed. Simple cubic phases were found to be favorable for the occurrence of superconductivity and the origin of this phenomenon was explored. The development of achieving high T_c in Nb_3Ge was used as an example to demonstrate how rapid quenching can significantly change the physical properties of an intermetallic compound or a solid solution without altering the crystal structure. From the study of amorphous phases, one concludes that the atomic-like characteristics play a more prominent role in the amorphous phase than in its crystalline counterpart, and that many electrical and magnetic properties of amorphous metallic solids can be described in terms of a distribution function for certain physical parameters.

REFERENCES

1. Duwez, Pol, Trans. ASM 60, 607 (1967).
2. Jones, H., Rep. Prog. Phys. 36, 1425 (1973); Jones, H. and C. Suryanarayana, J. Mater. Sci. 8, 705 (1973); and the references in these two review articles.

3. Gavaler, J.R., *J. Appl. Phys. Lett.* **23**, 480 (1973).
4. Luo, H.L. and W. Klement, Jr., *J. Chem. Phys.* **36**, 1870 (1962).
5. Giessen, B.C. in: "Developments in the Structural Chemistry of Alloy Phases", B.C. Giessen, Ed., Plenum Press, New York (1969), 227.
6. Turnbull, D., *Contemp. Phys.* **10**, 473 (1969).
7. Chopra, K.L., "Thin Film Phenomena," McGraw-Hill, New York (1969).
8. Tsuei, C.C., Huan-Chun Yen, and P. Duwez, *Phys. Lett.* **34A**, 80 (1971).
9. Giessen, B.C., U. Wolff, and N.J. Grant, *Trans. AIME* **242**, 597 (1968).
10. Jordan, C.B., *J. Chem. Phys.* **39**, 1613 (1963).
11. Luo, H.L., M.F. Merriam and D.C. Hamilton, *Science* **145**, 581 (1964).
12. Tsuei, C.C. and L.R. Newkirk, *Phys. Rev.* **183**, 619 (1969).
13. Bommel, H.E., A.J. Darnell, W.F. Libby, B.R. Tittmann, and A.J. Yencha, *Science* **141**, 714 (1963).
14. Darnell, A.J., A.J. Yencha, and W.F. Libby, *Science* **141**, 713 (1963).
15. Wittig, J. and B.T. Matthias, *Science* **160**, 994 (1968).
16. Johnson, W.L. and S.J. Poon, *J. Appl. Phys.* **45**, 3683 (1974).
17. Chen, W.Y.K. and C.C. Tsuei, *Phys. Rev.* **B5**, 901 (1972).
18. Newkirk, L.R. and C.C. Tsuei, *Phys. Rev.* **B4**, 2321 (1971).
19. Westbrook, J.H., "Intermetallic Compounds," Wiley, New York (1967), 587.
20. Weger, M., *J. Phys. Chem. Solids* **31**, 1621 (1970).
21. Tsuei, C.C. and W.L. Johnson, *Phys. Rev.* **B9**, 4742 (1974).
22. Testardi, L.R., J.H. Wernick, and W.A. Royer, *Solid State Comm.* **15** (1974); Testardi, L.R., R.L. Meek, J.M. Poate, W.A. Royer, A.R. Storm and J.H. Wernick, *Phys. Rev.* **B11**, 4304 (1975).
23. Pan, V.M., V.I. Latysheva, and E.A. Shishkin, "Physics and Metallurgy of Superconductors," ed. by E.M. Savitskii and V.V. Baron (Consultants Bureau, New York) [1970], 179.
24. Geller, S., *Acta Crystallogr.* **9**, 885 (1956).
25. Matthias, B.T., T.H. Geballe, R.H. Willens, E. Corenzwit, and G.W. Hull, Jr., *Phys. Rev.* **139**, A1505 (1965).
26. Gavaler, J.R., M.A. Janocko and C.K. Jones, *J. Appl. Phys.* **45**, 3009 (1974); Santhanam, A.T. and J.R. Gavaler, *J. Appl. Phys.* **46**, 3633 (1975).
27. The quantity $\Delta\rho$ is defined as the change in resistivity due to the addition of magnetic elements to the host alloy.
28. Hasegawa, R. and C.C. Tsuei, *Phys. Rev.* **B2**, 1631 (1970).
29. Kondo, J., *Solid State Physics*, ed. by F. Seitz, D. Turnbull and H. Ehrenreich, Vol. 23, p. 183, Academic Press, New York (1975).
30. Heeger, A.J., ibid, p. 283.
31. Tsuei, C.C. and R. Hasegawa, *Solid State Comm.* **7**, 1581 (1959).
32. Maitrepeirre, P.L., Ph.D. Thesis, CA Inst. Tech. (1969).
33. Hasegawa, R. and J. Dermon, *Phys. Lett.* **42A**, 407 (1973).
34. Buckel, W. and R. Hilsch, *Z. Physik* **138**, 109 (1954).
35. Barth, N., *Z. Physik* **142**, 58 (1955); Buckel, W. and R. Hilsch, *Z. Physik* **146**, 27 (1956).
36. Wühl, N., J.E. Jackson, and C.V. Briscoe, *Phys. Rev. Lett.* **20**, 1496 (1968).
37. Collver, M.M. and R.H. Hammond, *Phys. Rev. Lett.* **30**, 92 (1973).
38. Chencinski, N. and F.J. Cadieu, *J. Low Temp. Phys.* **16**, 507 (1974).
39. Johnson, W.L., S.J. Poon, and P. Duwez, *Phys. Rev.* **B11**, 150 (1975).

40. Anderson, P.W., <u>J. Phys. Chem. Solids</u> <u>11</u>, 26 (1959).
41. Hammond, R.H. and M.M. Collver, in: <u>Low Temperature Physics—LT13</u>, ed. by W.J. O'Sullivan and E.F. Hammel; Plenum Press, New York (1974), <u>3</u>, 532.
42. Kerker, G. and K.H. Bennemann, <u>Z. Physik</u> <u>264</u>, 15 (1973).
43. Hopfield, J.J., <u>Phys. Rev.</u> <u>186</u>, 443 (1969).
44. Collver, M.M., Lawrence Berkeley Laboratory Rpt. No. LBL-167 (1971).
45. Cargill, G.S. III, <u>Solid State Physics</u>, ed. by F. Seitz, D. Turnbull and H. Ehrenreich; Academic Press, New York (1975), <u>30</u>, 225.
46. Giessen, B.C. and C.N.J. Wagner in <u>Physics and Chemistry of Liquid Metals</u>, ed. by S.Z. Beer; Marcel Dekker, New York (1972), 633.
47. Tsuei, C.C. and H. Lilienthal [to be published in <u>Phys. Rev. B</u>].
48. Rhyne, J.J., J.H. Schelleng, and N.C. Koon, <u>Phys. Rev.</u> <u>B10</u>, 4672 (1974).
49. Sharon, T.E. and C.C. Tsuei, <u>Phys. Rev.</u> <u>B5</u>, 1047 (1972).
50. Handrich, K., Phys. <u>Stat. Sol</u> (<u>b</u>) <u>53</u>, K17 (1972).

MAGNETIZATION IN SOME IRON-BASE GLASSY ALLOYS

R. Hasegawa and R.C. O'Handley

Materials Research Center
Allied Chemical Corporation
Morristown, New Jersey

1.0 INTRODUCTION

Ferrous metallic glasses such as Fe-P-C exhibit relatively high Curie temperatures[1] ($T_c \sim 600K$) and high saturation magnetizations[2] ($\mu_s \sim 2.1 \mu_B$/ Fe atom) with low coercivity[3] ($H_c \sim 50$ mOe) and remarkable mechanical properties. Recent developments in material synthesis[4] and characterization[5,6] of low field magnetic properties have suggested the use of these alloys for magnetic applications. In addition to the low field properties, some iron-base glassy ferromagnets in a ribbon form were studied by means of Mössbauer[7] and ferromagnetic resonance techniques.[8] The purpose of the present work is to report thermomagnetization and x-ray data taken on METGLAS alloys #2826 ($Fe_{40}Ni_{40}P_{14}B_6$) and #2615 ($Fe_{80}P_{16}C_3B_1$) in an attempt to understand the magnetization and crystallization behavior of these alloys.

2.0 EXPERIMENTAL

Liquids having compositions $Fe_{40}Ni_{40}P_{14}B_6$ and $Fe_{80}P_{16}C_3B_1$ are rapidly quenched into continuous ribbons of width 2mm and thickness 35μm. X-ray diffraction confirms the glassy state of the ribbons. About $10 \sim 20$ mg of material was cut into squares (about 2x2mm) for the magnetization measurements made with a vibrating sample magnetometer. The temperature range covered was between 4.2 and 1050K. Several strips were also crystallized in vacuum separately for x-ray (CuKα and MoKα) analysis of the constituent phases.

3.0 RESULTS AND DISCUSSION

3.1 Fe$_{40}$Ni$_{40}$P$_{14}$B$_6$

A ferromagnetic resonance (FMR) technique has been used to determine the temperature dependence of the saturation magnetization and the uni-

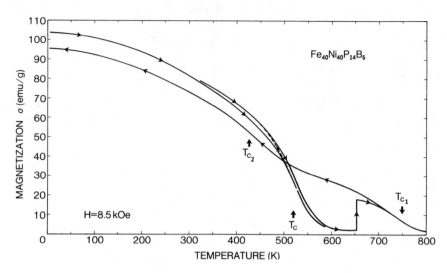

Figure 1. Magnetization at H = 8.5 kOe versus temperature for glassy $Fe_{40}Ni_{40}P_{14}B_6$ alloy. The arrows indicate the temperature cycle described in the text.

axial anisotropy energy. Those results indicated that the ferromagnetic Curie temperature (T_c) and the average crystallization temperature (T_{cr}) were about 525 and 655K respectively.[8] It has also been found[9,10] that thermal annealing at about 300°C significantly influences low field magnetic properties, such as coercivity and remanence, of this glassy alloy. With the knowledge of these findings, the magnetization behavior of the alloy was first studied as a function of temperature between 4.2 and 600K. The sample was then annealed at 573K without any external field for two hours and cooled down to 300K. The magnetization was measured during cooling and successive heating through T_c and T_{cr} and up to about 820K. The sample was then cooled to 4.2K. The magnetization behavior during these temperature cycles is shown in Figure 1. The magnetization at H = 8.5 kOe and T = 4.2K gives the saturation moment of 1.14 μ_B/transition metal atom. The magnetization in the vicinity of the Curie temperature during the initial heating (4.2–500K) is plotted against field in the lowest part of Figure 2. The Arrott method has been[11] found to be useful in determining the value of T_c for amorphous

460

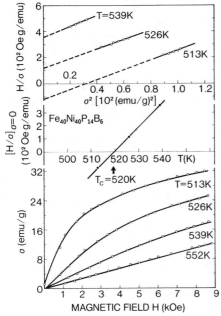

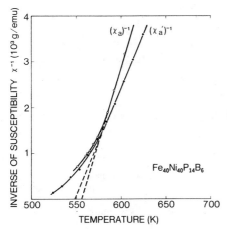

Figure 2. Magnetization behavior in the vicinity of Curie temperature of a $Fe_{40}Ni_{40}P_{14}B_6$ alloy. The upper portion of the figure illustrates the Arrott method of determining T_c.

ferromagnets, indicating that the thermodynamic equation of state near T_c is of the same form for glassy ferromagnets as it is in crystalline materials. The same method, therefore, was applied to the present case as shown in the upper part of Figure 2. The result shows that $T_c = 520K$, which is close to the value (525K) indicated by FMR data.[8] Above T_c the material behaves like a superparamagnet up to T_d ($\sim580K$), above which the susceptibility obeys the Curie-Weiss law $\chi_a = C/(T-\theta_p)$ where C is the Curie constant and θ_p is the paramagnetic Curie temperature. This behavior is shown in Figure 3. After annealing at 573K for two hours, the susceptibility behavior described above changes slightly as is indicated

Figure 3. Inverse of the susceptibilities above T_c as a function of temperature for a $Fe_{40}Ni_{40}P_{14}B_6$ alloy. The susceptibilities χ_a and χ_a' were measured respectively before and after annealing at 573K for 2 hours.

461

TABLE: Susceptibility Data for the Glassy $Fe_{40}Ni_{40}P_{14}B_6$ Alloy

As-quenched			Annealed at 573 K for 2 hours		
$X_a = C/(T-\theta_p)$			$X_a' = C'/(T-\theta_p')$		
C	θ_p	T_d	C'	θ_p'	T_d'
1.54×10^{-2} emu/g	556 K	580 K	1.99×10^{-2} emu/g	549 K	585 K

by X_a' in Figure 3. The Table (supra) summarizes the results of the susceptibilities X_a and X_a'. From the Curie constants C and C', we obtain the paramagnetic moments (gS) of 1.98 and 2.41 μ_B/atom respectively. The increase in gS may imply clustering of transition metal atoms during the annealing process, which may be responsible for the 20% increase of the saturation induction observed in ribbons annealed near 300°C.

As seen in Figure 1, the present alloy crystallizes at T = 653K above which the magnetization versus temperature is found to follow the Brillouin function of S = 1 with a Curie temperature of T_{c1} of 750K. This makes it possible to determine the magnetization contribution of the crystalline phase having $T_c = T_{c1}$, as shown in Figure 4. Upon cooling the sample from 820 to 4.2K, an increase of the magnetization is observed below ~500K indicating existence of an additional crystalline

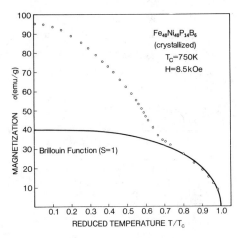

phase (Figures 1 and 4). By subtracting the magnetization contribution of the other crystalline phase (corresponding to the solid line in Figure 4) from the total magnetization and using the Arrott plot, it is found that the additional crystalline phase has the

Figure 4. Magnetization at H = 8.5 kOe (shown by open circles) plotted against T/T_c where T_c is the Curie temperature of one of the crystalline phases.

462

Curie temperature T_{c2} of 426K.

The x-ray diffraction data on the samples annealed at 700K for two hours show that the sample is 100% crystalline and contains two crystalline phases: fcc Fe_xNi_{1-x} and tetragonal $(Fe, Ni)_3P_{0.7}B_{0.3}$. The fcc phase has the lattice parameter a = 3.585 Å and is identified as $Fe_{0.55}Ni_{0.45}$ by using the published data of 'a' for the binary Fe-Ni system.[12] Furthermore, it has been reported[13] that the crystalline $Fe_{0.55}Ni_{0.45}$ alloy has $\mu_s = 1.73$ μ_B/atom and $T_c = 750$K which is identical to the T_c value of one of the ferromagnetic phases detected in the magnetic data (Figure 4). The fcc phase contributes about 40 emu/g to the total magnetization (see Figure 4) and thus amounts to about 24wt.% of the total crystallized sample. The tetragonal phase, on the other hand, has the lattice parameters a = 8.914 Å and c = 4.381 Å. Using the lattice parameter data[14] on Fe_3P, Ni_3P, and $Fe_3(P-B)$, the composition of the tetragonal phase is determined as $(Fe_{0.58}Ni_{0.42})_3P_{0.7}B_{0.3}$. This phase corresponds to the magnetic phase having $T_c = 426$K. If we estimate the saturation magnetization of this phase to be about 95 emu/g based on the published data for $(Fe-Ni)_3P$ alloys,[15] the magnetization contribution of about 56 emu/g toward T = 0 from the tetragonal phase corresponds to about 58wt.% of the total crystallized sample. This leaves an additional phase or phases of about 18% in weight. This residue is probably paramagnetic because no additional magnetic transition is detected, and $\sigma(8.5$ kOe$)$ at 4.2K for the crystallized sample is much smaller than that for the glassy sample. The paramagnetic residue probably has a composition close to $Ni_3P_{0.7}B_{0.3}$.

3.2 $Fe_{80}P_{16}C_3B_1$

The magnetization σ of 190.5 emu/g at H = 9 kOe and T = 4.2K obtained for the as-quenched alloy gives a saturation moment of $\mu_s = 2.13$ μ_B/Fe atom. The magnetization data at H = 9 kOe are shown in Figure 5 as a function of temperature. The value of T_c is determined by the Arrott plot to be 565K. In the vicinity of 600K, the glassy alloy starts to crystallize as evidenced by the increase of σ. A steady increase of σ lasts for about two hours at T = 605K (see Figure 5). When the sample is cooled down and reheated between 300 and 605K, the σ versus T curve is revers-

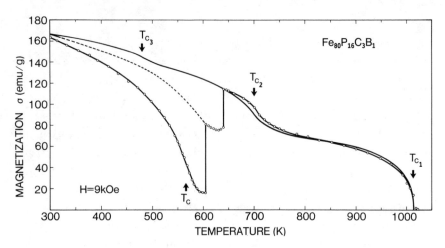

Figure 5. Magnetization at $H = kOe$ versus temperature for a $Fe_{80}P_{16}C_3B_1$ alloy. Data with open circles correspond to the heating cycle. The dashed line represents the thermally reversible region. The simple solid line is the cooling curve.

ible and does not show any indication of magnetic transitions (Figure 5). Above $T = 605K$, σ decreases with T to about 630K and starts to increase at this temperature, indicating the onset of a further crystallization process. At 635K, it takes about 2.5 hours to complete the crystallization. Further increase of the temperature reveals a magnetic transition at T_{c2} and finally the entire crystalline sample becomes paramagnetic at T_{c1} (Figure 5).

Upon cooling the sample from 1050K, an additional magnetic transition is observed at $T_{c3}(<T_{c2}<T_{c1})$. As in the previous section (3.1), a combination of the Brillouin function fitting of σ and the Arrott plots leads to the determination of these transition temperatures. It is found that $T_{c1} = 1013$, $T_{c2} = 700$, and $T_{c3} = 480K$. By comparing the x-ray data on the crystallized sample and published magnetic data, crystalline phases having these Curie temperatures are identified as α-Fe with about lat.% P ($T_c = 1018K$, $\sigma_s = 219$ emu/g)[13] Fe_3P ($T_c = 716K$, $\sigma_s = 155$ emu/g)[15] and Fe_3C ($T_c = 475K$, $\sigma = 173$ emu/g)[16] respectively. The saturation magnetization of 180 emu/g obtained for the crystallized sample at 4.2K is found to consist of about 100, 56, and 24 emu/g from α-Fe, Fe_3P and Fe_3C

phases respectively. These values further imply that the three phases amount to about 46, 36, and 24% in weight respectively in the crystallized sample.[17]

ACKNOWLEDGEMENT

The authors wish to thank L.A. Davis and J.J. Gilman for their interest in the present work. P. Harget contributed in the x-ray data analysis and M.O. Sullivan assisted in the magnetization measurements.

REFERENCES

1. Duwez, P. and S.C. H. Lin, J. Appl. Phys. 38, 4096 (1967).
2. Tsuei, C.C., G. Longworth, and S.C.H. Lin, Phys. Rev. 170, 603 (1968).
3. Hasegawa, R. [unpublished data].
4. Gilman, J.J., Physics Today 28, 46 (1975).
5. Egami, T., P.J. Flanders, and C.D. Graham, Jr., AIP Conf. Proc. No. 24, 697 (1975).
6. O'Handley, R.C., J. Appl. Phys. (November 1975)
7. Chien, C.-L., and R. Hasegawa, Magnetism and Magnetic Materials Conf., Philadelphia (1975).
8. Hasegawa, R., ibid.
9. Luborsky, F., IEEE Trans. MAG [to be published].
10. Mendelsohn, L., G. Bretts, and E.A. Nesbitt [to be published].
11. Hasegawa, R., J. Appl. Phys. 41, 4096 (1970).
12. Owen, E.A. and A.H. Sully, Phil. Mag. 31, 314 (1941).
13. Bozorth, R.M., Ferromagnetism; Van Nostrand Co., Princeton, 111 (1951).
14. Rundqvist, S. and E. Larsson, Acta Chem. Scand. 13, 551 (1959); and Rundqvist, S., Acta Chem. Scand. 16, 1 (1962).
15. Gambino, R.J., T.R. McGuire, and Y. Nakamura, J. Appl. Phys. 38, 1253 (1967); and Meyer, A.P.J. and M.-C. Cadeville, J. Phys. Soc. Japan 17, 223 (1962).
16. Bernas, H., I.A. Campbell, and R. Fruchart, J. Phys. Chem. Solids 28, 17 (1967).
17. The weight percentages add up to 96% which is satisfactory, considering the difficulty of such analyses.

MAGNETIC ANNEALING OF AMORPHOUS ALLOY TOROIDS

F.E. Luborsky, R.O. McCary, and J.J. Becker

General Electric Co.
Corporate Research and Development
Schenectady, New York

INTRODUCTION

In a previous paper[1] we described the characteristics of an amorphous
alloy with nominal composition of $Ni_{40}Fe_{40}P_{14}B_6$. Changes in d-c magnetic
properties, resulting from stress relief and magnetic annealing, were
described for straight ribbon samples of this alloy. Stress relief of
toroids resulted in substantial improvements in a-c characteristics.
Properties comparable to the Permalloy alloys were achieved.

In this paper we describe the a-c and d-c magnetic characteristics
of annealed amorphous alloy toroids. We show that the properties are determined almost entirely by the magnetically induced anisotropy since
crystal and strain-magnetostriction anisotropy are absent. A wide range
of hysteresis behavior is achieved by changing the magnetic anneal, resulting in a variation in properties similar to those obtained in
Permalloy.[2]

EXPERIMENTAL METHODS AND RESULTS

Toroids were prepared by winding METGLAS[R] 2826 ribbon, with a nominal
50 micron thickness, into aluminum or tungsten cups with a hole through
their center. The springiness of the ribbon resulted in the toroid being pressed against the outer rim of the cup. The average toroid diameter was 1.3cm. Some toroids were wound with MgO powder as insulation
between turns and some without MgO (see Table). The tungsten was chosen
because its coefficient of thermal expansion is less than that of the
ribbon so that, on cooling from the stress relief temperature, no
stresses were exerted on the turns resulting from the contraction of
the cup. The aluminum has a larger coefficient of expansion and will
stress the ribbon when cooled. Fifty turns of wire with high temperature insultation were wound on the cup for the drive and sense winding

R = Registered Trade Name of Allied Chemical Corp.   467

TABLE. FABRICATION DETAILS FOR THE VARIOUS TOROIDS

SAMPLE	MgO INSULATION	NUMBER OF TURNS OF RIBBON	TOROID CUP MATERIAL
FL-7	Yes	5.67	Aluminum
FL-11	Yes	13.67	Aluminum
FL-15	Yes	13.55	Tungsten
FL-25	No	12.39	Tungsten
FL-26	No	12.40	Aluminum

for the magnetic anneal and test. These wires were connected to feed-throughs in a glass tube. The assembly was outgassed under vacuum and then sealed with half an atmosphere of pure dry nitrogen. The toroids were treated and tested as previously described,[1] without removing from the sealed container.

Toroids were stress-relieved by heating for two hours at temperatures between 290° and 360°C.[1] If annealed below 290°C, the sample is not fully stress-relieved resulting in strain-magnetostriction effects. If annealed above 360°C, crystallization starts, producing a phase with a high coercive force. The properties achieved after this stress-relief anneal were dependent only on the subsequent cooling or treatment schedule in the vicinity of the Curie temperature, T_c, of the alloy, the field environment used and the previous magnetic state of the toroid.

Typical remanence ratios, M_r/M_s, after cooling from above T_c are shown in Figure 1. Temperatures were monitored by a thermocouple on the outside of the glass envelope but inside a thick, insulated sleeving. The rate through T_c is reported, but the cooling rate was maintained at roughly the same value down to about 200°C, i.e., to about 40° below T_c. Below this temperature changes in direction of K_u occur slowly enough to be insignificant.[3] Note that this alloy requires a rate of less than about 1 deg/min to develop maximum remanence during cooling in a satu-

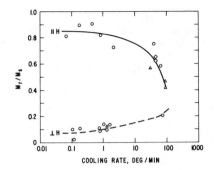

Figure 1. Effect of cooling rate on the remanence-to-saturation ratio for cooling toroids in a parallel, ∥, or transverse, ⊥, field from above T_C. Stress-relief anneal at 302°C, Δ; 325°C, ◊; and 356°C, o.

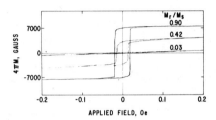

Figure 2. Typical d-c hysteresis loops at low, intermediate and high M_r/M_S ratios. From field cooling toroids FL-25 after stress-relief anneal of 2 hours at 355°C.

rating circumferential field, ∥H, or to develop minimum remanence during cooling in a field perpendicular to the plane of the toroid, ⊥H. Typical d-c hysteresis loops at the two extremes and at an intermediate value of M_r/M_S are shown in Figure 2. The intermediate value was obtained by a rapid cooling in ∥H. These loops are very similar to the loops of 65-Permalloy annealed in various ways.[2] The maximum induced anisotropy is assumed to develop in both extremes. From these two hysteresis loops K_u can be calculated in two ways. If we assume a rotation process, as described previously,[1] we extrapolate the initial slope of the ⊥H magnetization curve to saturation. This value is K_u and is 2.75 Oe. The second method measures the area between the ∥H and ⊥H magnetization curves; this gives a value of 2.8 Oe. These are in good agreement with the value of 2.6 Oe (800 ergs/cm³) calculated from measurements on straight ribbons of the same samples.[1]

At high cooling rates, greater than 90 deg/min, we assume that K_u becomes negligible. The hysteresis loop close to this case is shown in Figure 2 by the loop with M_r/M_S = 0.42. Approximately the same value is

469

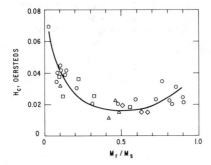

Figure 3. Dependence of coercive force on remanence-to-saturation ratio as in Figure 1. Data shown with ▢ were obtained by annealing below T_C in $\parallel$H from low M_r/M_S state.

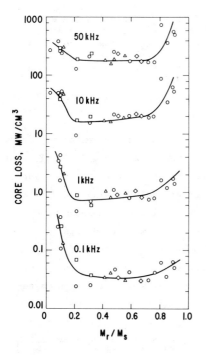

Figure 4. Core loss as a function of remanence-to-saturation ratio measured at various frequencies for B = 1000G for the same samples as in Figure 1.

obtained even without the field applied. Its shape is determined only by any residual anisotropies present and by wall nucleation. Various values of M_r/M_S were also obtained by annealing in a $\parallel$ or $\perp$ field at temperatures below T_C. The resultant properties appear to depend only on the final value of M_r/M_S; not on its previous history.

Typical trends in d-c coercive force, a-c core losses and permeability are shown in Figures 3-5. Most of the data were obtained from one toroid (FL-25) by cooling in either a $\parallel$ or $\perp$ field at various rates as shown in Figure 1. There appeared to be no trend with continued magnetic re-anneals. Using different materials for the cup, or using MgO insulation, as summarized in the Table, did

470

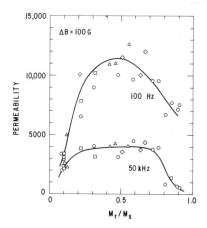

Figure 5. Permeability as a function of remanence-to-saturation ratio at ΔB = 1000G for the same samples as in Figure 4.

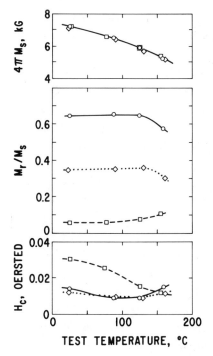

Figure 6. Effect of test temperature on d-c properties of toroid heat-treated to different values of M_r/M_s.

not significantly alter the properties achieved after stress relief. The different stress-relief temperatures, from 300° to 360°C, also did not produce significant differences (Figures 1, 3, and 4). The total losses in all cases were proportional to $f^a B^b$ where f is the frequency, B the maximum induction, and a and b are constants. The values of a and b varied linearly from 1.3 to 1.6 and from 2.1 to 1.8 respectively as M_r/M_s varied from 0.05 to 0.90. The values are in the ranges expected[2]; i.e., $1<a<2$ and $1.6<b<3$. The changes with M_r/M_s reflect changes in the

471

Figure 7. Effect of test
temperature on a-c
properties of the toroid
in Figure 6.

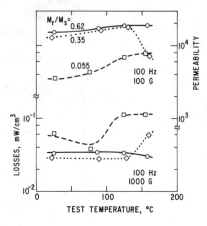

number and configuration of
the domain walls.

APPLICATIONS
_____

This wide range of d-c and
a-c properties provides characteristics suitable for different types of applications.
Samples with high M_r/M_s are
particularly suited to devices such as switch cores, high gain magnetic
amplifiers, and low-frequency inverters, where the square loop characteristic is needed. Samples with intermediate M_r/M_s values and thus
minimum a-c losses, are best applied in high frequency inverters or
transformers where minimum losses are important. This range is also useful in signal devices such as fluxmeters and ground fault detectors
where the highest permeability is desired. Samples with M_r/M_s near zero
are interesting for applications in which constant permeability is desired over a wide range of $4\pi M$ as in filter chokes or loading coils and
some specialized transducers.

Two important parameters of the material, of importance in applications, are the stability and the temperature coefficient of properties.
Stability is evaluated in a "worst case" situation for this alloy in a
separate paper.[3] The worst case was taken as aging in a d-c saturating
circumferential field beginning from the low M_r/M_s state. Under these
conditions the life time is controlled by magnetic annealing, not by
recrystallization. At 140°C it was found that 10% reorientation of the
magnetic anisotropy took only 17 hours; extrapolated to 100°C it would
take 1000 hours. Thus, application of the very low M_r/M_s state does not
now appear reasonable for this alloy unless the temperature can be kept
below about 70°C. However, improved stability might well be achieved by

472

changing the preparation conditions or alloy composition. Application of the high M_r/M_s state is quite reasonable since the domain orientation during application, as well as at rest, is in the direction of the induced anisotropy.

The temperature coefficients of both d-c and a-c properties were measured up to about only 160°C, to avoid aging effects during measurement. Typical results are shown in Figures 6 and 7 for FL-26 heat-treated to different M_r/M_s values. Although not shown, the properties measured at room temperature, after the last measurements at temperature, were generally significantly different from those in the original room temperature measurement. This was also true for tests made on FL-25 using a tungsten core box, as well as for FL-7 made from the aluminum box but using MgO insulation. It is assumed that this change is caused by a change in stress on the toroid. In general, the properties vary with temperature less than for ferrites and about the same as for the Permalloys.

CONCLUSIONS

A wide range of M_r/M_s can be achieved in stress relieved amorphous tape toroids by a magnetic anneal. The magnetically induced anisotropy is the major anisotropy present. The value of K_u = 2.8 Oe, calculated from the hysteresis loops of the toroids, agrees with the value of 2.6 Oe calculated from measurements on straight samples. The H_c, losses, permeability, and the dependence of losses on B and f, all vary with M_r/M_s. The stability of toroids with low M_r/M_s appear to be marginal for application at present, but the stability of toroids with high M_r/M_s is satisfactory. The temperature dependence of properties, up to 150°C, appear excellent.

ACKNOWLEDGEMENTS

The experimental work of B.J. Drummond is greatly appreciated.

REFERENCES

1. Luborsky, F.E., J.J. Becker, R.O. McCary, IEEE Trans. on Magnetics, MAG-11, 1644 (1975).
2. Bozorth, R.M., Ferromagnetism, Fig. 11-19, Chapters 5, 11, and 17. van Nostrand, New York (1951).
3. Luborsky, F.E., Proc. 21st Ann. Conf. on Magnetism and Magn. Mat., AIP Conf. Proc. No. 29, 209 (1976).

STRESS CORROSION CRACKING OF AMORPHOUS IRON BASE ALLOYS

A. Kawashima, K. Hashimoto, and T. Masumoto

The Research Institute for Iron, Steel, and Other Metals
Tohoku University
Sendai, Japan

To evaluate the potential for practical applications of the extremely corrosion-resistant amorphous Fe-Cr-Ni-P-C alloys,[1-3] their susceptibility to stress corrosion cracking (SCC) was examined.

EXPERIMENTAL PROCEDURES

Ribbon-shaped amorphous $FeCr_{7.5}Ni_{23}P_{13}C_7$ and $FeCr_7Ni_{20}P_{14}C_6$ alloys, where subscripts denote atomic percent, were prepared by roller quenching. Solutions used were H_2SO_4-NaCl and HCl in which crystalline stainless steels suffer SCC at room temperature.[4,5] After specimens were installed in a tensile test machine, specimens were exposed to the solutions for 2 hours without stress. Subsequently, SCC tests were conducted at constant strain rates of 2.2×10^{-6} to 5.6×10^{-4} sec^{-1} and various applied potentials. The amount of hydrogen in the specimens was measured by argon gas carrier fusion-gas chromatography.[6,7]

RESULTS AND DISCUSSION

Typical stress-strain curves in air and solutions are shown in Figure 1. The fracture stresses at anodic, cathodic and corrosion potentials in 5 N H_2SO_4 - 0.1N NaCl are lower than in air. The susceptibility to cracking of the alloy, expressed by the ratio of the fracture stress in solution $\sigma_{f.soln}$ to that in air $\sigma_{f.air}$, is shown as a function of applied potential in Figure 2. In the potential region less noble than the hydrogen evolution potential, the SCC susceptibility increases with decreasing potential in all solutions with and without Cl^- ions. All fracture surfaces that failed at cathodic potentials showed brittle patterns. When the potential is increased, the fracture stresses in 2N and 5N H_2SO_4 approach that in air at about the corrosion potential (E_{corr}). However, when the potential is raised to a value higher

475

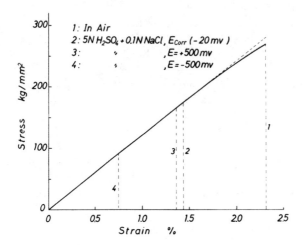

Figure 1. Stress–strain curves of amorphous $FeCr_{7.5}Ni_{23}P_{13}C_7$ alloy in air and solutions at strain rate of 4.2×10^{-6} sec^{-1}.

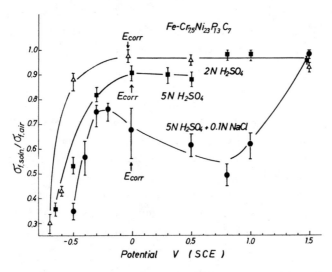

Figure 2. Variation of susceptibility to cracking of amorphous $FeCr_{7.5}$ $Ni_{23}P_{13}C_7$ as a function of applied potential in various solutions at a strain rate of 5.6×10^{-6} sec^{-1}.

Figure 3. Fractograph of amorphous $FeCr_{7.5}Ni_{23}P_{13}C_7$ alloy fractured at a strain rate of 5.6×10^{-6} sec^{-1}.

than the hydrogen evolution potential in 5N H_2SO_4-0.1N NaCl, the fracture stress decreases in the passive region including the corrosion potential. Fracture surfaces in the passive region exhibited a mixture of granular and vein-like patterns, an example of which is shown in Figure 3. There is a crack origin as indicated by an arrow. In the area showing a granular pattern the crack propagates into the specimen and a vein-like pattern appears in the vicinity of the surface and far from the crack origin. Under open circuit condition (E_{corr}) the fracture stresses in distilled water, neutral NaCl, 2N and 5N H_2SO_4, 2N H_2SO_4 - 0.5N NaCl and 0.1N HCl were almost the same as that in air. However, fracture stresses in 5N H_2SO_4 with 0.1 N or more NaCl and in 0.5 N and more concentrated HCl decreased. Therefore, embrittlement of the alloys in the passive region occurs in relatively strong acidic solutions with a certain amount of Cl^- ions. In the transpassive region the fracture stress recovers to that in air with increasing potential, and a vein-like pattern was observed on the fracture surface at 1.5 V. The decrease in SCC susceptibility in the transpassive region is associated with general corrosion.

To investigate the cause of embrittlement of cathodically polarized alloys, specimens were examined under various conditions after cathodic polarization for 1h at 12 mA/cm^2 in 2 N H_2SO_4 as shown in Figure 4. The fracture stress of a hydrogenized specimen as measured by a tensile test 5 min after cathodic polarization (curve 1) is less

477

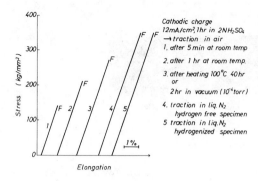

Figure 4. Stress-elongation curves of amorphous $FeCr_7Ni_{20}P_{14}C_6$ alloy treated differently after cathodic polarization for 1h at 12 mA/cm^2 in 2N H_2SO_4. Strain rate: 5.6 x 10^{-6} sec^{-1}. Symbol F denotes fracture of specimen.

TABLE . Amount of Hydrogen Absorbed (ppm)[9]

ENVIRONMENT		UNSTRESSED		STRESSED	
		E_{corr}	−500mV	E_{corr}	+500mV
As received	18.0				
5N H_2SO_4				37.2	37.8
5N H_2SO_4 +0.5N NaCl		25.4	138.3	43.3	52.4
5N H_2SO_4 +3N NaCl		28.4			

than one half of that in air (270 kg/mm^2). Standing for 1h in air at room temperature (curve 2) increases the fracture stress as compared with curve 1. After aging, a hydrogenized specimen in oil at 100°C or in vacuum (curve 3), the fracture stress is the same as that in air and the fractograph showed a vein-like pattern. Curves 4 and 5 exhibit no effect of cathodic polarization on the fracture stress in liquid nitrogen. As shown in the Table (supra), the amount of hydrogen absorbed by cathodic polarization at −500mV is one order of magnitude higher than that in the as-received "hydrogen free" specimens. Consequently, embrittlement of cathodically polarized amorphous alloys can be attributed to hydrogen embrittlement due to diffusible hydrogen absorbed.

478

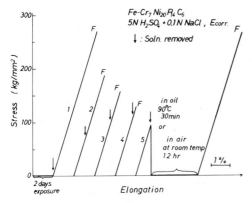

Figure 5. Stress-elongation curves of amorphous FeCr$_7$Ni$_{20}$ P$_{14}$C$_6$ alloy in 5N H$_2$SO$_4$-0.1N NaCl at corrosion potential with a strain rate of 5.6 x 10^{-6} sec^{-1}. Symbol F denotes fracture of specimen.

Embrittlement in the passive region, which includes the corrosion potential, was studied under various conditions as shown in Figure 5 (supra), where arrows indicate the time when the solution was drained from the electrolytic cell and the traction test continued further in air after the specimen was rinsed. No decrease of the fracture stress is observed in tensile tests in air after immersion for 2 days without stress in 5N H$_2$SO$_4$-0.1N NaCl, 2N H$_2$SO$_4$-3N NaCl or 3N HCl (curve 1). Therefore, the fracture stress of the alloys is not affected by prior immersion in corrodants. However, the fracture strength decreases with increasing time of traction in solution (curves 2, 3, and 4). Curve 5 was obtained by the following procedure: the cross-head of the tensile test machine was returned to the original position after a certain time of traction in the solution and the solution was drained from the cell. After the specimen was rinsed and held in oil at 90°C for 30min or in air at room temperature for 12h, tensile tests were resumed in air. The fracture stresses of the specimens thus treated were almost the same as that in air, and their fractographs showed vein-like patterns, an example of which is shown in Figure 6. The results obtained at +500mV were similar to those in Figure 5. As shown in the Table, the amounts of hydrogen absorbed during the tensile test are about 38ppm in 5N H$_2$SO$_4$ where no SCC occurred and about 47ppm in 5N H$_2$SO$_4$-0.5N NaCl in which SCC took place. The susceptibility to SCC increases with an increasing amount of hydrogen absorbed.

Figure 6. Fractograph of the amorphous FeCr7Ni20P14C6 specimen of experiment No. 5 in Figure 5.

Consequently, cracking can be ascribed to hydrogen embrittlement not only in the potential region lower than the hydrogen evolution potential but also in the higher potential region. If tensile stress is applied in relatively strong acidic solutions with a certain amount of Cl⁻ ions under anodic polarization, a sharp notch will be produced on specimens by stress assisted corrosion or by stress concentration on preexisting surface defects. Since film-free sites were highly reactive,[8] the solution-chemical condition for reduction of hydrogen ions, including enrichment in Cl⁻ ions, will occur on this site which results in penetration of hydrogen into the amorphous alloys. The region far from the crack origin and that in the vicinity of the surface where hydrogen can be released from the specimen by anodic polarization will not be embrittled, and hence these regions of the fracture surface showed a vein-like pattern similar to that in a hydrogen-free specimen.

CONCLUSIONS

SCC tests for amorphous Fe-Cr-Ni-P-C alloys were performed in H_2SO_4-NaCl and HCl solutions. The following conclusions can be drawn:

[1] Under cathodic polarization hydrogen embrittlement occurs in acidic solutions regardless of Cl⁻ concentration.

[2] In the passive potential region, which include the corrosion potential, no embrittlement was detected during

480

traction in acidic solutions with lower concentration of Cl⁻ ions and by neutral NaCl solution.

[3] If relatively strong acidic solutions contain a certain amount of Cl⁻ ions, hydrogen embrittlement occurs during traction even in the passive potential region. However, exposure of specimens to the above solutions prior to traction in air did not decrease the fracture strength of the alloys.

REFERENCES AND NOTES

1. Naka, M., K. Hashimoto, and T. Masumoto: *J. Japan Inst. Metals* 38 835 [in Japanese] (1974).
2. Naka, M., K. Hashimoto, and T. Masumoto, *Corrosion* 32 (1976).
3. Hashimoto, K. and T. Masumoto, Proceedings of the Second International Conference on Rapidly Quenched Metals (§II): MIT, Cambridge, Massachusetts, USA (1975); *J. Mat. Sci. Eng.* (1976).
4. Acello, S.J. and N.D. Greene, *Corrosion* 18, 286 (1962).
5. Bianchi, G., F. Mazza, and S. Torchio, *Corr. Sci.* 13, 165 (1973).
6. Goto, H., S. Ikeda, and M. Hosoya: *J. Japan Inst. Metals* 28, 764 [in Japanese] (1964).
7. Kawashima, A., K. Hashimoto, and S. Shimodaira, *ibid.* 38, 1046 (1974).
8. Hashimoto, K., K. Osada, T. Masumoto, and S. Shimodaira, *Corr. Sci.* 16 (1976).
9. Unstressed specimens were exposed for 2h at corrosion potential and then for 1h at −500mV. Traction of specimens was performed up to the stress of about 110 kg/mm^2 for about 1.5h in solutions.

INVESTIGATION OF THE CORROSION BEHAVIOR
OF SPLAT-COOLED ALUMINUM-IRON ALLOYS
WITH SMALL ADDITIONS OF MAGNESIUM, MANGANESE, AND CHROMIUM

G. Faninger, D. Merz, and H. Winter

Battelle-Institut, e.V.
Frankfurt-on-Main

INTRODUCTION

The main cost factor in the construction of seawater desalination
plants is due to the vast amount of nonferrous tubes for condensers and
heat exchangers; the tube material must withstand the severe corrosive
attack of seawater and brine for extended periods of time at elevated
temperatures; furthermore, the tube material should have a good heat
conductivity and a sufficient elevated-temperature strength. The copper-
nickel alloys of the 90/10 and 70/30 types have proved their durability
and have become standard in many desalination plants. It is believed
that the higher costs of these alloys more than compensate for the long-
term savings in maintenance and replacement.

Aluminum and its alloys are used to a much lesser extent in marine
service, particularly in superstructures of ships. In applications where
the material is completely immersed, even low concentrations of heavy
metals such as copper, nickel, and iron must be avoided. Otherwise, pre-
cipitations of these metals will cause intensive pitting by local gal-
vanic action. Small additions of manganese and chromium may improve the
corrosion resistance[1,2], nevertheless the values required for long-time
service in seawater barely are reached. Furthermore, the mechanical
strength at temperatures above 100°C, needed for a good erosion resistance, is not sufficient.

On the basis of a novel method for the rapid solidification of metal
melts, developed at Battelle-Frankfurt, alloys of aluminum with relatively high contents of iron (6-8wt.%) were developed which showed
drastically improved strength values at temperatures of up to more than
300°C.[3-6] This is due to a submicroscopic dispersion of metastable intermetallic compounds in the aluminum matrix. Surprisingly, it turned

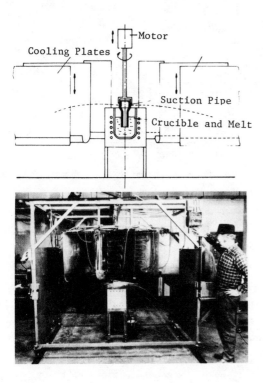

Cooling Plates

Motor

Suction Pipe

Crucible and Melt

Figure 1. Schematic diagram and photograph of the pilot plant for the production of splat foils.

Figure 2. Microstructure of extruded splat foils produced under different splat-cooling conditions (a,b) with different extrusion temperatures (b,c).

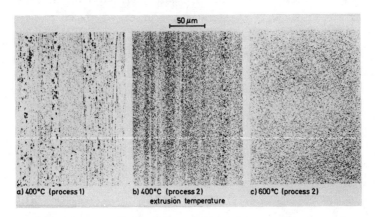

50 μm

a) 400°C (process 1) b) 400°C (process 2) c) 600°C (process 2)
extrusion temperature

484

out that the corrosion resistance in seawater of this new type of alloy surpasses the values of the commonly used aluminum alloys. [5,6] The good combination of elevated-temperature strength, corrosion resistance, heat conductivity and low-priced starting materials is of interest for a future application in seawater desalination plants.

EXPERIMENTAL PROCEDURE

Figure 1 shows the pilot plant used for the production of splat foils from aluminum-iron alloy melts. A rotating graphite cylinder is immersed into the melt contained in an induction-heated crucible; from the perforated rim of the graphite cylinder, droplets of the melt are thrown continuously by centrifugal forces onto water-cooled copper plates under a small angle; they rapidly solidify into elongated splat foils with an estimated solidification rate of 10^6 °C/sec.

The splat foils with a thickness of 10 to 50μm were cold-compacted under a pressure of 200 MN/m^2 into cylinders with a diameter of 70mm; after annealing for 2 hours at 400°C they were extruded at the same temperature into profiles with rectangular, circular or hollow cross section and an extrusion ratio between 1:25 and 1:38; the extrusion pressure varied between 14 and 18 MN/m^2 and the extrusion speed between 12 and 35 m/min.

RESULTS

The microstructure of the resulting extrusions is shown in Figure 2; the first picture shows an extrusion from splat foils produced in air; only a part of the cross-sectional area has the typical splat-cooled structure, the rest shows an intermediate structure with coarse precipitates. It was concluded that the oxidation of the surface of the free-flying droplets in air reduces the solidification rate. Therefore, argon was blown on the outlets of the melt in the rim of the rotating graphite cylinder (Figure 1); the argon shield was very efficient and resulted in a much better microstructure of the extrusions as is revealed in the second and third pictures of Figure 2.

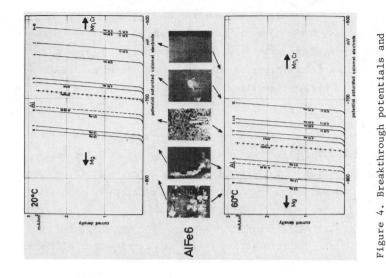

Figure 4. Breakthrough potentials and samples of extrusions made from splat foils of the alloy AℓFe6 with different additions and exposed to artificial seawater for 2 months at 150°C.

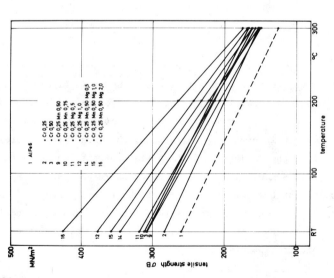

Figure 3. Temperature dependence of the tensile strength of extrusions made from splat foils of the alloy AℓFe6 with different additions.

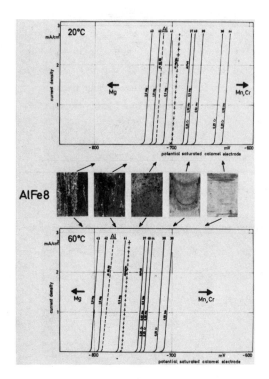

Figure 5. Breakthrough potentials and samples of extrusions made from splat foils of the alloy AℓFe₈ with different additions and exposed to artificial seawater for 2 months at 150°C.

Figure 3 shows the tensile strength as a function of temperature for an Al-6wt.% Fe alloy with and without various additions of chromium, manganese, and magnesium. These additions raise the strength values of the base alloy. It can be noticed that the tensile strength drops linearly with increasing temperature; this is a typical sign for a dispersion hardening effect.

Figure 4 shows the breakthrough potentials in artificial seawater for the base alloy Al-6wt.% Fe with various additions of Cr, Mn, and Mg for temperatures of 20 and 60°C; one may notice the beneficial effect of small additions of Cr and Mn; the same is true for the surface corrosion, as can be seen in the accompanying pictures.

Figure 5 (supra) shows the breakthrough potentials in artificial seawater for the base alloy Al-8wt.% Fe with various additions of Cr, Mn, and Mg for temperatures of 20 and 60°C; here too, the beneficial effect of small additions of Cr and Mn becomes evident.

CONCLUSIONS

The rapid solidification of aluminum alloys with 6 and 8 weight percent iron in a newly devised pilot plant leads to splat foils which can be

487

processed into ductile and corrosion-resistant extrusions with good elevated-temperature strength. Small additions of chromium and manganese have a beneficial effect. The use of an argon gas shield for the emerging droplets greatly improves the properties of the resulting alloys. If mass production can be accomplished, condenser and heat exchanger tubes made from these alloys may have a wide field of application, especially in seawater desalination plants.

REFERENCES

1. Verink, E.D. and P.F. George: Materials Protection and Performance 12, No. 5, 26-30 (1973).
2. Bonewitz, R.A.: Corrosion 29, No. 6, 215-222 (1973).
3. Battelle report on research project T/0204/92420/91430 BMVg, December (1970).
4. Battelle report on research project T/0240/12420/11035 BMVg, September (1972).
5. Battelle report on research project CV 31-71 BMFT, IV B 5, November (1972).
6. Ahlborn, H. and D. Merz: Aluminium 47, 671-677 (1971).

SATURATION MAGNETOSTRICTION AND VOLUME MAGNETOSTRICTION OF AMORPHOUS RIBBONS BASED ON Fe-Ni and Fe-Co

K.I. Arai, N. Tsuya, M. Yamada

Research Institute of Electrical Communication,
Tohoku University, Sendai

H. Shirae, H. Fujimori, H. Saito,
T. Masumoto

The Research Institute for Iron,
Steel, and Other Metals
Tohoku University, Sendai

INTRODUCTION

Recently, it has become possible to produce metallic amorphous materials in the form of long ribbons with reasonably uniform cross-sections by a centrifugal solidification and a roller quenching technique.[1,2] Concerning the mechanical properties of the materials, it has been known that the yield and fracture stresses are extremely high.[3]

As to magnetic properties, it is well-established that amorphous metals can be ferromagnetic if the corresponding crystalline alloys (without the metalloids) are ferromagnetic; the saturation magnetization and Curie temperatures of the amorphous materials are somewhat reduced.[4] Since an amorphous material should have no macroscopic magnetic anisotropy, the magnetostriction would play a very important role in the magnetization process. The magnetostriction constants were measured to some extent by Tsuya, et al. for an amorphous Fe ribbon[5,6] and by Egami, et al. for amorphous Fe-Ni ribbons.[7]

We have extended to greater details the measurement of the magnetostriction of amorphous Fe ribbons, including the amorphous Fe-Ni system and Fe-Co system which exhibit low coercive force and high permeability. In this paper, we report first the results of measurements of the saturation magnetostriction and forced magnetostriction of amorphous ribbons of the compositions

$Fe_{0.80-x}Ni_xP_{0.13}C_{0.07}$ $\quad (0 \leq x \leq 0.4)$,

$(Fe_{1-x}Co_x)_{0.80}P_{0.13}C_{0.07}$ $\quad (0 \leq x \leq 0.7)$, and

$(Fe_{1-x}Co_x)_{0.75}Si_{0.15}B_{0.10}$ $\quad (0.75 \leq x \leq 1.0)$

with the magnetic field applied in various directions of the ribbons by using a three terminal capacitance method and a dilatometric method. Next, we report the temperature dependence of the magnetostriction in these ribbons between liquid nitrogen and room temperature.

EXPERIMENTAL PROCEDURE

The amorphous ribbons of these compositions used in this experiment were prepared by the centrifugal solidification technique as reported previously.[1] The structure of these ribbons was confirmed by an x-ray study to be amorphous. All of the ribbons were about 0.6mm x 25μm in cross-section.

The three-terminal capacitance method and a modified dilatometric method of the suspension type were employed for measuring the magnetostriction. To prepare the sample in the capacitance method, each ribbon was shaped into ten square platelets about 1.5 x 0.6mm^2 in size, and they were piled up in the form of slab using the adhesive No. 3000 (Cemedine Co. Ltd.).

The structure of the capacitance cell used in this method was reported in the previous paper.[6] In this method, a change in dimension of the sample caused by the magnetostriction is observed as a change in capacitance across two electrodes, one of which moves with a change in dimension of the sample.

The relations between the direction of the applied magnetic field and that of the observation of the magnetostriction are shown in the Table, in which the principle axes of the ribbon are taken as the rectangular coordinate system (x,y,z) with the x-axis parallel to the di-

TABLE. Directions of the Magnetic Field and of the Observation of the Magnetostriction

case	magnetic field	observation
1	(xy) plane	x-axis
2	(xy) plane	y-axis
3	(xz) plane	x-axis
4	(yz) plane	y-axis

490

rection of the length of the ribbon, z-axis perpendicular to its surface and y-axis perpendicular to both x- and z-axes.

The dilatometric method was used in measuring the low field dependence and the detailed compositional dependence of the long samples in
the vicinity of the composition with $Fe_{0.05}Co_{0.70}Si_{0.15}B_{0.10}$.

RESULTS AND DISCUSSION

Taking θ as the angle between the direction of the applied magnetic
field and that of the observation of the magnetostriction, the magnetostriction changes as $\cos^2\theta$ for all cases in every sample used in this
experiment and shows a maximum when the direction of the applied field
is parallel to that of the observation. When the magnetic field is applied parallel to that of the observation. When the magnetic field is applied parallel and perpendicular to the direction of the observation,
the field dependence of the magnetostriction $\Delta\ell(0°)$ / ℓ and $\Delta\ell(90°)$ / ℓ
of $Fe_{0.65}Co_{0.15}P_{0.13}C_{0.07}$ at room temperature is shown in Figure 1 for
cases 1 and 3. In this figure, the field dependence of the magnetostriction $\Delta\ell(0°)$ / ℓ in case 1 is quite similar to that in case 3. The positive magnetostriction $\Delta\ell(0°)$ / ℓ is saturated at about 1.5KOe and then
increases linearly with the increase of the applied field. The magnetostriction $\Delta\ell(90°)$ / ℓ in case 1 is negative, showing a minimum at about 4KOe
and increasing linearly with the increase of the applied field. However,
the magnetostriction $\Delta\ell(90°)/\ell$ in case 3 is positive at a low magnetic
field, turning negative at about 3KOe and then increasing linearly with
the increase of the field above 12KOe. Such an anomalous field dependence
of the transversal magnetostriction is found in all compositions of the
Fe-Ni system and Fe-Co system as far as case 3 is concerned and seems
to give information important in accounting for the behavior of the
magnetic domains in these materials.

For an isotropic material such as a polycrystalline material, the
magnetostriction $\Delta\ell(\theta)/\ell$ can be given as a function of $\cos^2\theta$ as follows:

$$\Delta\ell(\theta) / \ell = \frac{3}{2} \lambda s \left[\cos^2\theta - \frac{1}{3}\right],$$

where λ_s is the saturation magnetostriction constant. On the assumption
that the above equation holds for an amorphous material, we can calculate the magnetostriction constants λ_s of the sample under examination

491

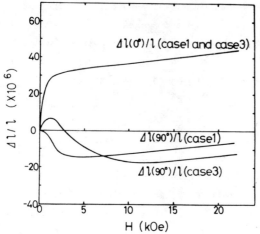

Figure 1. Field dependence
of the magnetostriction
$\Delta\ell(0°) / \ell$ and $\Delta\ell(90°) / \ell$
of $Fe_{0.65}Co_{0.15}P_{0.13}C_{0.07}$
in cases 1 and 3.

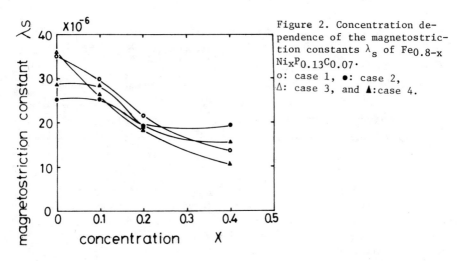

Figure 2. Concentration dependence of the magnetostric-
tion constants λ_s of $Fe_{0.8-x}$
$Ni_xP_{0.13}C_{0.07}$.
o: case 1, ●: case 2,
△: case 3, and ▲:case 4.

from the values $\left[\Delta\ell(0°) / \ell - \Delta\ell(90°) /\ell\right]$ with the correction for the
shape effects.

   The results for $Fe_{0.80-x}Ni_xP_{0.13}C_{0.07}$ measured in the magnetic field
at 20KOe by the capacitance method are shown in Figure 2. From this
figure, we find that the magnetostriction constants of each amorphous
ribbon depend on the directions of the magnetic field, hence the amor-

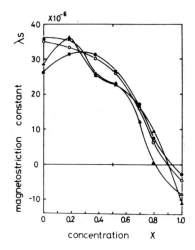

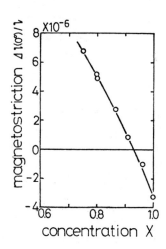

Fig.3 Concentration dependence of the magnetostriction
constants λ_S of Fe-Co system
O:case 1, ●:case 2, Δ:case 3
and ▲ :case 4

Fig.4 Concentration dependence of the longitutidal
magnetostriction $\Delta\ell(0°)/\ell$ of
$(Fe_{1-x}Co_x)_{0.75}Si_{0.15}B_{0.10}$
$(0.75\leq x\leq1.0)$.

phous ribbons are not completely magnetically isotropic but partly anisotropic. The difference between the anisotropic parts of the magnetostriction constants for $Fe_{0.80}P_{0.13}Co_{0.07}$ is about 11×10^{-6}, and the
average magnetostriction constant λ_s is 31×10^{-6}. The average magnetostriction constant decreases monotonously with the increase of Ni concentration and is 15×10^{-6} for $Fe_{0.40}Ni_{0.40}P_{0.13}Co_{0.07}$.

The concentration dependence of the magnetostriction constants λ_s
of $(Fe_{1-x}Co_x)_{0.80}P_{0.13}Co_{0.07}$ $(0\leq x \leq 0.7)$ and $(Fe_{1-x}Co_x)_{0.75}Si_{0.15}B_{0.10}$
$(0.75 \leq x \leq 1.0)$ is shown in Figure 3. Figure 4 shows the results for
$(Fe_{1-x}Co_x)_{0.75}Si_{0.15}B_{0.10}$ $(0.75 \leq x \leq 1)$ obtained by the dilatometric
method. As shown in Figure 3, the magnetostriction in the Fe-Co system
is also not completely isotropic, but shows a slight dependence on the
directions of the magnetic field. As shown in Figure 4, the longitudinal magnetostriction $\Delta\ell(0°) / \ell$ is zero at or about the composition
$Fe_{0.047}Co_{0.703}Si_{0.15}B_{0.10}$. This fact accounts for the remarkably low
coercive force H_c of 0.005 Oe and high permeability μ_m of 1,200,000

which were observed for the same composition. Since the amorphous alloys have no crystalline anisotropy, it is not surprising that H_c is small when the magnetostriction is small in the Co-rich amorphous materials. On the other hand, a greater value of H_c in other compositions can be understood on the basis of the magnetoelastic effect considering the internal stress distribution and a large magnetostriction. Perhaps the most significant aspect of this consideration is to point out the fact that a similar concentration dependence is seen in both $H_c(x)$ and $\lambda_s(x) / I_s(x)$, where $I_s(x)$ is the saturation magnetization.

The forced magnetostriction constants $\delta\omega / \delta H$ of the Fe-Ni and Fe-Co systems are obtained from the field dependence of the saturation magnetostriction. The constant is found to be $21 \times 10^{-10}/Oe$ for $Fe_{0.80}P_{0.13}Co_{0.07}$; it decreases monotonically with the increase of Co or Ni concentration.

The magnetostriction constants λ_s of $Fe_{0.80}P_{0.13}Co_{0.07}$ have broad maximums near 130K in cases 1, 2, and 3 and then decrease with the increase of the temperature. On the other hand, in case 4, the magnetostriction constant has a broad minimum near 200K. The temperature dependences of the magnetostriction constants λ_s of the Fe-Ni system in case 1 have maxima near 90K in the composition with $x = 0.2$, 0.4 and decrease monotonically with the increase of the temperature.

ACKNOWLEDGEMENTS

We would like to thank Prof. Y. Shiraga of Fukui University for his valuable help in the course of the measurement of the magnetostriction. The present work has been partially supported both by the Grant in Aid of the Ministry of Education and by Kurata Funds.

REFERENCES

1. Masumoto, T. and R. Maddin, Act metallurgica 19, 725 (1971).
2. Chen, H.S. and C.E. Miller, Rev. Sci. Instrum. 41, 1237 (1970).
3. Chen, H.S. and D.E. Polk, J. Non-crystalline Solids 15, 174 (1974).
4. Sinha, A.K., J. Appl. Phys. 42, 338 (1971).
5. Tsuya, N., K.I. Arai, Y. Shiraga, & T. Masumoto, Phys. Lett 51A, 121 (1975).
6. Tsuya, N., K.I. Arai, Y. Shiraga and T. Masumoto, to be published in: Phys. Status Solidi (a), 31 (1975).
7. Egami, T., D.J. Flanders, and C.D. Graham, Jr., 20th Annual Conf. on 3M, San Francisco (1974).
8. Fujimori, H., K.I. Arai, H. Shirae, H. Saito, T. Masumoto, and N. Tsuya [unpublished].

494

INTERACTIONS AND ISOLATED IMPURITIES
IN SUPERSATURATED ZnFe ALLOYS

E. Babić, J.R. Cooper, B. Leontić, and M. Očko

Institute of Physics
University of Zagreb, Yugoslavia

INTRODUCTION

Although the occurrence of localized magnetic moments (LMM) in dilute alloys with transition metal impurities was found experimentally some 40 years ago,[1] a more detailed theoretical understanding of this phenomenon was achieved only recently.[2] The theory relates to systems which form LMMs and describes well the transition from an apparently nonmagnetic behaviour at low temperatures ($T \ll T_K$) to a magnetic one at high temperatures ($T > T_K$). It was experimentally found, however, that there are several systems which cannot be clearly characterized as magnetic or nonmagnetic. One of these is the ZnFe system which seems (together with a few other systems) to lie near the border between magnetic and nonmagnetic alloys. By investigating this alloy we may gain an understanding of the nature of the transition between magnetic and nonmagnetic systems.

The main reason why earlier investigations of ZnFe alloys were unable to give a reliable answer to this question was mainly the very low solid solubility of Fe in Zn. As a result, resistivity measurements of very dilute ZnFe alloys at low temperatures[3] seemed to indicate a magnetic behaviour while more recent magnetic susceptibility measurements[4] showed the opposite. Our aim was to clarify this situation by measuring the electrical resistivity and thermoelectric power of more concentrated ZnFe alloys in a wide temperature range.

EXPERIMENTAL

In order to investigate the impurity resistivity of ZnFe alloys over a broad temperature range, one has to subtract a rather large phonon contribution to the total resistivity. For this subtraction procedure to have sufficient accuracy, the impurity concentration should reach at least 0.1at.% Fe which is not possible without rapid quenching.

The starting materials were supplied by Koch-Light Lab. Ltd.; their levels of purity were 6N for Zn and 4N8 for Fe. The master alloys were prepared in evacuated and sealed quartz tubes heated to 700°C in a resistance furnace. The nominal concentrations of these alloys were 0.04, 0.1, 0.2, and 0.3 at.%Fe. Some ZnAg master alloys (1, 2 and 3 at.%Ag) and two ZnCo alloys (0.11 and 0.22at.%Co) were also prepared in a similar fashion and later used for the subtraction of the phonon resistivity in ZnFe alloys. The actual concentrations of all the alloys were determined by electron microprobe examination. The agreement between the nominal and actual concentrations was within 10%.

As the equilibrium solid solubility of Fe in Zn seems to be below 0.01at.%,[4] we used a rapidly rotating mill device[5] to quench our master alloys. The quenched samples were in the form of long strips up to 50cm long and 20-50μm thick. In all samples a texture was detected which seemed to be more pronounced in the more concentrated samples. The crystallographic planes (0001) were parallel to the plane of cooling, while no texture was detected in the direction of rolling. We selected the samples for resistivity and thermoelectric power (TEP) measurements by measuring the residual resistance ratio $RRR = R_{4.2}/(R_{273} - R_{4.2})$. These values and other data relevant to our samples are given in the Table. The methods we used for measuring the resistivities[6] and TEP[7] have been described elsewhere and we will not discuss them here at any length. The relative accuracy of the resistivity measurements was 1 part in 10^5 while the absolute one was about 1%. The corresponding figures for TEP measurements were 3 and 10% respectively.

RESULTS AND DISCUSSION

The linearity of the residual resistivity as a function of the concentration (vide Table) for all our samples indicates that they are good solid solutions. The slope $\rho_0/c = (15 \pm 1)$ μΩcm/at.% also agrees with that found for very dilute alloys.[8] While residual resistivity is a good check for the existence of solid solution it is usually not very sensitive to interimpurity interactions. These interactions show better

496

TABLE: Some Relevant Data
for <u>Zn</u>Fe Alloys

c	RRR	ρ_0	θ
0.04	0.13	0.71	120
0.04	0.14	0.76	135
0.04	0.15	0.82	125
0.11	0.31	1.69	99
0.11	0.32	1.74	97
0.11	0.33	1.80	85
0.2	0.605	3.10	61
0.2	0.68	3.7	63
0.3	1.00	5.45	52

c, concentration in at.%
RRR, residual resistance
ratio (see text)
ρ_0, residual resistivity
in $\mu\Omega$cm
θ, characteristic temperature deduced from the low
temperature (T ≲ 2K) re-
sistivity dependence

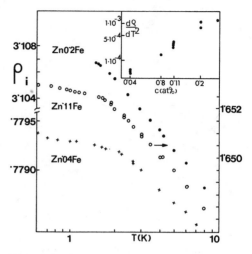

Figure 1. Low temperature impurity resis-
tivities ρ_i (in $\mu\Omega$cm) of 3 <u>Zn</u>Fe alloys.
Inset: $(d\rho/dT^2)$ (in $\mu\Omega$cm/K^2) versus concentration for T<2K.

in the temperature dependence of the resistivity at lowest temperatures.

In Figure 1 (supra) we show the impurity resistivities of three <u>Zn</u>Fe
alloys (with 0.04; 0.11, and 0.2at.%Fe) in the temperature interval 0.5
— 10K. These samples show a $-T^2$ saturation of the impurity resistivity
below 2K and a weak logarithmic decrease at higher temperatures. Using
(for T ≲ 2K) the empirical relation:

$$\rho_T = \rho_0 \left(1 - T^2/\theta^2\right) \tag{1}$$

we have evaluated the θ values given in the Table (supra). A rather
strong concentration dependence of these θ values as well as the relative narrowness of the temperature interval in which this $-T^2$ variation
was observed (T < 0.02θ) indicate that this behaviour is not typical of
isolated Fe impurities. In order to find out the origin of this $-T^2$ variation, we plot $[d\rho/d(T^2)]$ versus concentration c for T < 2K in the inset
of Figure 1. The approximate c^2 dependence indicates that this resistivity variation is caused by interacting Fe pairs having a much lower θ
value. This conclusion is also supported by a very weak Curie-Weiss

497

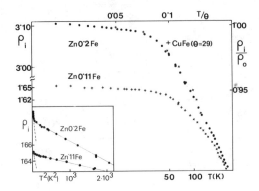

Figure 2. Impurity resis-
tivities ρ_i(in $\mu\Omega$cm) of two
ZnFe alloys versus tempera-
ture (logarithmic scale).
Together with the Zn 0.2Fe
alloy, data for a CuFe alloy[10] are plotted in the reduced scales (right and top
scale). Inset: the impurity
resistivities of the same
ZnFe alloys versus T^2.

variation of the magnetic
susceptibility of dilute
ZnFe alloys showing a de-
pendence on c^2 rather than
on c.[4] Due to a very low
characteristic temperature
of Fe pairs in Zn, their contribution to the impurity resistivity (i.e.,
its temperature dependence) is expected to be small at higher temperature.

In order to separate out the single impurity behaviour, we therefore extended our measurements to higher temperatures (T ≾ 270K). In
Figure 2 (supra), we show the impurity resistivities of two ZnFe alloys
(0.11 and 0.2at.%Fe) versus temperature (logarithmic scale). The sample with 0.04at.%Fe is not shown on this figure since the ratio of the
phonon resistivity to the impurity resistivity was too large for the
subtraction procedure to have sufficient accuracy in this case. Figure
2 is rather similar to Figure 1, i.e., there is a T^2 saturation below
40K and a rather strong logarithmic decrease of the impurity resistivity above 100K. In the inset the impurity resistivity at T < 50K
is shown plotted against T^2. The good linearity of these data (apart
from the points at the lowest temperatures which have already been
discussed) indicates that the empirical relation (1) is again obeyed
but with a higher characteristic temperature ($\theta \sim 300$K). However, this
T^2 variation is observed in the same temperature interval (T ≾ 0.1·θ)
as in some other similar systems (AlMn, AuV) and [d ρ/d(T^2)]

498

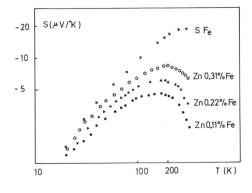

Figure 3. Thermoelectric powers of three Z̲n̲Fe alloys versus temperature. The upper curve is the extrapolated impurity thermopower SFe.

values for T < 40K are proportional to the concentration which clearly indicates single impurity behaviour. This high θ value explains the apparent nonmagnetic behaviour of the magnetic susceptibility[4] of very dilute Z̲n̲Fe alloys below 50K.

Our data enable us to verify a recent suggestion[9] that the impurity resistivities of alloys with different θ values behave similarly and that the normalized impurity resistivities ρ_{imp} / ρ_0 of alloys with different hosts but the same impurity should overlap when plotted against the reduced temperature T/θ. In Figure 2 we include (in a plot of this type) an earlier measurement[10] of a C̲u̲ 0.01at.%Fe alloy (denoted by crosses) together with our Z̲n̲ 0.2at.%Fe alloy. Although the θ values for these two alloys differ by more than a factor of ten, the data superimpose very well through the entire temperature interval. Since the C̲u̲Fe is a "classical" example of a magnetic alloy our resistivity results seem to indicate that no sharp transition exists between magnetic and nonmagnetic systems.

Finally, in Figure 3, we show the thermoelectric power data for three Z̲n̲Fe alloys (0.11, 0.22, and 0.31at.%Fe) in the temperature interval 1.5-300K. All three alloys show a linear variation of the thermopower (S) with temperature below 20K. The initial slope S/T is practically concentration-independent and equal to $-(0.16 \pm 0.01)$ $\mu V/K^2$ which is only about a factor of two higher than in A̲l̲Mn alloys. Also, the overall shape of the thermopower data for Z̲n̲Fe alloys is very similar to that for A̲l̲Mn alloys.[7] However, in contrast to A̲l̲Mn,

499

S is strongly concentration-dependent for $\underline{Zn}$Fe. Because of this, the Nordheim-Gorter rule has been used to find the impurity thermopower (S_{Fe}). These values are plotted in the same figure together with the measured values. Within the uncertainty arising from this extrapolation, S_{Fe} seems to become constant only above 200K. A comparison of the thermoelectric power in some other systems[7] indicates that the impurity thermopower reaches a broad maximum at temperatures around 0.8θ. The thermopower data lend further support to our value for the characteristic temperature of $\theta \sim 300K$. Although the rather large uncertainty in the θ value and the S_{Fe} extrapolation prevent us from carrying out of a more detailed comparison, our data also seem to support a recently found similarity in the behaviour of the thermopower[7] for alloys with very different characteristic temperatures and therefore indicate a smooth transition from magnetic to nonmagnetic systems.

CONCLUSION

Within the scope of our results for the impurity resistivity and thermopower, alloys in the $\underline{Zn}$Fe system seem to be weakly magnetic with a characteristic temperature $\theta \sim 300K$. This value is much higher than the value of $\theta \sim 80K$ obtained earlier for more dilute alloys,[8] and it agrees better with a recent magnetic susceptibility measurement.[4] Careful analysis of the impurity resistivity at the lowest temperatures shows that interacting Fe pairs give rise to a fortuitous characteristic temperature $\theta \sim 80K$. The new value of θ (300K) enabled us to compare this system with other alloys having different θ values. This comparison lends further support to the notion that the impurity resistivity[9] and thermopower[7] behave similarly in magnetic and nonmagnetic (or weakly magnetic) alloys.

This, of course, is in contrast to earlier ideas of a sharp transition between magnetic and nonmagnetic alloys.

ACKNOWLEDGEMENT

We thank Professors C. Rizzuto and B.R. Coles for useful suggestions, and Mr. M. Miljak for help in some experiments.
500

REFERENCES

1. De Has and De Boer, Physics 1, 609 (1933/1934).
2. Wilson, K.G., Proceedings of the Nobel Symposium XXIV, Uppsala, (1973).
3. Caplin, A.D., Phys. Lett. 26A, 46 (1967).
4. Bell, A.E. and A.D. Caplin, J. Phys. F. 5, 143 (1975).
5. Babić, E., E. Girt, R. Krsnik, and B. Leontić, J. Phys. E3, 1015 (1970).
6. Babić, E., M. Očko, and C. Rizzuto, J. Low Temp. Phys. 21, 243 (1975).
7. Cooper, J.R., Z. Vučić, and E. Babić, J. Phys. F. 4, 1489 (1974).
8. Ford, P.J., C. Rizzuto, and E. Salamoni, Phys. Rev. B6, 1851 (1972).
9. Rizzuto, C., E. Babić, and A.M. Stewart, J. Phys. F3, 825 (1973).
10. Star, W.M., F.B. Basters, and C. Van Barle, Proc. of LT11 2, 1252 (1968).

LOW TEMPERATURE DEVIATIONS FROM MATTHIESSEN'S RULE IN RAPIDLY QUENCHED ALUMINIUM ALLOYS

E. Babić, R. Krsnik, B. Leontić, and M. Očko

Institute of Physics
University of Zagreb, Yugoslavia

INTRODUCTION

A long time ago, Matthiessen and Vogt[1] found that, for most alloys, the temperature coefficient of resistivity (at room temperature) is equal to that of the pure host metal. Consequently the contributions to the resistivity from the impurities and the lattice vibrations of the host metal were simply regarded as being additive. This hypothesis was named Matthiessen's rule.

More recently, careful analysis of the resistivity of pure metals and alloys at low temperatures has shown that Matthiessen's rule (MR) is hardly ever obeyed and that the total resistivity of an alloy is always greater than the sum of the impurity and the host phonon contributions. At low temperatures, the additional term is much larger than the pure host phonon contribution and in rather pure metals may amount to a significant fraction of the impurity resistivity. This term, called deviation from Matthiessen's rule (DMR), is concentration dependent and may also have a different temperature dependence (T^3, T^4) than the pure phonon part (usually T^5).

A better understanding of this phenomenon is of great practical importance. Advanced technology requires large-scale applications of superconductivity. Of particular importance is the required quality of metals such as copper and aluminium which are necessary in any application of superconductivity; their low temperature electrical properties must be well known.

EXPERIMENTAL

In our investigation we used Al-alloys with transition metal impurities since they have residual resistivities[2] about a factor of ten greater than alloys with the same concentration of impurities of normal metals

503

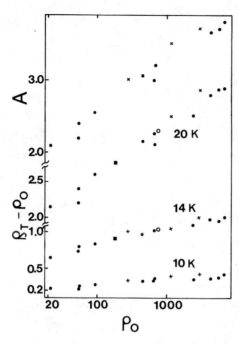

Figure 1. The results for ρ_T-ρ_o versus ρ_o (in nΩcm) at 10, 14, and 20K; •AlV alloys, ×AlTi alloys, ■AlFe and oAlNi samples. At the top, the coefficient A (10^{-3} Ωcm/k³) of the DMR T³-term versus ρ_o.

and since they also do not appear to form Guinier-Preston zones in aluminium. As the equilibrium solid solubilities for most of the transition metals in metals in aluminium are practically zero, we had to employ a rotating mill quenching device[3] for the production of samples. This quenching technique allows not only a large enhancement of the solid solubility but also produces samples of suitable (and controllable) shape. The samples were in the form of long strips with typical dimensions of 5-10cm length, 2-10mm width, and 20-60μm thickness. We checked the "as obtained" samples by measuring the residual resistance ratio RRR = $R_{4.2}$ / (R_{273} - $R_{4.2}$) which also gives the concentration of the impurities assuming them to be in solid solution. Simultaneously, the existence of solid solutions was verified by x-ray measurements; the actual concentrations were determined by electron microprobe analysis.

In order to prevent possible later precipitation in these metallurgically supersaturated alloys all samples were kept in liquid nitrogen prior to the measurements. Selected samples were cut to the desired shape, fitted with the potential and current leads and mounted in a special cryostat. The temperature was controlled by the amount of helium exchange gas and a resistance heater mounted on the copper plate supporting the samples. A calibrated Ge thermometer situated near the

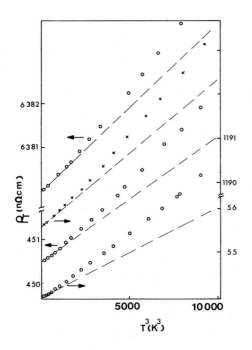

Figure 2. Total resistivities (ρ_T) of four alloys vs T^3: ○A1V alloys, ×A1Ti alloys. The dashed lines are extrapolations of the low temperature T^3 slopes.

samples was used to monitor the temperature. The resistances of the samples were measured with a relative accuracy of a few parts in 10^6 by means of a potentiometer. The absolute error in the resistivity was about 2% due to uncertainty in the geometrical shape factor.

RESULTS AND DISCUSSION

Previous investigations [4,5] of pure Al and its alloys indicated the existence of an unusual T^3 temperature dependence of the resistivity at the lowest temperatures which was related to the DMR. The concentration (or ρ_0) dependence of the temperature-dependent part of the resistivity ($\rho_T - \rho_0$) was less clear since the data due to different authors always covered narrow concentration intervals only; in particular, there was a lack of studies on very "impure" (high ρ_0) alloys (since, for aluminium, such alloys can only be produced by fast quenching). We investigated a number of A1V and some A1Ti, A1Ni, and A1Fe alloys with residual resistivities (ρ_0) ranging from 20–6500 nΩcm in the temperature interval 1.5 – 20K. Our aim was to find the detailed ρ_0 dependence of $\rho_T - \rho_0$ and the limiting low temperature resistivity variation.

In considering the concentration dependence of the DMR, we plot in the lower part of Figure 1 the values for $\rho_T - \rho_0$ versus ρ_0 (in a logarithmic scale) for all our samples at the three temperatures 10, 14, and

505

20K. At these temperatures, our data are consistent with a slow, approximately logarithmic varion of $\rho_T - \rho_0$ (or the DMR) with ρ_0. The results for the lowest temperature (10K) and for the most impure alloys ($\rho_0 > 1 \ \mu\Omega cm$) may seem to be varying more slowly, but these data are the least accurate of all due to the limited resolution of the potentiometer. This indicates that, for a more detailed investigation of very "impure" alloys, an even better resolution is required (10^7) which is impossible to achieve with conventional potentiometers. We can summarize that our measurements do not indicate a saturation of the DMR in very impure Al-alloys at these temperatures. It is, however, an open question whether saturation cannot take place at even lower temperatures where our resolution was not sufficient to establish isotherms of this type with any significant accuracy.

Next we investigated the temperature dependence of the resistivity. In Figure 2 we plot the total resistivities (ρ_T) of four samples versus T^3. We took only four samples (with very different ρ_0) but all other samples had the same behaviour. All these samples show some variation steeper than T^3 (dashed line). Since the T^3 temperature scale gives more weight to the high temperature points, it is not possible to find out more about this deviation by using this type of plot. Some conclusion can be reached, however, by looking at the relative behaviour of alloys with different ρ_0: it is seen that, although the entire difference $\rho_{20} - \rho_0$ is greater for the more impure alloys, the difference between the dashed lines (T^3) and the experimental points does not depend on ρ_0. This recalls a previous suggestion[4] that $\rho_T - \rho_0$ is the sum of a DMR term ($\sim T^3$ and increasing with ρ_0) and of an "ideal" resistivity term ($\sim T^5$ and independent of ρ_0).

In the upper part of Figure 3 we plot the data for two samples in a form suitable for a fit to $\rho_T = \rho_0 + AT^3 + BT^5$. Rather good linearity of our data (down to about 10K) seems to indicate simultaneous existence of T^3 and T^5 terms in the resistivity. We have also tried to fit the data for one of the samples to two simple polynomial expressions different from the one which we had used previously. As none of these expressions provides a good fit to our data in any significant temperature

506

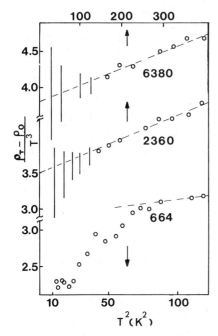

Figure 3. The $\rho_T-\rho_o / T^3$ data (in 10^{-13} Ωcm/K^3) for three A1V alloys versus T^2 [See text; the numbers below the points denote the residual resistivities in nΩcm]. Note that the bottom curve has an expanded temperature scale.

interval (Figure 4) we believe that they are indeed best represented by a combination of the T^3 and T^5 terms.

From the type of plot shown in Figure 3, we have extracted (for all samples) the coefficients A and B (corresponding to the T^3 and T^5 terms respectively). While B is independent of ρ_o and equal to (2.2 ± 0.3) · 10^{-6} Ωcm/K^5 (which is very close to the value derived from the Bloch-Grüneisen formula for pure Al), the coefficient A increases with ρ_o. The variation of A with ρ_o is shown in the upper part of Figure 1. We note two details: i) A is larger than B by at least a factor of 10^3 for all samples; ii) A has the same ρ_o dependence as $\rho_T - \rho_o$. The latter comes from the fact that A actually measures the DMR where it has a T^3 dependence. Again, it clearly indicates that the DMR does not saturate in our alloys in this temperature interval.

It is interesting to find out whether this ρ_o dependence of the DMR also continues to the lower temperatures (T < 10K). As mentioned before, at very low temperatures the resolution of about 5 parts in 10^6 was not sufficient for a more detailed investigation of very impure alloys. This is clearly indicated in the upper part of Figure 3 where the error bars at the lowest temperatures denote the errors produced by a change in the ρ_o value of ±5 · 10^{-6}. In order to gain some insight into the temperature and ρ_o dependence of the DMR below 10K we have performed ex-

507

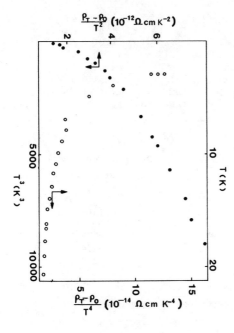

Figure 4. An <u>AlV</u> alloy (ρ_0=450nΩcm) fitted to $\rho_T = \rho_0 + CT^2 + DT^5$ and $\rho_T = \rho_0 + ET^4 + FT^5$ (open circles). A fit to $\rho_T = \rho_0 + GT^3 + MT^4$ was also attempted (not shown) but the result was less satisfactory than that in Figure 3.

tremely precise measurements ($\sim$1 part in 10^6) on a few somewhat less impure Al-alloys ($\rho_0 < 1\mu\Omega$cm). The data for one of these samples (ρ_0 = 664 nΩcm) are shown in the lower part of Figure 3. It can be seen that the coefficient B (of the T^5 term) seems to increase significantly below 10K while A is considerably reduced. The other samples which we have measured with the highest accuracy showed a rather similar behaviour and the ρ_0 variation of A below 10K was also considerably less pronounced than that in Figure 1.

CONCLUSIONS

Within the scope of our results, the theoretical models[6] which try to explain the DMR in polyvalent metals by the anisotropy of both the impurity and phonon scattering rates over the Fermi surface seem to be qualitatively correct. They predict a rather strong variation of the DMR with ρ_0 in the "transitional" regime similar to the one observed in Figure 1. However, they also predict an apparent saturation of DMR at high ρ_0 values (which should occur within two decades in ρ_0), in contrast to our results. As far as the temperature dependence is concerned, the same model predicts a nearly T^3 variation in the temperature interval 10-20K which is in fair agreement with our results. At very low temperatures the enhancement of the T^5 - term in the resistiv-

508

ity by more than a factor of ten (with respect to the Bloch-Grüneisen value) is predicted.[7] Although the experimental results (see the lowest curve in Figure 3) indicate an increase in the value of B below 10K, this increase is considerably smaller than the one predicted. Finally, the prediction that the $\rho_T - \rho_0$ variation with ρ_0 should saturate at very low temperatures seems also to be qualitatively correct since our A values show considerably weaker ρ_0 dependence below 10K (although for only a few samples measured). Further investigations[8] may clarify these differences.

ACKNOWLEDGEMENT

We acknowledge useful discussions with Drs. A.D. Caplin and J.A. Campbell and Professors J.W. Wilkins and C. Rizzuto. We also thank M. Miljak for help in some experiments.

REFERENCES

1. Matthiessen, A. and C. Vogt, Ann. Phys. 122, 19 (1864).
2. Boato, G., M. Bugo, and C. Rizzuto, Nuovo Cim. 4, 226 (1967).
3. Babić, E., E. Girt, R. Krsnik, and B. Leontić, J. Phys. E.3, 1014 (1970).
4. Caplin, A.D. and C. Rizzuto, J. Phys. C, 3, L117 (1971).
5. Krsnik, R., E. Babić, and C. Rizzuto, Solid St. Commun. 12, 891 (1973).
6. Dosdale, T. and G.J. Morgan, J. Phys. F., 4, 402 (1974).
7. Lawrence, W.E. and J.W. Wilkins, Phys. Rev. B, 6, 4466 (1972).
8. Babić, E., R. Krsnik, and M. Očko (1975 [to be published].

THE DOMAIN STRUCTURE OF FERROMAGNETIC METALLIC GLASSES

H.J. Leamy, S.D. Ferris, D.C. Joy,
R.C. Sherwood, E.M. Gyorgy, and H.S. Chen

Bell Laboratories, Murray Hill, New Jersey

INTRODUCTION

Ferromagnetic metallic glasses were produced in the form of splat quenched droplets nine years ago[1]. Recently however, methods for the production of continuous, glassy ribbons have been developed, and applied to the production of glassy ferromagnets. We have employed both roller [2] and centrifugal spin [3] quenching techniques to produce such specimens, and have studied their ferromagnetic properties and domain structure. This paper contains a summary account of our experience in the study of domain structures in these alloys by scanning and transmission electron microscopy (SEM and TEM, respectively).

EXPERIMENTAL METHODS

Roller quenched alloys, which were the subject of our initial report [4], are produced by directing an 0.5mm stream of molten alloy between a pair of rapidly rotating steel rollers as indicated schematically in Fig. 1a. Upon contact with the rollers, the outer layer of the stream solidifies and is compressed. The remaining liquid flows laterally from between the solidified crust and is quenched by the rollers to produce a 2mm by 40µm ribbon. Cooling rates of $\sim 10^6$ K/sec. are achieved and crystal-free material is reproducibly obtained. The samples contain a deformed central zone that corresponds to the original liquid stream. This central zone is made easily visible by electropolishing, indicating that cold flow has occurred during the quenching operation [5]. The central zone is flanked by smoother regions that correspond to the lateral liquid flow. Spin quenched samples are produced by directing the molten alloy stream against the inside surface of a rapidly rotating drum. In this process, no cold flow is occasioned by mechanical contact with a second roller, and the liquid flow is controlled by centrifugal and surface tension forces. Typical ribbon dimensions produced by this technique are 1mm by 25µm.

The domain structure of the glassy ribbons was examined by techniques that rely upon the alteration of electron trajectories within the material by the internal magnetic induction, $\vec{B}$. In the TEM case, well known Lorentz microscopy techniques [6,7] were applied to ribbons that had been thinned by ion milling. In order to examine thick specimens, the SEM technique of Fathers, et al. [8,9,10], was utilized.

511

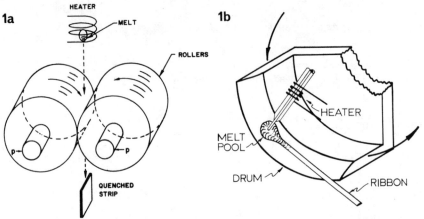

Fig. 1: Schematic representation of: (1a) the roller quenching and (1b), the centrifugal spin quenching process.

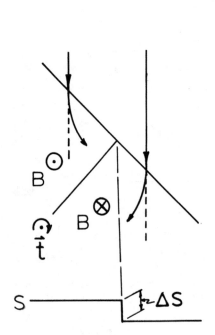

Fig. 2: A schematic illustration of domain contrast formation in the SEM.

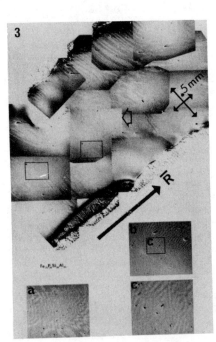

Fig. 3: SEM domain contrast in a roller quenched, $Fe_{75}P_{16}Si_{06}Al_{03}$ ribbon. $\underline{t}$ is perpendicular to $\underline{R}$.

In this technique, electron trajectories within samples tilted about $\bar{B}$ are deflected either towards or away from the surface as illustrated in Figure 2. The number of electrons that escape from the sample and appear as the backscattered electron signal increases or decreases, depending upon the sign of the deflection force: $\bar{F} = e\bar{v} \times \bar{B}$. Observable contrast between domains of different $\bar{B}$ orientation thus occurs when the component of $\bar{B}$ parallel to the tilt axis, $\bar{t}$, is large enough to produce a change ΔS in the backscattered signal, S, that is distinguishable from noise or other fluctuations in S. $\Delta S/S$ for metallic glass specimens imaged at 50kV is always less than 0.5% and is easily masked by surface topography. For this reason, samples for SEM observation were always polished to produce a flat, strain free surface. Since only the component of $\bar{B}$ parallel to $\bar{t}$ and perpendicular to the electron beam contributes to ΔS, the orientation of the in-plane component of $\bar{B}$ may be determined by rotation of the tilted sample about its surface normal.

RESULTS

The domain structure revealed by SEM examination of an $Fe_{75}P_{16}Si_{06}Al_{03}$ sample and shown in Fig. 3 is typical of all positively magnetostrictive, roller quenched, amorphous ribbons. Maximum domain contrast is observed when $\bar{R}$, the ribbon axis, is perpendicular to $\bar{t}$. This indicates that $\bar{B}$ is predominantly directed perpendicular to $\bar{R}$; i.e., is directed across the ribbon. Furthermore, $\bar{B}$ is predominantly uniaxial and most of the domain walls examined by specimen rotation were 180° walls. The complex maze domains shown in Fig. 3, for example, are all magnetized perpendicular to $\bar{R}$ in antiparallel directions. Other regions, such as that shown arrowed in Fig. 3a, exhibit maze structures in which $\bar{B}$ is differently oriented. As will be discussed later, the maze domain pattern is indicative of material for which $\bar{B}$ does not lie within the plane of the ribbon. The domain pattern shown in Fig. 3 exhibits a correspondence with the flow produced during roller quenching. The maze patterns occur primarily in the strained central zone of the ribbon and are flanked by domains of much larger size that run perpendicular to $\bar{B}$. The maze domains, which are reminiscent of patterns observed in materials which have been strained during polishing [11], vanish upon stress relief annealing of the material. The domain pattern in a stress relieved sample is shown in Fig. 4. Here, the predominant orientation of $\bar{B}$ is again perpendicular to $\bar{R}$, although the central zone contains regions within which $\bar{B}$ is parallel to $\bar{R}$. The domain pattern observed in a roller quenched, nonmagnetostrictive ($\lambda = 0$) alloy [12]: $(Fe_{04}Co_{96})_{75}P_{16}B_{06}Al_{03}$, is similar to that shown in Fig. 4. We conclude therefore, that the maze patterns reflect the presence of a local anisotropy of magnetostrictive origin.

513

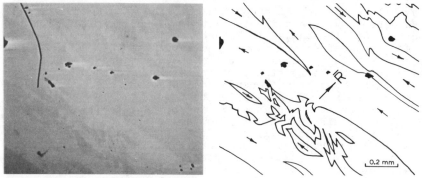

Fig. 4: The domain structure of a roller quenched, $Fe_{.75}P_{.16}B_{.06}Al_{.03}$ ribbon after a 15 min., $400°$ C stress relief anneal.

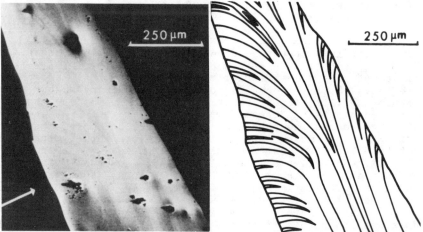

Fig. 5: The domain structure of a centrifugal spin quenched, $\lambda = 0$ alloy. $\bar{t}$ is perpendicular to $\bar{R}$.

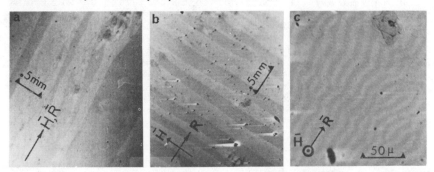

Fig. 6: The domain structure of a roller quenched, $\lambda = 0$ alloy after annealing for 45 min. at $270°$ C in a saturating field with: (6a) $\bar{H}$ parallel to $\bar{R}$, (6b) $\bar{H}$ perpendicular to $\bar{R}$, and (6c) $\bar{H}$ perpendicular to the ribbon plane.

The anisotropy evident in the absence of strain or magnetostriction (Fig. 4) however, must have its origin in the atomic scale structure of the glass, since magnetostatic (shape)
effects would favor the formation of a domain pattern in
which $\bar{B}$ is parallel to $\bar{R}$. This anisotropy is thus similar
to the "pair ordering" anisotropy commonly observed in crystalline materials [13]. Its occurrence implies a lack of
complete structural isotropy in the glassy ribbon. The
similarity of the domain structures described above and the
flow processes envisioned for roller quenching suggests that
this structural anisotropy is formed during flow of the molten alloy just prior to solidification. This hypothesis is
strengthened by results obtained for a centrifugal spin
quenched, $\lambda = 0$ alloy specimen. In this quenching process,
a pool of melt is formed at the point of impingement of the
stream and the rotating wheel. The wheel drags the melt out
of the pool tangentially in the direction of $\bar{R}$, while surface tension tends to pull the melt towards the center of
the ribbon [3]. The resultant domain structure appears to
reflect this flow behavior, as Fig. 5 illustrates. Here, $\bar{B}$
is parallel to $\bar{R}$ in the center of the ribbon and perpendicular to $\bar{R}$ at the ribbon edges.

As might be expected, magnetic anisotropy of structural
origin is susceptible to alteration by metallurgical processing. Magnetic field annealing, for example, is effective
in these glasses [14-16]. The domain patterns obtained aft-
er annealing a $\lambda = 0$ sample in fields parallel to $\bar{R}$ and perpendicular to the ribbon plane are shown in Fig. 6. These
changes in the orientation of the anisotropy, $\bar{K}_u$, are accompanied by the expected modification of the magnetization
behavior of the ribbon [14]. The pattern produced by annealing with the magnetic field, $\bar{H}$ perpendicular to the ribbon plane is of particular interest. As Fig. 6c shows,
this treatment yields a maze pattern that is similar in most
respects to that observed in as quenched, magnetostrictive
glasses. The maze pattern in both cases is observable by
the SEM techniques described above because the in plane component of $\bar{B}$ is large. SEM contrast due to stray field deflection of low energy electrons [17,18] moreover, is not
observed, indicating the absence of stray fields at the sample surface. This, together with the strong maze contrast
visible in Figs. 3 and 6(c), suggests that the maze domains
are closure domains that have formed in order to reduce the
magnetostatic energy of the underlying structure, which is
magnetized perpendicular to the ribbon plane.

Deformation processing, a technique commonly employed
to alter the properties of crystalline ferromagnets, also
affects the domain structure of metallic glasses. Although
the structural rearrangements that accompany cold flow in
metallic glasses are not understood in detail, flow is known
to proceed inhomogeneously at high stress levels with little

515

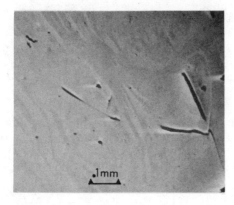

Fig. 7: The domain structure of a roller quenched, $\lambda = 0$ ribbon after a 20% reduction in thickness by cold rolling. Note that the predominant orientation of $\bar{B}$ is perpendicular to $\bar{R}$.

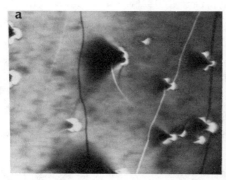

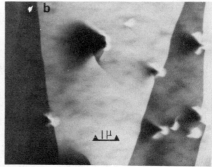

Fig. 8: TEM micrographs of domains in a roller quenched $Fe_{75}P_{16}B_{06}Al_{03}$ alloy. Fig. 8a is a Fresnel micrograph while Fig. 8b is a Foucault micrograph of the same area. Slight motion of the domain walls between exposures is appparent. Fig. 8c shows the approximate orientation of $\bar{B}$ within the material. The shaded areas are hillocks left by ion beam thinning.

accompanying work hardening [19]. Within the localized deformation bands, therefore, we might well expect alteration of the local anisotropy and a consequent refinement of the domain structure.

This expectation is confirmed by Fig. 7, which shows the domain structure of a cold rolled, λ = 0 alloy. The serrations in the domain structure are morphologically quite similar to the deformation bands observed on the surface of the unpolished sample.

Because the glasses described here possess no magnetocrystalline anisotropy, they generally exhibit excellent soft magnetic properties [20,21]. The magnetization processes in these materials are dominated by residual anisotropies of magnetostrictive or structural origin. Furthermore, because metallic glasses are capable of supporting large stresses, the magnetostrictive component of $\overline{K}$ may be quite appreciable. The glasses that possess the largest magnetization, moreover, are magnetostrictive; e.g., for $Fe_{75}P_{16}B_{06}Al_{03}$, λ = 20x10^{-6}. We have therefore examined the magnetic microstructure of this alloy by TEM without stress relief annealing. A typical result is shown in Fig. 8. Fig. 8 contains a domain pattern characteristic of uniaxial anisotropy. The domains visible in these micrographs are pinned by surface hillocks that were produced during ion milling. They are separated by 180° walls that occasionally branch into lower angle walls. Fig. 3 and Fig. 5 indicate that the orientation of $\overline{K}$ is nonuniform on a macroscopic scale. Fig. 8 however, indicates that within the microscopic field of view shown, there exists no detectable spatial variation of the easy axis of magnetization. Specifically, no ripple contrast like that observed in inhomogeneous or polycrystalline thin films is observed. Finally, the domain wall structure observed at the tips of the uniaxial domains in Fig. 8 is of interest. The lack of wall closure at the domain tip has been observed previously in crystalline materials and has been analysed in detail by Torok [22]. It is indicative of weakly uniaxial material.

CONCLUSION

Discussion of the microstructural features presented here is incomplete without consideration of the corresponding magnetic properties and magnetization behavior. We refer the reader to recent reviews [20,21] in which this information as well as a more extensive bibliography may be found.

REFERENCES

1. Duwez, P. and S.C.H. Lin, J. Appl. Phys. 38, 4096 (1967).
2. Chen, H.S. and C.E. Miller, Rev. Sci. Inst. 41, 1237 (1970).
3. Chen, H.S. and C.E. Miller, Mat. Res. Bull. 11, 49 (1976).
4. Leamy, H.J., S.D. Ferris, G. Norman, D.C. Joy, R.C. Sherwood, E.M. Gyorgy, and H.S. Chen, Appl. Phys. Lett. 26, 259 (1975).
5. Pampillo, C.A., Scripta Met. 6, 915 (1972).
6. Hale, M.E., H. W. Fuller, and H. Rubinstein, J. Appl. Phys. 30, 789 (1959).
7. Hirsch, P.B., A. Howie, R.B. Nicholson, D.W. Pashley, and M.J. Whelan, Electron Microscopy of Thin Crystals, Butterworths, London, 388-414 (1965).
8. Fathers, D.J., J.P. Jakubovics, and D.C. Joy, Phil. Mag. 27, 765 (1973).
9. Fathers, D.J., J.P. Jakubovics, D.C. Joy, D.E. Newbury, H. Yakowitz, Phys. Stat. Sol. A20, 535 (1973).
10. Fathers, D.J., J.P. Jakubovics, D.C. Joy, D.E. Newbury, and H. Yakowitz, Phys. Stat. Sol. A22, 609 (1974).
11. Vide, e.g., D.J. Craik and R.S. Tebble, Ferromagnetism and Ferromagnetic Domains, Wiley, New York (1965).
12. Sherwood, R.C., E.M. Gyorgy, H.S. Chen, S.D. Ferris, G. Norman, and H.J. Leamy, AIP Conf. Proc. 24, 745 (1975).
13. Vide, e.g., S. Chikazumi and S. Charap, Physics of Magnetism, Wiley, New York (1964).
14. Chen, H.S., S.D. Ferris, E.M. Gyorgy, H.J. Leamy, and R.C. Sherwood, Appl. Phys. Lett. 26, 405 (1975).
15. Egami, T. and P.J. Flanders [to be published in: the AIP Conf. Proc. series: Proc. of the 1975 Conf. on Magnetism and Magnetic Materials, Philadelphia, December (1975)].
16. ibid., F.E. Luborsky.
17. Banbury, J.R. and W.C. Nixon, J. Sci. Inst. 44, 889 (1967).
18. Joy, D.C. and J.P. Jakubovics, Phil. Mag. 17, 61 (1973).
19. Leamy, H.J., H.S. Chen, and T.T. Wang, Met. Trans. 3, 699 (1972).
20. Egami, T., P.J. Flanders, and C.D. Graham, Jr., AIP Conf. Proc. 24, 697 (1974).
21. Gyorgy, E.M., H.J. Leamy, R.C. Sherwood, and H.S. Chen, [to be published in the AIP Conf. Proc. series: 1975 Conf. on Magnetism and Magnetic Materials, Philadelphia, December (1975)].
22. Torok, E.J., AIP Conf. Proc. 24, 753 (1975).

STRUCTURE AND SUPERCONDUCTIVITY
OF METASTABLE PHASES
IN LIQUID QUENCHED Zr-Rh ALLOYS

Kazumasa Togano and Kyōji Tachikawa
National Research Institute for Metals, Tokyo

INTRODUCTION

The Zr-Rh alloy system has attracted interest, since the intermetallic compound Zr_2Rh shows a high superconducting transition temperature
of 11.2K despite its noncubic structure and the absence of V_a-group elements.[1] In previous work,[2] the authors have investigated the superconducting critical fields and critical currents of Zr-Rh alloys prepared
by conventional arc melting and reported that the upper critical field,
H_{c2}, of Zr_2Rh is as high as 80KOe at 4.2K. In this study, we have
quenched Zr-Rh alloys containing 0-36at.%Rh from liquid state and have
obtained new metastable phases, i.e., supersaturated solid solution of
β-Zr and noncrystalline phase. In this paper we report on the structure
and superconducting properties of these new metastable phases.

SPECIMEN PREPARATION AND STRUCTURE STUDY

Zr-Rh alloys containing 0-36at.%Rh were prepared from 99.9% pure
zirconium flakes and 99.9% pure rhodium powder by arc melting on a
water-cooled copper hearth in an argon atmosphere. Small blocks of
several hundred milligrams were cut from the buttons and rapidly quenched from the liquid state by an apparatus adapted to an arc furnace. The
apparatus used is similar to that described by Ohring and Haldipur.[3]
Quenched specimens have the shape of thin foils, about 1.5cm in diameter and 50μm in thickness. The quenching rate for this method has been
reported to be of the order of 5×10^5 °C/sec.[3] The actual composition
of the quenched foils was determined by chemical analysis.

The structure of the specimen was studied by x-ray diffraction, electron diffraction and electron microscopy. The identification of phases
and the determination of the lattice parameters of β-Zr were carried out
with an x-ray diffractometer. In transmission electron microscopy,

519

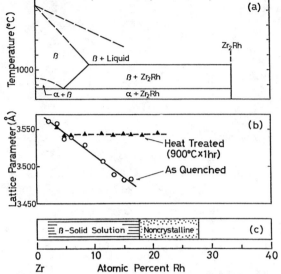

Figure 1. (a) Equilibrium phase diagram of the Zr–Rh alloy system[1]; (b) Lattice parameter of β-Zr solid solution (bcc); (c) Concentration ranges in which metastable phases are obtained by liquid quenching.

thinner foil regions which were found near the edge of the foil were examined without further thinning. Micrographs and selected area diffraction patterns were taken on a JEM150 unit operating at 150KV. The Zr-rich end of the Zr–Rh equilibrium phase diagram reported by Zegler[1] is shown in Figure 1(a) (supra). This diagram suggests the existence of a low-melting-point eutectic at which the liquid is in equilibrium with the intermetallic compound Zr_2Rh and body-centered cubic β-Zr solid solution. β-Zr has a maximum solubility of about 8at.%Rh. Our preliminary study on the metallographic structure of arc-melted ingots has revealed that the eutectic point lies at about 23at.%Rh.

For the 2.5-16at.%Rh alloys quenched from the liquid, only the peaks of retained β-Zr were observed in the x-ray diffraction pattern. On the plot of lattice parameter versus concentration shown in Figure 1(b), the lattice parameter decreases linearly with increasing Rh concentration beyond the equilibrium solubility limit of 8at.%Rh. These results indicate that the solubility limit of Rh in β-Zr solid solution can be significantly increased by rapid quenching from the liquid state. The decrease in the lattice parameter can be attributed to the smaller size of the Rh atom and agrees fairly well with an assumed Vegard's law line which is calculated from the extrapolated lattice parameter of pure β-Zr and the atomic size of Rh corrected according to the coordination change.

520

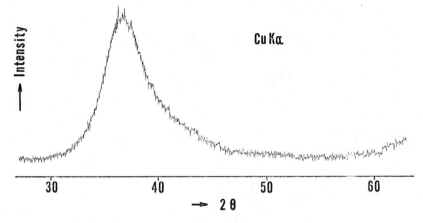

Figure 2. X-ray diffraction pattern of the liquid quenched Zr-23at.%Rh alloy.

The x-ray diffraction analysis of the liquid-quenched alloys containing 18-26at.%Rh yields diffraction patterns with broad maxima which indicate the absence of long-range periodicity in the atomic arrangement. The first broad peak is found at $2\theta = 36.5 - 36.9°$ and the peak width at half-maximum height is $\Delta(2\theta) = 4.6 - 5.0°$, using Cu-Kα radiation. Figure 2 (supra) shows the x-ray diffraction pattern of the liquid quenched Zr-23at.%Rh specimen. The transmission and the election graphs of these specimens reveal no diffraction contrast the electron diffraction patterns do no have sharp Bragg-type diffraction maxima. The designation "noncrystalline" was chosen for this structure,[4] since the detailed radial distribution analysis required for the characterization of the alloy as possessing true "amorphousness" was not carried out in this work.

Figure 1(c) shows the composition ranges in which the metastable phases are obtained. For pure Zr, the β-α transition cannot be suppressed even by liquid quenching. For alloys containing more than 28at.%Rh, the peaks for the equilibrium phases are obtained in the diffraction pattern.

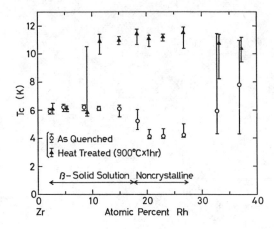

Figure 3 (supra). Superconducting transition temperatures of liquid quenched and heat-treated Zr-Rh alloys.

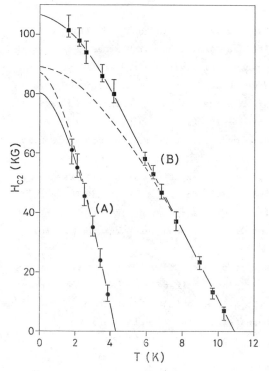

Figure 4 (left). Temperature dependences of upper critical field for the noncrystalline Zr-23at.%Rh alloy (A) and the Zr_2Rh compound (B). The dashed lines were calculated using the WHH theory, ignoring the paramagnetic limiting and spin-orbit scattering effects ($\alpha=0$ and $\lambda_{so}=0$).

The superconducting transition temperature, T_c, was measured by a resistive method. T_c was taken as the temperature at which one-half of the normal-state resistance of the specimen was restored. In Figure 3, the T_c values of the rapidly quenched alloys are plotted as a function of the Rh concentration. The T_c of the β-Zr solid solution is about 6K and almost independent of the Rh concentration even in the supersaturated region. The noncrystalline phase shows a sharp transition with a midpoint of 4.1K, though superconductivity persists up to a higher temperature (a few percent of the whole transition). This residual superconductivity is probably due to a very small amount of retained crystalline phases included in the noncrystalline matrix, which could not be detected by x-ray diffraction. Since it was reported that the T_c of amorphous transition-metal alloys obtained by vapor quenching shows a maximum at an electron-to-atom ratio e/a of about 7,[5] a higher T_c might be expected upon addition of elements with larger e/a. The T_c of the specimens heat-treated for 1h at 900°C after liquid quenching is also shown in Figure 3 (previous page). The T_c of alloys containing more than 8at.%Rh is significantly increased by the formation of the equilibrium Zr_2Rh phase which has a higher T_c of about 11K. The broad superconducting transition for the liquid quenched Zr-33at.%Rh alloy is probably due to atomic disorder in the Zr_2Rh compound produced by quenching. Atomic ordering by annealing improves the T_c of the Zr-33at.%Rh alloy.

UPPER CRITICAL FIELD

An upper critical field measurement for the noncrystalline Zr-23at.%Rh specimen was performed by the resistive method using a superconducting solenoid. In this measurement, current was supplied parallel to the magnetic field. The upper critical field, H_{c2}, was defined as the midpoint of the transition measured at a current density of 10 A/cm^2. The measurement was also carried out for the intermetallic Zr_2Rh compound. The magnetic field was applied perpendicular to the columnar axis.[2]

The results are shown in Figure 4. For the noncrystalline Zr-23at.%
Rh alloy, the upper critical field at 0K, obtained by extrapolation of
experimental values by the function $H_{c2}(t) = H_{c2}(0)[1-at^2+bt^4]$ is 80.1
KG, where $t = T/T_c$. The upper critical field gradient at the transition
temperature, $-(dH_{c2}(T)/dT)_{T=T_c}$ is 31 KG/K and much larger than that of
the Zr_2Rh compound. This gradient gives the temperature dependence of
$H_{c2}(T)$ for type II superconductors in the dirty limit, according to the
theory proposed by Maki[6] and Werthamer, et al. (WHH).[7] The dashed line
in curve A is the calculated plot ignoring both the paramagnetic
limiting effect and the spin-orbit scattering effect, and thus
it gives a value of H_{c2}^* (0) of 87 KG. The paramagnetic limitation parameter is given by $\alpha = 2H_{c2}*(0)/H_p(0)$, where $H_p(0)$ is the
zero-temperature upper critical field limit given by Clogston's equation, $H_p(0) = 1.84 \times 10^4 T_c$. Thus $\alpha = 1.6$ is obtained for the noncrystalline Zr-23at.%Rh alloy. (If complete paramagnetic suppression is assumed, a very low $H_{c2}(0)$ of about 50KG results.) From $_{c2}(0) = 80.1KG$ and
$\alpha = 1.6$, a spin-orbit scattering frequency parameter $\lambda_{so} = 8$ is obtained
using the WHH theory. This relatively large λ_{so} seems to be related to
the large normal state resistivity of the noncrystalline Zr-23at.%Rh alloy. The presence of a large spin-orbit scattering effect in the non-
crystalline state has also been predicted for Au-La alloys and a few
transition metal alloys.[8,9]

ANNEALING BEHAVIOR

The annealing behavior of the metastable phases was investigated for a
few specimens by continuous observation in the hot stage of an electron microscope, electrical resistivity change and isochronal dependence of T_c on annealing temperature. The resistivity change of the supersaturated solid solution in the Zr-11at.%Rh alloy indicates that the
transformation occurs between 320 and 500°C. T_c rapidly increases during the transformation due to the precipitation of the equilibrium Zr_2Rh
phase.[2] The transformation of the noncrystalline phase can be divided into two stages. In the first stage, the noncrystalline phase transforms
to a crystalline transition phase with hexagonal structure. The transition phase shows a superconducting transition at about 6.5K. In the
second stage, the equilibrium α-Zr phase and the Zr_2Rh compound are
524

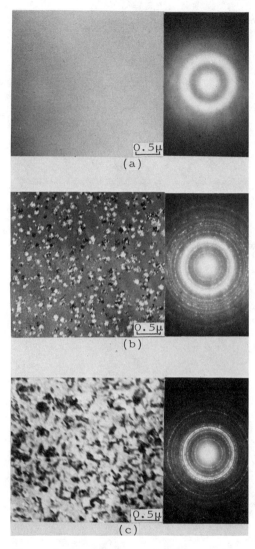

Figure 5. Transmission electron micrographs and electron diffraction patterns of a liquid-quenched Zr-23at.%Rh alloy. (a) As liquid quenched. (b) Partially crystallized on heating at 450°C. (c) Fully crystallized on further heating at 450°C.

formed. The formation temperature of the transition phase and that of the final equilibrium phases are about 400 and 500°C, respectively.

Figure 5(a) shows the transmission electron micrograph and the corresponding electron diffraction pattern of a liquid quenched Zr-23at.%Rh specimen. Transformation of the noncrystalline state to a crystalline state was observed within the microscope by heating the Zr-23 at.%Rh specimen to and at about 450°C. As can be seen in Figure 5(b), large numbers of crystalline nuclei precipitate homogeneously and then grow at the expense of the noncrystalline matrix. In the electron diffraction pattern corresponding to this intermediate state relatively sharp crystalline peaks are superimposed on the broad noncrystalline peak. After further heating, the crystallization is complete and an equiaxed polycrystalline structure and sharp diffraction rings are obtained as shown in Figure 5(c).

ACKNOWLEDGEMENTS

The authors are grateful to Dr. H. Wada of NRIM for many helpful discussions and to Mr. Y. Shimura for his assistance in this research. They are also indebted to Prof. S. Tanuma of University of Tokyo for making the Nb_3Sn superconducting solenoid available to them.

REFERENCES

1. Zegler, S.T., J. Phys. Chem. Solids 26,1347 (1965).
2. Togano, K. and K. Tachikawa, J. Less-Common Met. 33,275 (1973).
3. Ohring, M. and A. Haldipur, Rev. Sci. Instrum. 42,530 (1971).
4. Revcolevschi, A. and N.J. Grant, Metall. Trans. 3,1545 (1972).
5. Collver, M.M. and R.H. Hammond, Phys. Rev. Lett. 30,92 (1973).
6. Maki, K., Phys. Rev. 148,362 (1966).
7. Werthamer, N.R., E. Helfand, and P.C. Hohenberg, Phys. Rev. 147,295 (1966).
8. Johnson, W.L., S.J. Poon, and P. Duwez, Phys. Rev. 11, 150 (1975).
9. Johnson, W.L. and S.J. Poon, IEEE Trans. on Magnetics, MAG-11,189, (March 1975).

A

540

MICROSCOPIC OBSERVATION OF FRACTURE BEHAVIOR IN A
$Ni_{55}Pd_{35}P_{10}$ METALLIC GLASS - S. Takayama and R. Maddin

AC LOSSES AND TEMPERATURE-DEPENDENT PROPERTIES OF AN AMOR-
PHOUS MAGNETIC ALLOY - W.M. Swift and K. Foster

THE IMPURITY RESISTIVITY AND INTERACTIONS IN SUPERSATURATED
AlMn AND AlCr ALLOYS - A. Hamzić, E. Babić, and B. Leontić

SPECIFIC HEAT AND TUNNELING MEASUREMENTS ON AMORPHOUS INDIUM
AND THALLIUM FILMS - S. Ewert, A. Comberg, H. Wühl

PRELIMINARY EXPERIMENTS FOR THE DEVELOPMENT OF DUCTILE SUPER-
CONDUCTING ALLOYS ON THE BASE OF COPPER BY RAPID SOLIDIFICA-
TION FROM THE MELT - G. Faninger, D. Merz, and H. Winter

SOFT FERROMAGNETIC PROPERTIES OF SOME AMORPHOUS ALLOYS -
H. Fujimori, Y. Obi, T. Masumoto, and H. Saito

EXTREMELY HIGH CORROSION RESISTIVITY OF CHROMIUM-BEARING
AMORPHOUS IRON ALLOYS - K. Hashimoto and T. Masumoto

METALLIC STATE OF Si IN RAPIDLY QUENCHED Si-NOBLE METAL
ALLOYS - A. Hiraki, M. Iwami, A. Shimizu, K. Shuto

FURTHER EVIDENCE FOR THE $T^{-\frac{1}{2}}$ SINGULARITY IN AMORPHOUS KONDO
ALLOYS - R. Hasegawa

SEMI-CONDUCTION PROPERTIES OF AMORPHOUS V_2O_5 PREPARED BY
SPLAT-COOLING - J. Livage and R. Collongues

MICROCRYSTALLINE/AMORPHOUS IRON ALLOY FILMS FOR CORROSION-
RESISTANT COATINGS - W.B. Nowak

[57]Fe MÖSSBAUER MEASUREMENTS IN Zn-Fe ALLOYS - J.M. Williams,
J.B. Dunlop, and B. Leontić